CHINA KNOWLEDGE SERIES

ANCIENT CHINA'S TECHNOLOGY AND SCIENCE

Compiled by the Institute of
the History of Natural Sciences,
Chinese Academy of Sciences

FOREIGN LANGUAGES PRESS
BEIJING

First edition   1983

ISBN   0-8351-1001-X

Published by the Foreign Languages Press,
24 Baiwanzhuang Road, Beijing, China

Printed by the Foreign Languages Printing House,
19 West Chegongzhuang Road, Beijing, China

Distributed by China Publications Centre (Guoji Shudian),
P.O. Box 399, Beijing, China

Printed in the People's Republic of China

Armillary (*right*) and equatorial torquetum (*below*), both made in the 15th century and now preserved at the Purple Mountain Observatory in Nanjing.

Reconstructed model of Zhang Heng's seismograph.

Reconstructed model of the south-pointing carriage by Wang Zhenduo.

Reconstructed model of the odometer by Wang Zhenduo.

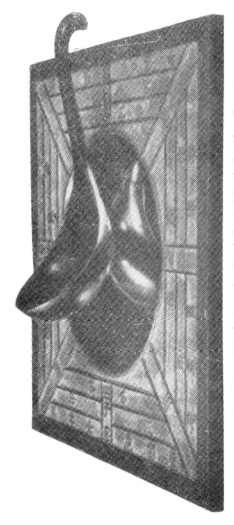

Reconstructed model of *sinan* (south-pointing ladle) of the Han Dynasty.

The earliest printed literature in existence — *Diamond Sutra* — bearing the date of the 9th year of the reign of Xiantong (A.D. 868), Tang Dynasty *(nearest left column)*.

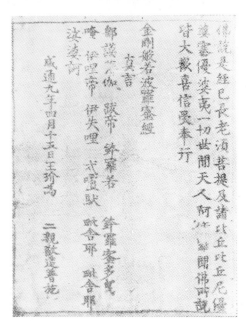

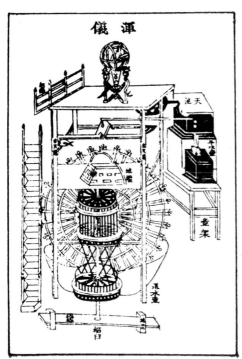

The water-driven astronomical clock tower as shown in Su Song's book *Xin Yi Xiang Fa Yao* (*New Design for an Armillary Clock*).

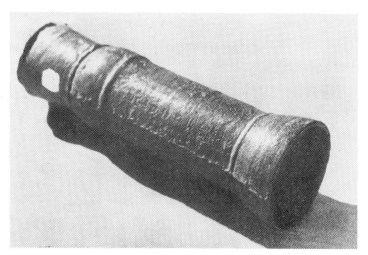

A bronze cannon cast in 1332.

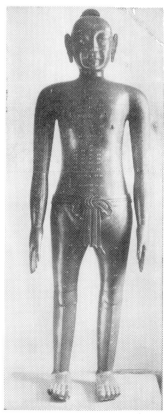

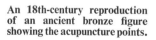

An 18th-century reproduction of an ancient bronze figure showing the acupuncture points.

The Great Wall snakes along mountain ridges.

Zhaozhou Bridge in Hebei Province.

CONTENTS

Preface

For 2,000 years the Chinese people had a most remarkable record in science and technology. Then from two to three centuries ago a decline set in. Commenting on ancient Chinese technological discoveries and inventions, Dr. Joseph Needham writes in the preface to his great work *Science and Civilisation in China* that they were "often far in advance (as we shall have little difficulty in showing) of contemporary Europe, especially up to the 15th century".[1]

Scientific and technological achievements are without exception most vividly reflected in various activities of mankind. An original idea in science or a technological invention is no achievement unless it is verified in human practice and becomes a motive force of history. We are pleased to be able to say that all the achievements dealt with in this book qualify according to the above definition, that they are achievements in the true sense of the word.

Except for a few short periods, China has been a political entity from ancient times. The Chinese nation has stood firm on earth for the past 4,000 years and has been steadily prospering. A major reason for this is that China has achieved brilliantly in science and technology as outlined in part in this book.

The impact of China's science and technology on the rest of the world has been great. Early in the Han Dynasty, from 138 B.C., Zhang Qian in the capacity of a diplomatic envoy blazed the trail which was the "Silk Road" leading to Middle and West Asian countries. And there were other early Chinese explorers who carried China's advanced culture and science abroad, and returned with cultural achievements from other lands. There was the Buddhist priest Jian Zhen of the Tang Dynasty, who in the early years of the 8th

---

1 J. Needham, *Science and Civilisation in China*, Vol. 1, p. 4.

century braved the hazardous voyage to Japan several times and finally succeeded in landing there. There was also Zheng He of the Ming Dynasty, who between 1405 and 1433 headed an impressive fleet on seven voyages to Southeast Asia and Africa.

In the 17th century, Western science and technology began flowing into China via the Jesuit missionaries. Some 200 years later, towards the end of the Qing Dynasty, the feudal rulers who had panicked before imperialist gun-boats suddenly turned from xenophobia to blind worship of anything foreign. This latter type of delusion infected certain influential people, who advocated "wholesale Westernization" even after the patriotic May 4th Movement of 1919. China was submerged in Western science and technology at the cost of almost total obliteration of her own fine traditions.

The Chinese Communist Party and the People's Government rehabilitated China's national traditions in science and technology after liberation in 1949. Today, in a new drive for socialist modernization, these traditions are carried forward alongside a quest for world-wide advanced theories and practices.

Investigation and excavation of cultural relics are meticulously done since these are of national concern. The many objects unearthed in the past 30 years are publicly exhibited and studied throughout the country, broadening and deepening people's knowledge of their cultural past. Ancient palaces, temples, pagodas and bridges — architectural wonders that have survived historical vicissitudes — and other sites are now kept in a good state of preservation. These and the unearthed articles are most eloquent witnesses to ancient Chinese scientific and technological achievement in the formation of one of the oldest civilizations of mankind.

Remarkable among them is a pair of jade burial suits sewn with gold thread which were found at Mancheng of Hebei Province in the tombs constructed in 113 B.C. for the princely house of Liu Sheng. Also the body of man buried 2,100 years ago at Jiangling of Hubei Province under the reign of the Emperor Wen Di of the Western Han Dynasty, the body and internal organs being better preserved than those of his contemporary the Marchioness of Dai at Mawangdui near Changsha in Hunan Province. Other ancient wonders are the large-scale construction site dating back to the Western Zhou Dynasty 3,000 years ago in Shaanxi Province, the remains of Epang Palace of the First Emperor of Qin (Qin Shi Huang) at Xianyang which was burned by Xiang Yü the Conqueror near the end of the 3rd century B.C., and the Qin and Han ship-building site recently unearthed at Guangzhou.

These cultural relics inspire utmost admiration for the ancient artisans' superb skill, which matches or even excels the best of today. Handicraft production is necessarily conditioned by the materials used, whether bronze, iron, jade, gold, silver or hardwood. The procuring and refining of these materials demanded skills in metallurgy, mining, mechanical engineering and even chemical engineering which warrant great pride today.

Science and technology thrived in the cultural upsurge after the Qin and Han dynasties. In astronomy, mathematics, physics, chemistry, meteorology, seismology and related sciences, China was once centuries in advance of the West. Zu Chongzhi's calculated ratio between the circumference of a circle and its diameter was not far off *pi*. In technology China has contributed to the world the compass, gunpowder, paper-making and printing. Originality is shown in many engineering works too. The Chinese people have through the ages shown resourcefulness and intelligence in science and technology as well as in politics, economics, military affairs, literature and art.

The creativeness and ability of the Chinese people were long handicapped by feudal obscurantism, however, especially in science and technology. The talents of many outstanding scientists, master technicians and artisans were buried in oblivion. Though nameless, these labouring people deserve their place in history. The fruits of their labour have come down to us.

Time cannot detract from their contributions. A look at modern achievements in science and technology often reveals footprints of predecessors who closely approached modern aims. The new is in fact brought into existence by sorting through the old. The facade of the new mansions may be different, but the materials used are invariably from familiar sources. Even science and technology from the West sometimes reveal certain Chinese traditional vestiges.   A fitting example of ancient technology derived from long practice, forming a fine tradition and made to serve the present is the Zhaozhou Bridge built 1,300 years ago in Hebei Province.   Its open-spandrel arch construction has been improved by the use of reinforced concrete in a cross-curved arch, and the bridge is still used today.

Turning past achievement into a new motive force, Chinese scientists and technologists believe that China now can and will reach or even surpass certain advanced world levels in the not too distant future.

Mao Yisheng

Records of Astronomical Events

Chen Xiaozhong

As one of the first countries to have started astronomical researches, China has in possession a store of written records of celestial events dating back to 40 centuries ago. Some of the records, systematically covering a wide range, remain valuable to modern astronomy. The present article, however, will be restricted to the documents with respect to sunspots, comets, meteors and novae.

"Dark Gaseous Masses" in the Sun

The ancient Chinese were earnest experimenters and industrious observers. They persisted in giving detailed descriptions of events on the sun, the source of light and heat. Chinese historical writings provide a rich source of such accurate records. A report of sunspots, the earliest known in the world, appeared in *Han Shu* (*History of the Han Dynasty*). It states that in the third month of the year 28 B.C., "the sun bore a yellowish colour when it rose and a mass of dark gas was observed at its centre". This account carries all the necessary information with respect to the time of occurrence as well as the precise location of the spots.

Actually, Chinese sunspot reporting started at a still earlier time. The book *Huai Nan Zi* (*The Book of the Prince of Huai Nan*) which was compiled around 140 B.C. told about the appearance of "a perching crow in the centre of the sun", which was nothing but a sunspot in the shape of a black bird. Later in the fourth month of 43 B.C., according to *History of the Han Dynasty*, "a black object the size of a pellet lay aslant over the sun near its edge".

Sunspots, or dark regions on the solar surface, are subject to constant changes arising as a result of the powerful movements involving enormous masses of matter there. Ancient Chinese astronomers made keen observations of the sunspots' durations, which ranged from no more than a day to half a year in a few particular cases, and such observations were recorded. A paragraph in *History of the Han Dynasty* states that in the first month of A.D. 188 "the sun wore the colour of orange and there was seen in its centre a dark gaseous mass in the shape of a flying magpie, which remained until several months later". And, "black spots as big as a plum", according to *Song Shi* (*History of the Song Dynasty*), appeared on March 12, 1131 and "did not vanish until three days later".

Sunspots remain for a time and change in shape during their course. A small black point near the edge of the sun at first, a sunspot grows and splits into two separate groups with a great number of tiny dots between them. Such processes did not go unobserved by the ancient Chinese. The source cited above gave a vivid description of the phenomenon involving huge masses of sunspots when it stated that on May 2, 1112 "in the centre of the sun there were dark spots the size of chestnuts, sometimes two and then three".

Relying solely on the naked eye, the ancient Chinese were able to watch the sun only when it was dimmed by clouds or haze, or when it was close to the horizon and enshrouded in mist. Otherwise they could only look at its image reflected by oil in a basin. But for all these restrictions documentary accounts of sunspots amounted to more than 100 within a period of some 1,600 years up to the 17th century. Chinese historical writings used such terms as "coins", "chestnuts", "flying magpies", etc. to depict the form of sunspots, and such expressions as "remained for several months" and "did not vanish until three days later" to show their respective durations.

As sunspots vary in size, time of occurrence and duration, only the large and conspicuous ones could be detected in the absence of the telescope. Thus the ancient records gave the number of the spots as no more than two or three and their durations as no less than several days or months. Modern observations provide convincing explanations for this trend. And a description like "the sun became crimson and dim" was nothing but a truthful and scientific representation of the environments in which the particular observation was made.

Chinese astronomical observations have won the praise of scientists abroad. George Ellery Hale, an American astronomer, pointed out that the ancient Chinese made industrious and accurate astronomical observations. He said that the Chinese were roughly 20 centuries ahead of the Westerners in sunspot observation. Hale also noted the consecutiveness of the reports, which he called reliable sources.

The 11-year cycle of sunspots was discovered in 1843 by the German scientist H.S. Schwabe. Analysis of the ancient Chinese records leads to the same conclusion. The Yunnan Observatory in 1975 compiled a complete catalogue of Chinese sunspot records covering the period from 43 B.C. to A.D. 1638. A cycle of 10.6 $\pm$0.43 years was found through an analysis of the total 106 occurrences, together with two longer cycles of 62 and 250 years respectively. This is an important discovery in the study of ancient Chinese records and in the researches into the laws of sunspots.

The concurrence of the climaxes of auroras with those of sunspots was also mentioned in ancient Chinese writings. The above-mentioned observatory in an analysis made in July 1977 with the help of ancient Chinese data found that the same 11-year cycle applies to both sunspots and auroras and that the cycle is not a phenomenon confined to the last three centuries. This discovery, instrumental to the solution of a series of geophysical and astronomical problems, gives another evidence to the value of Chinese sunspot accounts.

Comets

Descriptions of the comets appeared very early in China. *Chun Qiu (Spring and Autumn Annals)*, a historical record, noted that in the seventh month in the autumn of 613 B.C. a comet was seen travelling into the Great Dipper. This was the world's earliest report of Halley's Comet. Later, an entry in *Shi Ji* (*Records of the Historian*) gave the year though not the precise date of its reappearance in 467 B.C. This comet with an average cycle of 76 years is the largest and brightest one to be seen. Within a period of more than 2,500 years from the 7th century B.C. to the beginning of the present century, Chinese astronomers have kept written accounts of the appearance of Halley's Comet on 31 occasions. The description of 12 B.C. given in *History of the Han Dynasty* provided the best detail. It stated:

On the 12th of the seventh month in the first year of the reign of Yanyuan of the Yuan Dynasty a star first appeared in the Dongjing[1] constellation, then travelled past Wuzhuhou[2] and the region north of Heshu[3]. It was seen successively in Xuanyuan[4] and the enclosure Taiwei[5] and was six degrees behind the sun when it appeared in the east in the morning. On the 13th it was seen in the west after sunset. Its tail was again in the enclosure Ziwei.[6] It headed south at the same longitudes as Tajiao [Arcturus] and Nieti [Muphrid]. It slowed down when it reached the enclosure Tianshi[7] with its tail in the centre of that group of stars. Ten days later it flew westward and after 56 days it disappeared along with Canglong[8] of the ecliptic.

---

[1] Dongjing, a lunar mansion, with μ Geminorum as its determinative star.

[2] Wuzhuhou, five stars which are θ, λ, ι, γ, and φ Geminorum.

[3] Heshu, a group of six stars which are λ, β and ε Canis Majoris and β, α and δ Geminorum.

[4] Xuanyuan, a group of stars in Leo and Leo Minor.

[5] Taiwei, an "enclosure" of stars in Leo, Virgo, Coma Berenices, Canes Venatici and Ursa Major.

[6] Ziwei, an "enclosure" of stars most of which are the circumpolar stars.

[7] Tianshi, an "enclosure" of stars in Hercules, Ophiuchus, Serpens, Bootes and Coronas Borealis.

[8] Canglong, the Blue Dragon. This region includes seven mansions: Jiao, Kang, Di, Fang, Xin, Wei, Qi. It covers the Virgo, Libre, Scorpius and part of Sagittarius.

The passage, in concise and vivid language, gave lively descriptions of the route, apparent velocity and time of appearance and disappearance of this gigantic comet. Other Chinese historical writings also contain fairly detailed accounts of its appearance.

By 1910 Chinese comet reports amounted to no fewer than 500 and they were by no means confined to Halley's. The comet, providing a splendid sight, was by far not the only one that showed such glory. A Chinese historical article gave the information that in the year 676 a comet appeared in Gemini, heading for the region near Castor and was "as long as three *chi*".[1] It travelled to the northeast and "became three *zhang*[2] in length". It swept over the region near λ and μ Ursae Majoris, and headed for the region around θ Ursae Majoris. This passage, characteristic of all Chinese reports, gave not only a vivid picture of the comet itself, but also its exact locations and the name of every big star along its route.

Furthermore, the ancient Chinese tried to explain the origin of the comet's tail. *Jin Shu* (*History of the Jin Dynasty*) was evidently correct when it stated:

When a comet appears in the morning, its tail points towards the west, and when it appears in the evening, its tail points to the east. Whether south or north of the sun, a comet always throws its tail away from the sun, and the tail varies in brightness and length. That is because the comet itself does not shine but merely reflects the sunlight.

The ancient Chinese also reported the splitting of comets. In the 10th month of A.D. 896, according to *Xin Tang Shu* (*New History of the Tang Dynasty*), three "travelling stars, one bigger than the other two, were seen" between Equuleus and Pegasus. "They travelled together eastwards, now close to each other and then

---

[1] According to researches by Yin Shitong of the Beijing Planetarium, a *chi* used in astrometry during the Tang Dynasty (608-907) was equivalent to 24.525 centimetres.

[2] 1 *zhang* = 10 *chi*.

farther apart, as if engaged in a fight.   Three days later the big star vanished and then the other two."   In this elaborate description the "travelling stars" were certainly a disintegrating comet.

Though sometimes laying emphasis on astrology, the ancient Chinese comet observers made great contributions to later researches by their persistent work and systematic written information.   Often European researchers refer to Chinese documents in studying the orbits and cycles of the comets, the early information about Halley's Comet being but one such report.

Early in the 20th century the British astronomers Andrew Claude de la Chrois Crommelin and Herbert Philip Cowell compared their own calculations of Halley's Comet's perihelion and cycle with Chinese data of 240 B.C. and found general conformity between the two sets of figures.

In recent years the American scientist Joseph L. Brady also made use of ancient Chinese information in his researches on the comet's movement from 1682 to the present century.   His studies centred round the comet's anticipated reappearance in 1986 and the possible existence of a 10th major planet affecting its orbit.   The Chinese astronomer Jiang Tao (T. Kiang) working at Dunsink Observatory, Ireland also studied the comet's motion.   He had published his article "The Past Orbit of Halley's Comet" in *Memoirs of the Royal Astronomical Society*, Vol. 76, Part 2, 1972.   Another Chinese astronomer, Zhang Yuzhe (Y.C. Chang), head of the Purple Mountain Observatory in Nanjing, China had published in 1978 his article "The Evolution of the Orbit of Halley's Comet, Its Trend and Ancient History".   He cited data from Chinese historical writings and used computers to determine the orbital elements of the comet's reappearances.   One of the results of the comparisons he made of those data was a clue to the answer to certain problems in Chinese chronology.   The modern significance of ancient Chinese data concerning comets is obvious.

The French astronomer F. Baldet concluded in the fifties when studying the orbits of 1,428 comets that the Chinese records of comets were among the world's best.

Meteorite Showers

The earliest records of meteorite showers were again Chinese. A detailed account in *Zuo Zhuan* (*Zuoqiu Ming's Chronicles*), a book of history, stated that at midnight of a certain day of the fourth month of 687 B.C. "stars disappeared and meteors dropped in a shower". This is the world's earliest written mention of the Lyrids.

In Chinese writings such reports totalled roughly 180, among which nine were about the Lyrids, a dozen about the Perseids and seven about the Leonids. These data will play an important role in studies on the evolution of the traces of meteoric swarms.

Meteorite showers are spectacular sights. The ancient Chinese described them vividly, as exemplified here in depicting the Lyrids:

> In the third month . . . of the fifth year of Daming [461], the moon eclipsed the stars in the Xuanyuan constellation . . . millions upon millions of shooting stars of various sizes and lengths travelled together westward and the scene remained until dawn. (*Song Shu* [*History of the Liu Song Dynasty*])

The Perseids were, however, no less impressive:

> On the night of the first of the fifth month of the second year of Kaiyuan [714], a stream of shooting stars headed for the northwest. In it were both bigger stars the size of pots or measures and innumerable smaller ones. The stream flowed through the Pole and all the stars in the sky shook. The sight persisted until daybreak. (*Xin Tang Shu* [*New History of the Tang Dynasty*])

The fact that when meteors dropped on the earth they became stony or iron meteorites was also noted by the ancient Chinese. *Records of the Historian* written in 200 B.C. contains the statement "a fallen meteor is a stone". And Shen Kuo of the 11th century went further to point out that of some meteorites the major element was iron. In his *Meng Xi Bi Tan* (*Dream Stream Essays*) appears:

> One evening [in 1064] a sound like thunderbolt was heard in Changzhou. It was from a big star shooting through the

southeastern sky, equalling the moon in size. Later another explosion was heard and the star was in the southwest. With a deafening noise it fell into the garden of a certain Xu family in Yixing County, Jiangsu Province. The blaze was seen even from a distance. . . . At the scene a very deep hole as big as a drinking cup was discovered. People looked into the hole and found the star in it, still shining. Then it darkened but the heat remained. After a long time people dug and found a round rock at the depth of three *chi*. It was the size of one's fist and felt warm. There was a slight bulge on one side of it. It was the colour of iron and was as heavy.

The oldest meteorite existing in China today is the one at Long-chuan, Sichuan Province. Having fallen some time before the 17th century, it was excavated in 1716. It weighs 58.5 kilogrammes and is preserved in the Chengdu Geological Institute.

"Guest Star"

In ancient Chinese writings, the term "guest star" applied casually to a comet, but in most cases to a nova, because the latter resembles a guest in its behaviour — unpretentious, even invisible at first. Then it becomes thousands to millions of times brighter (if it is a nova) and even hundreds of millions of times brighter (if it is a supernova), after which it darkens gradually and in a few years or over a decade becomes as dim as at first.

Records about novae were found in the tortoise shell inscriptions of the 16th century B.C. The earliest systematic written record known dates to the 2nd century B.C. In the fifth month of 134 B.C., according to *History of the Han Dynasty*, "a guest star appeared in the lunar mansion of Fang", i.e., the head of Scorpion. The same nova is mentioned in the historical writings of various countries, but the Chinese version is the only one that named the month of its appearance and gave its celestial location. The 19th century French astronomer Jean-Baptiste Biot based his placing of this nova at the head of his compendium of novae on the Chinese reference.

A total of roughly 90 novae were mentioned in Chinese writings from the 15th century B.C. to A.D. 1700, the most impressive of which was the nova seen in 1054 near ρ Tauri, which dimmed and disappeared two years later. A paragraph in *Song Hui Yao* (*Administrative Statutes of the Song Dynasty*) noted that in 1056 "the astronomers reported the vanishing of a guest star and interpreted it as a sign foretelling the departure of a guest". The star first appeared in 1054, rising in the region of ρ Tauri "in the east and was seen even in the daylight. It resembled Taibai [Venus], giving white rays. Its brightness persisted for 23 days".

A late 18th-century observer discovered with a telescope a nebula in the shape of a crab near ρ Tauri and named it Crab Nebula. In 1921 the nebula was found to be expanding. Calculations based on the rate of its expansion revealed that the nebulosity was formed nine centuries before and that it was the product of a supernova outburst. Crab Nebula is the source of both light and radio pulsations and emits X-rays and gamma rays. All these radiations have the very short, stable pulsation period of approximately 0.033 seconds. Comprehensive research studies resulted in the consensus that this nebula is the remnants of the nucleus of a supernova that went through an outburst, or a neutron star. This process, the last stage in the evolution of stars, had been only a theoretical hypothesis until it found support in this Chinese written evidence.

The nova that appeared in Cassiopeia in 1572 was visible to the naked eye at noon at the height of its brightness. *Ming Shi Lu* (*Veritable Records of the Ming Dynasty*) states:

> On the third of the 10th month . . . a guest star as big as a bullet appeared in the northeast. . . . After 19 days, it grew to the size of a cup, emitting rays of orange colour. . . . During the 10th month it was even seen at noon.

Another nova, the one of 1604, equalled Venus in brightness. On October 10 that year, according to *Ming Shi* (*History of the Ming Dynasty*), "in the southwestern sky a star like a big bullet was seen" at the tail of Scorpion. "It was no more to be seen the next month." On February 3, 1605 it reappeared and was "in the south-eastern sky but the same mansion. It remained there until the eighth month of the next year".

World astronomers are highly interested in China's ancient documents on novae and supernovae at the present time when radio astronomy is rapidly developing. They seek enlightenment on the significance of supernovae to research on the radio sources in the Galaxy. During the 1950s Chinese astronomers began working to provide systematic collections of ancient literature relevant to novae and supernovae. Their work is highly appreciated by scientists abroad. Of the dozen ancient Chinese reports on supernovae, eight or nine are found to correspond to radio sources. Such information merits being called the ancient Chinese observers' brilliant success and significant contribution to modern research workers.

Astrometry and Astrometric Instruments

Bo Shuren

Astrometry is the science which, for practical and scientific purposes, studies the means to locate the heavenly bodies or to ascertain the specific time when a given heavenly body is to come into a particular position. It also includes knowledge of the instruments and implements used in such studies as well as how to operate these devices. Astrometry has been an important branch of astronomy in China since very ancient times when there was the office of *Huozheng* (Observer of Antares) responsible for the announcement of seasons according to the rise and set of that star.

The Star Catalogue of Shi Shen —
One of the World's Earliest

Of the series of star catalogues compiled in China, that compiled by Shi Shen, an astronomer who flourished during the 4th century B.C., was the first and is therefore important.

Shi was the author of a very valuable eight-volume book entitled *Tian Wen* (*Astronomy*), generally called by later astronomers *Shi Shi Xing Jing* (*Shi's Classic on Stars*). Only fragments of it are extant as quotations in an 8th-century astronomical writing entitled *Kai Yuan Zhan Jing* (*Kaiyuan Classic on Astrology*). Study of these fragments made it possible to compile Shi's catalogue listing the equatorial co-ordinates of a total of 115 stars[1] including the determina-

---

[1] Numbers given to the stars listed in the *Kaiyuan Classic on Astrology* suggest a total of 121 stars, though the names of six have been lost.

tive star for each of the 28 lunar mansions and other stars.

In this catalogue the equatorial co-ordinates fall into two separate categories. The first, relating to the determinative star of each of the lunar mansions, includes *judu* and *qujidu*. *Judu* is the distance between two determinative stars in terms of degrees on the celestial equator, while *qujidu* or the North Pole distance is a star's complement of declination. The second form of co-ordinates relates to the stars

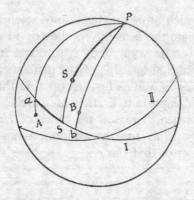

Diagram showing the ancient Chinese equatorial co-ordinates

I   equatorial circle
II   horizon circle
P   North Pole
A and B determinative stars of two neigh-
     bouring lunar mansions
ab   *judu* of the determinative star A
S   a given celestial body
as   *ruxiudu* of S
SP   North Pole distance of S

other than the determinative stars of the lunar mansions. That includes a heavenly body's *ruxiudu* or angular distance to the determinative star of the lunar mansion to which the body belongs and, again, its North Pole distance.

Either of the above-mentioned forms is in conformity with the equatorial co-ordinates system widely used in modern astronomy.

There have been slow variations in the equatorial co-ordinates of stars owing to the annual precession. Comparisons between the ancient and modern co-ordinates will therefore, through calculations based on the law of precession, reveal the year when the co-ordinates were set. Such a study of the data provided by Shi's catalogue showed that some of them had actually been compiled in the 4th century B.C. while the rest were readjusted during the A.D. 2nd century.

The earliest Greek catalogue was made by the astronomer Hipparchus in the 2nd century B.C. Two other Greek astronomers before Hipparchus did try to ascertain the relative positions of a number of stars, but that was no earlier than the 3rd century B.C.

Shi's catalogue served as a basis for astrometry in the later generations. His data for the distance between each two lunar mansions' determinative stars continued to be useful in observing the motion of the sun, the moon and the planets. They were an important series of data essential to Chinese calendrical calculations.

Star Maps

A star map is a visual record of astronomical observations and a helpful instrument for the identification and location of stars. It is as indispensable in astronomy as a map in geography.

China has a long history of making star maps. Scientific star maps showing the definite position of stars, besides those drawings which were merely descriptive or partial, started to appear as early as the period before the 1st century B.C.

Chinese star mapping originated in an ancient diagram of a dome constructed on the basis of a "theory of hemispherical dome", or the *gaitian* theory.[1] Bearing some resemblance to a revolving star map used in teaching astronomy today, the whole set of the dome had a complete star map as its background. According to *Zhou Bei Suan Jing* (*Arithmetical Classic of the Gnomon and Circular Paths*) written in the 1st century B.C., there were seven concentric circles with an equal distance between each described as the "diagram of seven declination circles". The celestial North Pole was the common centre. The smallest of the circles corresponded to the modern summer solstitial colure, the circle in the middle corresponded to the celestial equator, and the largest to the winter solstitial colure. Between the colures was another circle touching each, and

---

[1] The *gaitian* theory which embodied an ancient school of Chinese astronomy held that the heavens were a hemispherical dome, like a basin turned upside down, with the Great Dipper and the Pole Star at its centre around which the sun, moon and stars were believed to revolve.

that referred to the ecliptic, in the proximaty of which there were the stars of the lunar mansions as well as other stars.   When one put the diagram behind a piece of thin silk bearing a black circle outlining the region to be observed and turned the background map counterclockwise, then he could see different pictures of the sky within a day or a year.

Since the A.D. 2nd century, with the *gaitian* theory outdated, the diagram of the hemispherical dome gradually fell into oblivion. Yet its background map survived and developed into a star map which became an important independent document in astronomy.

One development was the ever greater number of stars inscribed on it.   Information contained in *Han Shu* (*History of the Han Dynasty*) showed that by the beginning of the A.D. 1st century there were inscribed on that map as many as 118 star groups, each bearing a particular name and consisting of one or more stars which totalled 783.

Chen Zhuo, Royal Astronomer of the Kingdom of Wu, outshone his predecessors with his own original star map.   He made comparisons of the stars named by Shi, Gan and Wu Xian, three schools of astronomers of the 4th century B.C., and produced a complete map showing 1,464 stars in 283 groups. His work was highly appreciated as a model by later astronomers.

Another development was in the technique of star mapping. The system of seven declination circles was replaced by three concentric circles, the smallest of which was called the circle of perpetual apparition or inner circle.   Its radius was equivalent to the latitude of the place of observation. Stars within this circle were visible the year round. The celestial equator was represented, as it had been in other maps, by another circle in the middle. The outer circle, that of perpetual occultation, was set as the limit beyond which no star ever rose above the horizon. The distance between either the inner or the outer circle and the equator was the same. The system, first mentioned by Cai Yong of the 1st century as the "Official Star Map" in his *Yue Ling Zhang Ju* (*Notes to the Monthly Ordinances*), typified the familiar form of ancient Chinese star mapping.

The oldest of such star maps extant were found on a pair of

stone tablets reclaimed from two excavated tombs of a 10th century king of the state of Wu Yue and his concubine. The tablets are inscribed with maps of the stars in the lunar mansions. Each of these highly valuable maps shows about 180 stars fairly accurately located.

The planisphere-on-stone of Suzhou, Jiangsu Province, another illustration of this system known to many astronomers in the world, was drawn in 1190 and committed to stone in 1247. Though smaller than the star maps on stone mentioned above, this planisphere is nonetheless 83 centimetres in diameter. The Milky Way is marked out and there are the longitudes of the determinative star of each lunar mansion drawn between the inner and outer circles. Owing to the erosion of the stone, scientists disagree on the exact number of stars shown. A recent count, necessarily inaccurate, gives the total as 1,434.

Researchers affirmed that the Suzhou planisphere was based on observations made between 1078 and 1085. With its unprecedented coverage of stars it is one of China's most significant archaeological findings as it serves as reliable information on ancient Chinese stellar knowledge.

The hemispherical maps may also be called planispheres because they were based on polar co-ordinates. In such maps a star's North Pole distance can be readily measured, and the same is true of a star's angular distance to the determinative star of a given lunar mansion. Such maps give the accurate relative positions of the cir-cumpolar stars, while one of their failures is that there is less accuracy in this field with stars farther away from the North Pole.

Another technique of ancient Chinese star mapping was based on rectangular co-ordinates. It used the longitudes as the horizontal axes and the latitudes as the vertical axes. Provided the same unit is employed for measurement, the map would be rectangular with the base twice as long as the height, thus it was called a "horizontal map". The stellar map made by Gao Wenhong before the early 7th century and mentioned in *Sui Shu* (*History of the Sui Dynasty*) exempli-fies this model.

In such a map the inexactness with respect to the relative position of stars near the celestial equator is negligible, but the disproportion remained great for the circumpolar regions. Though the variance in the South Pole region posed no serious problem for the Chinese, the North Pole region data meant very much. A new method was then adopted. It was to draw a planisphere for the celestial region within the inner circle and a crosswise chart for the zone between the outer and the inner circles. This method is largely in line with the theories of modern star mapping.

The oldest set of maps of such a combined type extant is found in Dunhuang. Drawn perhaps in the early Tang Dynasty (618-907), it shows over 1,350 stars. The circumpolar stars are placed in a planisphere, the other stars in 12 separate "crosswise" diagrams arranged one after another along the celestial equator following the order in the apparent annual motion of the sun. The set agrees with the information from "Yue Ling" ("Monthly Ordinances") in *Li Ji* (*Record of Rites*) instead of the actualities in the early Tang time on the ecliptic position of the sun and the name of the star to occur at the zenith in the evening or morning given for each month. This shows that the set was a mere reproduction of an earlier chart. It was nevertheless the world's earliest comprehensive star map.

One of the features of the Dunhuang star maps is the lack of co-ordinate lines — either the inner circle in the planisphere or the equator and longitudes in 12 separate crosswise diagrams. The map fails in precision, but this failure suggests that the chart was a copy. Illegally removed from the country by the imperialist Aurel Stein in 1907, the map is now in the possession of the British Museum, London.

Later, the astronomer Su Song (1020-1101), in his *Xin Yi Xiang Fa Yao* (*New Design for an Armillary Clock*), gave his own map which was based on the combination of a planisphere and two horizontal maps and was more accurate. His planisphere was called *Map of the Stars in the Enclosure Ziwei*, and his two horizontal maps, *Diagrams of the Stars of Inner and Outer Groups of the Northeast* and *Diagrams of the Stars of Inner and Outer Groups of the Southwest*, showed the sky of the two semesters — from the autumn equinox to the spring equinox and vice versa. These charts were based on observations made from 1078 to 1085.

Another remedy for the failure arising from the imprecise representation of the relative position of stars was to divide at the equator the celestial sphere into two equal portions, one having the North Pole and the other the South Pole as its centre. Such a model was also found in the above-mentioned work by Su Song. That book, written between 1094 and 1096, was supposed to explain the water-driven astronomical clock tower constructed under his supervision. The charts were nothing but reproductions of the globe itself. Thus Su and his colleagues developed two separate systems of star mapping when they tried to project the globe on a plane.

Measuring the Meridian

The length of each degree of the meridian as a basic factor in geography, geodesy and astronomy was sought as early as the 3rd and 1st centuries B.C. by Greek astronomers, who failed however to take field measurements but relied solely on the evaluations made by caravans or ships.

Over-all survey of the meridian was made in the 8th century by Chinese astronomers, the work being carried out on the initiative of Yi Xing in 724 as a part of his effort to make a new calendar.

Besides the setting of the years, months and days, ancient Chinese calendar makers took as their task the prediction of the solar and lunar eclipses, the fixing of the 24 solar terms and the forecasting of the length of days. Satisfactory fulfilment of these tasks demanded the determination of the latitude of the place where observations were to be made. Astronomical surveys for this purpose were made in the 8th century in different places in China when astronomers were working to make a new calendar applicable for the entire country.

A total of 12 locations were selected. The items for observation included the altitude of the Pole Star and the length of the shadow of an 8-*chi*[1] gnomon at noon on the days of both equinoxes

---

[1] 1 astronomical *chi* = 24.525 cm. in the Tang Dynasty (618-907).

and both solstices.

The series of observations made by Nangong Yue and his collaborators in four towns in the region which is now part of Henan Province was more significant than were the observations made by other teams. In addition to the altitudes of the Pole Star and the lengths of shadow, they measured the distance between each two of the four localities, Baima, Xunyi, Fugou and Shangcai, which were with negligible variations on the same longitude.

Basing himself on the data from field observations, Yi Xing found that the difference between the lengths of the shadow of the gnomons erected respectively in Baima and Shangcai which stood 526 *li* plus 270 *bu*[2] away from each other was a little more than two *cun*.[3] Thus he disproved the long-held supposition that the length of the shadow varied by one *cun* for every 1,000 *li*. The astronomer He Chengtian as early as A.D. 442 published his refutation of that tradition, yet held that between any two places where the same length of shadow was observed the distance must be identical. This proposition was wrong in suggesting that the earth was flat. Liu Zhuo of the early 7th century noted the variety of the proportion between the length of shadow and the distance between two places, and Li Chunfeng of the same century also recognized the inconstance of this proportion. But it was Yi Xing who confirmed the discoveries by his predecessors, relinquished the concept of the lengths of shadow and used the polar altitude data instead. Simple calculations on the figures secured in field surveys produced the proportion that the North Pole would be one Chinese degree[4] higher for a place 351 *li* plus 80 *bu* to the north of another, and that was nothing but the figure for a Chinese degree on the meridian. In terms of modern measurement, the value is 129.22 kilometres for a meridional degree, with an error of 13.9 per cent when compared with the modern figure of 111.2 kilometres.

With such an error, Yi Xing's work nevertheless represented the world's first field survey of the meridian. It opened the way to

2   1 *li* = 300 *bu*, and 1 *bu* = 5 *chi*.
3   1 *cun* = 1/10 *chi*.
4   365.25 Chinese degree = 360°.

further researches on the earth by field surveying, put an end to the traditional misconceptions and for the first time combined the measurement of latitudes with geographical distance, thus paving the way for a more up-to-date calendar and laying the cornerstone for future astronomical geodesy.

Yi Xing's endeavour convinced himself of the truth that a law discovered through researches in a limited space or scope becomes a fallacy if it is blindly extended without limit. The concept of a shadow one *cun* longer 1,000 *li* further north was wrong because it arbitrarily applied a theory abstracted from plane surveying within a limited scope to broader areas or even to infinite space. Yi Xing made a great stride forward in science by his experiments which meant a refutation of that fallacy. He thus exemplified the necessity of the spherical concept in geodesy in greater scopes.

Nautical Astronomy

Chinese sailors who were seafaring thousands of years ago developed a whole set of nautical techniques among which astronomical navigation had a very important place.

A passage in *Huai Nan Zi* (*Book of the Prince of Huai Nan*), an encyclopaedic book compiled in the 2nd century B.C., reads: "Sailors who have lost their bearings will locate the North Pole when they find the Great Dipper" — obviously a conclusion drawn from very long experience.

Chinese sea-going ships visited the Indian Ocean coasts and islands in southern Asia as early as the 3rd century B.C. The outstanding Buddhist Fa Xian (c. 337-422) made his return trip from India and Sri Lanka by ship in 412. He wrote in a narrative of his voyage:

In the boundless sea we could not tell the east from the west. We sailed on by observing the sun, the moon and the stars. We did not know where we were when we had been driven by wind on cloudy or rainy days. . . . It was only when the sky opened that we recovered our bearings and found the correct route to follow.

Flourishing foreign trade, supported by growing production, encouraged the development in Chinese nautical techniques. The introduction of the compass and other instruments into navigation and nautical astronomy gave rise to significant advance in the science of seafaring.

The earliest extant document containing specific data on celestial navigation is *Zheng He Hang Hai Tu* (*Charts of Zheng He's Voyages*), an early 15th-century work included in *Wu Bei Zhi* (*Treatise on Armament Technology*) compiled by Mao Yuanyi in the 16th century.

The collection consists of a 20-page nautical chart and four star maps. By the chart we know that Zheng relied solely on the compass when he embarked at the port of Liujiagang near Suzhou, Jiangsu Province, left the mouth of the Changjiang (Yangzi River), sailed along the Chinese coastal provinces of Zhejiang and Fujian and the Chinese islands in the South China Sea to reach the northern end of Sumatra. Further in his voyage to Ceylon (modern Sri Lanka) he began to use astronomical navigation. Greater emphasis was laid on the observation of stars in co-ordination with the use of the compass when he sailed from Ceylon to the south and west Asian and east African coasts. The chart gave the altitudes of the Pole Star and another eight stars in Ursa Minor observed at 64 localities along his voyage. These star maps, called Star-Aiming Charts for Crossing the Oceans, contained the data on the positions and altitudes of many stars as seen in different locations in the Indian Ocean.

These charts show that the ancient sailors used the technique known as "star-aiming". The instrument for this technique, a set of star-aiming plates, is described by the 16th-century scholar Li Xu in his *Jie An Lao Ren Man Bi* (*Essays by the Old Man of Jie An*). The set consisted of 12 square ebony plates, the largest of which measured to 12 digits. Each side was 24 centimetres long. The smallest measured only one digit. Each plate was bigger than the last, and the difference between each two side lengths was equal and constant. There was an ivory tablet with a cleft on each angle, the clefts standing for half a digit, an eighth, a quarter and three quarters respectively. A quarter meant a quarter of a digit.

Recent studies by the Research Group on Nautical Astronomy sponsored by four institutions including the Beijing Observatory and South China Teachers College show how the set was employed. The observer chooses a plate, holds it in such a manner that its plane lies vertical to the level of the sea, its top side aligned with the heavenly body to be observed and its bottom coinciding with the horizon. If the object is out of contact, the observer chooses a larger plate and reads out the altitude of the star by the graduation on one side of the plate. Attached to the lower part of the plate is a string with which the observer keeps the plate at a constant distance from his eyes. He usually chooses the time when the required star is on the celestial meridian. The following equation shows the relation between the object's horizontal altitude (A), its declination (D) and the geographical latitude (L) of the place where the observation is made.

$$L = 90° + D - A$$

The value 90° + D for a given star can be recognized as constant within a fairly long period of time. The observer can locate the latitude of his ship's position simply by determining a star's altitude in terms of digits and quarters when it is at its meridian.

It is interesting to note that the Research Group on Nautical Astronomy found that the measure digit was employed by Chinese astronomical observers since the 2nd century B.C. A quotation in *Kaiyuan Classic on Astrology* from a book entitled *Wu Xian Zhan* (*Astrology of Wu Xian*) written in that century gave the difference between the latitudes of Venus and the moon as 5 digits which, in terms of modern measurement, is 9.4°, while the value of the digit for centuries remained at 1.9°, a figure congruent with that in *Charts of Zheng He's Voyages*.

Ancient Chinese nautical astronomy with star-aiming as one of its typical techniques played a great role in guaranteeing safety and accuracy in ocean-going navigation and was thus another notable contribution to world civilization.

The Armillary Sphere and Equatorial Torquetum

The most popular instrument by which the ancient Chinese astronomers obtained the celestial co-ordinates was the armillary sphere. In contrast to the ancient Greek armillary, the Chinese instrument stressed the equatorial co-ordinates and was from the very beginning equatorial while the Greek instrument was ecliptical in nature. There are no written or material traces of the origin of the Chinese armillary. Scientists who studied data from Shi's star catalogue compiled no later than the 4th century B.C. concluded that an armillary had been used in obtaining these figures.

In all probability the earliest Chinese armillary consisted of four rings arranged in two nests. The outer nest had three intersecting concentric rings equal in diameter. One ring parallelled the natural celestial equatorial plane and the other the horizontal plane. The third was a double ring with an interspace of about an inch between the two members which lay parallel to the celestial meridional plane. The inner nest had only one double ring whose centre also coincided with that of the rings in the outer nest. A bar serving as the diameter of this ring and the axle around which it spun was fixed on both ends to the third ring in the outer nest, parallel with the axis of the earth's rotation, that is, the double ring parallel to the meridional plane. The circle in the interspace between the two members of the double ring in the inner nest would thus stand for a declination. A movable sighting tube was attached to this double ring. The observer turned this ring and trained the sighting tube on his object so as to read the star's North Pole distance figure on the ring, which was shown by the sighting tube's specific position. The right ascension figure could be read out on the equator ring in the outer nest. In order to know a star's distance from the nearest star of the lunar-mansion determinative stars, that is, its *ruxiudu*, the observer could turn the inner declination ring to a position so that that determinative star could be seen through the space between its two members, read out the figure for its right ascension and calculate the difference between this value and that of the right ascension of the object. The earliest description of such a model appears in *Sui Shu* (*History of the Sui Dynasty*) in the book's account of the armillary constructed by the Royal Astronomer Kong Ting in 323. Here the armillary is said to be based on an ancient model. Elaborate study of the parts of this instrument as described in that book showed that each part was essential and indispensable to any useful equatorial armillary. We can say then that the early armillaries, including the one built by Luoxia Hong in the late 2nd century B.C., were similar in structure.

Within the subsequent roughly 10 centuries the armillary underwent development from the simple to more complicated forms, and then was simplified. But, besides the celestial sphere's diurnal equatorial motion, the sun has its annual ecliptical motion and the moon its motion along its own path. Equatorial co-ordinates were obviously inadequate for the researches on these motions. Amateur astronomers of the 1st century B.C. and the early 1st century A.D. modified the traditional armillary by adding an ecliptical ring to it. In 633 Royal Astronomer Li Chunfeng of the Tang Dynasty, commissioned by the Emperor Tai Zong, invented the ring for the lunar path as another addition.

As the ecliptic and the lunar paths have their daily motion along with the celestial sphere's diurnal rotation, the ring symbolizing these paths should be correspondingly movable. There are, however, no markings on the natural celestial sphere, a lack that poses great difficulties for an observer in adjusting his ecliptic ring to the real ecliptic. As a remedial supplement Li Chunfeng constructed the *sanchenyi* (component of the three arrangers of time), a set of three intersecting rings parallel to the ecliptic, the lunar path and the equator respectively. These rings were placed between the outer nest and the right ascension ring in the inner nest and could be put into circumpolar revolutions. The equator ring of the *sanchenyi* bore marks symbolizing the equatorial position of the determinative star of every lunar mansion. When the observer turned this ring and adjusted it to the determinative stars, the other two rings in this set would automatically come into correspondence with the ecliptic and lunar paths respectively. Moreover, 249 pairs of holes were bored into the ecliptic ring for the adjustment of the lunar path ring.   As the node of the ecliptic and lunar paths recessed along the ecliptic, the observer could move the lunar path ring and fix it to another pair of holes on the ecliptic ring at the end of every nodical month, making the lunar path ring correspond better with the natural lunar path itself.

Li Chunfeng's armillary and that constructed by Yi Xing, Liang Lingzan and their colleagues 90 years later — an armillary with an ecliptically mounted celestial latitude ring attached thereon — marked the climax in the trend towards complex construction.   The latter was a complicated instrument consisting of as many as seven different rings.   The result was the inevitable narrowing of the scope for the observer who looked through the sighting tube attached to the inner nest.   Besides, because the position of the ecliptic or the lunar path naturally changed, the observer found he was already behind the time even though he had readjusted his armillary with the help of the *sanchenyi* when he started using his sighting tube, and this made his observation inaccurate.   With the advance in mathematics, astronomers came to master the conversion between the two sets of co-ordinates — the equatorial for one and the ecliptic and lunar path co-ordinates for the other.   The period from the 10th to the 13th century saw further improvement in the technique of conversion. More and more people came to realize that it would be much better to convert an equatorial figure which was readily readable into an ecliptic figure or one pertaining to the lunar path than to take the pains to try to find out these figures directly with the help of the instruments.   A prominent astronomer who flourished in the 11th century, Shen Kuo, pioneered in this new and superior approach. He simplified the armillary by first rejecting the lunar path ring, the use of which had been more difficult and given less precise results than that of the other parts.   The process of simplification thus started agreed in theory with modern astronometrical techniques based on the conversion of different co-ordinates.

In the three years from 1276, Guo Shoujing made a breakthrough in the process of simplifying the armillary.   Discarding the ecliptic ring and the double meridian ring in the outer nest, he detached the horizon ring from the instrument and constructed a separate set, called *liyunyi*, for the determination of the horizontal co-ordinates. He converted the equator ring in the set of *sanchenyi* into a member of the equator double ring in the outer nest. Attached to this ring were four rollers to allow the easy rotation of the rings of the outer nest. He placed the equator double ring to the south of the axle of the right-ascension ring and supported the North Pole axle with two sets of frames. The ring for the right ascension was now high above all other parts and free from any interference. This new device, the equatorial torquetum, was simple in structure and termed a simple instrument.

The equatorial torquetum was unprecedented in precision — 0.05°. On either end of its sighting tube was a line through the centre which was vertical to the plane of the right-ascension ring. This was virtually a forerunner of the cross hairs in telescopes. The equatorial ring fixed at the southern end of the mobile axle, a device initiated by Guo Shoujing for this instrument, is now widely accepted by modern telescope makers. Johan Ludvig Emil Dreyer, the Danish astronomer who compiled *New General Catalogue of Nebulae and Clusters of Stars*, pointed out in his commentary on the significance of Guo's equatorial torquetum: " . . . the Chinese people often came into possession of great inventions many centuries before the Western nations enjoyed them."

An armillary and an equatorial torquetum built between 1437 and 1442 preserved at the Purple Mountain Observatory in Nanjing are material evidence of the splendid achievements of ancient Chinese astronomers.

Water-Driven Astronomical Clock Tower

A tracking device is indispensable for any modern equatorial telescope. An invention by a group led by the astronomer Su Song approximately nine centuries ago might be considered as a forerunner of such an apparatus. This was a water-driven astronomical clock tower, a development of an earlier water-driven celestial globe.

The first Chinese celestial globe existed no later than the 1st century B.C. In the 2nd century A.D. the celebrated astronomer Zhang Heng, after observing common water-driven devices in use, built a mobile celestial globe operated by water whose flow was regulated by a clepsydra. His water-operated globe revolved in correspondence with the diurnal motion of the natural celestial sphere.

Yi Xing and Liang Lingzan of the 8th century added a clock to that device, a combination that served not only as an astronomical instrument but also as a chronometer that underwent constant improvement through several centuries. It might be considered an early model of the modern clock.

The best representative of such instruments was the water-driven astronomical clock tower designed and made in 1088 by Han Gonglian under the sponsorship of his superior Su Song, then Minister of Personnel. Fortunately the directions to this installation titled *Xin Yi Xiang Fa Yao* (*New Design for an Armillary Clock*) and written by Su himself are extant for our comprehensive study of this complicated machine. Wang Zhenduo in the 1950s did extensive research on it and led a group of Museum of Chinese History workers who produced a copy of it. Their clock tower, one fifth the size of its 900-year-old model, now stands in that Museum.

Han's water-driven astronomical clock tower was a wooden building 12 metres in height. On the top platform was an armillary placed in a chamber with a removable roof which resembled the dome of a modern observatory. The armillary was connected by gears with the driving machinery of the whole installation, which enabled it to follow the diurnal motion of the natural celestial sphere. When the observer aimed the sighting tube at the sun, the mechanical motion of the armillary would keep the sun in the visual field for a fairly long time. Primitive and crude as it was in comparison with the modern tracing device, this mechanism was nevertheless in a sense a significant chronometric forerunner, highly appreciated by later generations.

The first floor of the building housed the celestial globe, revolving mechanically to follow the spinning of the natural celestial sphere. The observer in this building knew the actual position of each constellation from this mobile sphere. Below this chamber were the driving wheel, the clepsydras and the apparatus for announcing the time. The driving wheel, more than three metres in diameter, was furnished with water scoops along its tyre. When water from the clepsydras filled one of the tanks, the wheel turned for imbalance. A mechanical device regulated the motion of the wheel so that it ceased to turn until the next tank was filled. The role of this device was comparable to that of a clock's escapement. The transmission system passed on the mobile force from the wheel to the time-announcing device, the celestial globe and the armillary.

The time announcer was a five-storey building constituting the southern panel of the lowest part of the whole installation. The first storey produced sound. A wooden puppet in the middle door would beat his drum at every *ke* and the one in the left door would ring his bell at every *shi*,[1] and when the latter puppet struck a bigger bell it would be the middle of a *shi*. The 24 puppets in the second-floor chamber would appear one after another at the door, holding a tablet with an inscription announcing each *shi*. The third-floor box roomed 96 puppets who would announce the *ke* except for the quarters, halves and three-quarters by showing their own tablets in turn. The puppet in the fourth-storey chamber would play his string instrument at each *geng* and *chou*[2] of the night. The name of the *geng* and *chou* were shown by each of the other 25 puppets with his own tablet.

---

[1] In ancient Chinese chronometry, a day was divided into 12 *shichen* each of which consisted of two *xiaoshi* or hours. The *shichen* are named after the earthly branches of the sexagenary system. Thus the mean of the *shichen zi* falls on midnight and that of the *shichen wu* at noon.

Ancient Chinese chronometrists divided a day into 100 *ke* in addition to the system of *shichen*. *Ke* Zero falls on midnight and *Ke* 50 at noon, while the beginning of four *ke* among the total coincided with the means of *shichen*. Only the beginning of 96 *ke* needed therefore to be announced.

[2] A *geng* was 1/5 the length of a night, and a *chou* was 1/5 the length of a *geng*. The length of every *geng* and every *chou* was thus subject to the seasonal change in the length of the night itself. Adjustments had to be made in the mechanism of the clock to suit different seasons.

China's ancient water-driven astronomical clock tower was further solid evidence of high mechanical technique. As the escapement is a mechanical clock's key apparatus, foreign scholars today highly appreciate these water-driven clocks, especially Han's installation, and call them, as does Joseph Needham of Britain, the possible ancestor of the astronomical clocks of medieval Europe. Moreover, as an astronomical instrument, this installation with its mechanical dome and especially its tracing device, was a convincing manifestation of the creative ability of the ancient Chinese scientists and technicians.

Chinese Calendars

Chen Jiujin

Man's practice taught him that to measure time by the regular changes observable in Nature, changes which directly affected his productive activities, was a convenient way. In this light, the tropical year showing the annual seasonal changes, the synodical month based on the change of the phase of the moon and the solar day which varied in length became suitable chronometrical units. A calendar based on all these factors is generally known as a solar-lunar calendar, while the solar calendar takes into account only the changes in the tropical year and the lunar calendar inflexibly fixes every year at twelve synodical months.

The history of Chinese calendrical science is a long one. The earliest calendars, only fragmentary evidences of which are extant, are as yet subjects for further archaeological research. The systematic calendar, originating in the quarter-remainder calendar which was used in the 3rd and 2nd centuries B.C., subsequently underwent more than 100 reforms (see appendix, p. 49) that brought it to a very high degree of perfection. Traditionally solar-lunar in nature, Chinese ancient calendars were not limited to the arrangement of the years, months and days. They included the prediction of the apparent motion of the sun, the moon and the five visible planets, forecast of solar and lunar eclipses and definition of solar terms. Calendrical reforms meant specifically developing new theories, finding accurate astronomical data, and improving the technique of calculation. Chinese traditional calendrical science occupies an important place in the history of world astronomy. The following is an introduction to a few of the major aspects of that branch of the world's calendrical science.

33

Researches on the Sun's Apparent Motion

As the earth rotates round an axis not vertical to the plane of its orbit and there is an angle of 23.5° between the ecliptic and the equator, the sun's horizontal altitude in a given place changes regularly and that results in the cyclical alteration in the local weather. Researches on the apparent motion of the sun are therefore necessary for calendar makers. These researches were usually made in either of two different ways. One was to measure the shadow of a gnomon at noon and, basing on such measurement, to announce the seasons and determine the tropical year. The other way was to study the annual variation in the sun's apparent velocity and ascertain the value of the annual alteration of the winter-solstice point by observing the sun's position against the background of the stars with the help of such instruments as the armillary.

Ascertaining the Winter Solstice
and Tropical Year

Precise announcement of the seasons was impossible without accurate identification of the time of the winter solstice. Because several precise figures relating to the winter solstice were necessary for the ascertainment of the tropical year, ancient Chinese calendar designers attached great importance to observations relevant to its exact time. The earliest data on the length of the gnomon's shadow on the day of the winter solstice known to us date from the period 655-522 B.C.

Theoretically, data on two successive winter solstices should suffice for the computation of the tropical year, but this was not so in practice. The reason is that the data obtained by simply observing the alternation in the shadow's length could be inaccurate by up to several days. Moreover, the solstice did not necessarily occur at noon when the length of the gnomon's shadow was read. To correct this error the ancient calendar makers used the data on as many winter solstices as possible.

In the 5th century B.C. China adopted the quarter-remainder calendar — a calendar whose tropical year was simply 365.25 days. This was the most accurate calendar in the world at that time. It provided seven intercalated months, or 235 lunations, for every 19 years, while the synodical month was set at 29.53085 days, a figure of appreciable accuracy for that time.

Ever more elaborate calendars were aimed at as human society and science developed. In time the quarter-remainder calendar was found not in keeping with the practical astronomical observations. The Taichu Calendar of 104 B.C. and the revised quarter-remainder calendar of A.D. 85 represented attempted revisions, but as time went by the scientists realized that frequent revision of the calendar was not a practical long-range solution. Deeper probes into calendar-making theory led Liu Hong at the beginning of the 3rd century to realize that the quarter-remainder calendar's inaccurate figure for the tropical year was responsible for the errors. Liu Hong made some changes by shortening the year and a better calendar was the result.

From the 2nd century B.C. the gnomon of eight chi[1] had been used to measure the shadow on the day of the winter solstice. The reading was, however, far from satisfactory. Zu Chongzhi, who flourished between A.D. 430 and 510 and was among the astronomers who strove for a more accurate value of the tropical year, improved the technique of observation. What had stood in the way of the observers was the absence of conspicuous difference in the daily length of the shadow that could be easily measured throughout the day of the winter solstice and the impossibility of ascertaining the precise hour of the solstice. Zu found a solution to these problems. He extended his shadow-measuring to a time-span of 24 days or so to obtain a mean value. The trend that greater changes in the length of shadow could be observed in days more remote from the winter solstice contributed to the success of Zu's new technique.

---

[1] 1.84 metres.

His Daming Calendar set the tropical year at 365.2429 days, a figure whose accuracy was not challenged until 1064 with the appearance of the Mingtian Calendar which marked the beginning of the practice of making observations at more and different localities.

Guo Shoujing, a 13th-century scientist who made important contributions in mathematics and astronomy and also invented new instruments, made a device called the shadow-definer which made possible a more clear-cut view of the margin of the shadow. It focused the sunlight through a pinhole "no bigger than a rice grain" at a point on the gnomon plate. This device opened the way for the use of longer gnomons for higher accuracy. The Tower for the Measurement of the Shadow now standing in Dengfeng County, Henan Province is, so to speak, a gigantic gnomon. It was built by this same Guo with stone and bricks. With a height of 13.33 metres it was four times taller than gnomons in use at that time. Basing on his own elaborate observations and using the data of the six large-scale observations so far made since 462 when Zu Chongzhi's Daming Calendar was inaugurated, Guo Shoujing set the tropical year at 365.2425 days, confirming the researches by Yang Zhongfu who designed the Tongtian Calendar in 1199.

Following Guo, the astronomer Xing Yunlu (1573-1620) built a gnomon 20 metres high and obtained the figure of 365.242190 days for a tropical year, the most accurate in the world at that time and inferior to the figure obtained by modern computation by no more than 0.000027 of a day.

Locating the Point of the Winter Solstice and Determining the Precession

The point of the winter solstice means the relative position of the sun against the background of the stars on the day of the solstice. Modern astronomy indicates this position by the right ascension and the declination of the sun. Ancient Chinese astronomers used the datum for the difference between the right ascensions of the sun and the determinative star of a given lunar mansion.

Between the 5th and 3rd century B.C. when the quarter-remainder calendar was in use, the winter solstice point was set at the entrance to the lunar mansion Qianniu or the region approaching Giedi (β Capricorni). The Zhuanxu Calendar of 221 B.C. did the same, and this may be accepted as China's earliest such datum obtained through actual observations.

As it was impossible to ascertain the sun's relative position by direct observation, the ancient astronomers resorted to indirect means. They identified the day of the winter solstice, ascertained its midnight by using the clepsydra and found out the distance of the star at the zenith to the nearest lunar-mansion determinative star. They could thus tell the location of the sun, which stood directly opposite that star. The data obtained were inevitably inaccurate because the clepsydra was hardly a satisfactory chronometer.

The Chinese astronomers were ignorant of the precession until the 3rd century. Before that they thought the sun made an exact round trip along its orbit in the celestial sphere between winter solstices. That was why they fixed the tropical year at 365.25 days and divided the natural celestial sphere into as many degrees. As far as they knew the winter solstice point was fixed, and the notion that the point was in the region approaching Giedi remained. Makers of the Taichu Calendar of 104 B.C. virtually based themselves on the same data. In A.D. 7 Liu Yin noted, if vaguely, the wavering of the winter solstice point. It was in A.D. 85 that Jia Kui announced unequivocally that the point was 20.25° away from φ Sagittarii. Though still unaware of the precession, those astronomers observed the existence of shifting of the winter solstice point. Jiang Ji of the 5th century evolved the brilliant idea of locating the sun by calculation based on the data concerning the moon's motion as observed in lunar eclipses. Proceeding from such information an astronomer could ascertain the sun's position on the day of the winter solstice. Jiang's fairly precise researches set the winter solstice point at 17° to φ Sagittarii.

Owing to the precession, the winter solstice point has an annual movement to the west by the rate of 50.0 seconds or a degree within a period of 71 years and 8 months and, in terms of ancient Chinese measures, a degree within every 70.64 years.

Around A.D. 330 Yu Xi, after comparing the data about the winter solstice point so far accumulated, for the first time affirmed the existence of the variation of the point. Having realized that the sun's "celestial motion" was different from its "yearly revolution", he proposed the approach of "dealing with the heavens as the heavens and the year as the year". He was the first to use the term "precession" and set its rate at one degree in every 50 tropical years. Later than Hipparchus by nearly 450 years, Yu arrived at a figure far more precise than the Greek astronomer's, which was one degree for a century.

Not long after Yu, the notion of precession came to be practically applied in calendar making. He Chengtian who flourished around the year 450 also made researches on the precession and set its rate at one degree in every 100 years, though he did not make use of this value when he designed the Yuanjia Calendar. Zu Chongzhi pioneered in applying the precession figure in calendar formulation. He compared his data on the winter solstice point (15° from φ Sagittarii) with Jiang's and ascertained the rate of precession as one degree in every 45 years plus 11 months. Though this figure was hardly accurate, Zu became an innovator in calendar making by introducing the concept of the precession into this science. Liu Zhuo, maker of the calendar of 604, updated the rate and set it at one degree for every 75 tropical years which should be considered of high precision at that time. Liu's figure continued to be used until 1199 when the designers of the new Tongtian Calendar adopted the more accurate figure of one degree for every 66 tropical years and 8 months.

Discovery of the Fluctuations in the Velocity of the Sun's Ecliptic Revolution

There is a slight eccentricity in the earth's orbit that causes the fluctuations in the sun's apparent motion along the ecliptic. Until the 4th century Chinese astronomers failed to detect these fluctuations for lack of adequate instruments. Due to this and also to their ignorance of the precession, they set a year at 365.25 days and accordingly measured the celestial sphere by as many degrees. That would mean that the sun travels exactly one degree each day. The span between every two solar terms was given as a constant 15.2 days. Later astronomers called such an approach the method of "mean solar terms" or "constant solar terms". After many years' observation in the latter part of the 6th century, Zhang Zixin discovered the asymmetry in the sun's apparent motion and noted that "the sun travels at a lower velocity after the spring equinox and faster after the autumn equinox". This statement largely agreed with the facts at a time when the winter solstice was only 10° behind the perigee of the ecliptic. By the 13th century these two points coincided.

Zhang's discovery was soon taken over by the calendar makers. Liu Zhuo of the 7th century and his collaborators proposed the system of 24 terms for the sun's entire journey in a tropical year. The time for the sun to traverse the distance of a term varies because of the fluctuations in its velocity. Liu thus set the length of a term at 14.718 days near the winter solstice and 15.732 days near the summer solstice. He noted that from the autumn equinox there were 88 days to the winter solstice and from the spring equinox there were 93 days to the summer solstice. His figures were, however, inaccurate and it was left to Yi Xing of the 8th century to make adequate corrections. This Buddhist monk who designed the Dayan Calendar noted that the sun's apparent velocity was the highest in the fortnight approaching the winter solstice — a notion in conformity with the facts at that time when the earth's perihelion was 9° ahead of the winter solstice. His calendar gave the total length of six terms from the winter solstice to the spring equinox, or the time for the sun to make a quadrant along the ecliptic as 88.89 days and the length of the following six terms, or the time for the sun to make the next quadrant, as 91.73 days. The figures given for the other half of the year around the autumn equinox were identical to these respectively.

Guo Shoujing, who made the calendar of the year 1281, affirmed that the winter solstice was the time when the sun travelled at the highest speed. His data were highly accurate, as the earth's perihelion then was less than one degree behind the winter solstice. Guo, basing himself on actual astronomical observations, also noted that each quarter after the third day following the autumn equinox was 88.91 days in length and the corresponding figure for the other quarter was given as 93.71 days.

These facts lead us to the conclusion that there had been ever more accurate calculations of the sun's ecliptic motion and the fluctuation in its apparent velocity from A.D. 728 when Yi Xing published his Dayan Calendar. The length of each of the 24 solar terms, however, had remained to be considered as constant until the 17th century, except in the computation on the sun's actual progress along the ecliptic and on the nodes joining the ecliptic and the lunar path.

Researches on the Motion of the Moon

Researches on the motion of the moon were given an important place in ancient Chinese astronomy because they served as the basis on which calendar makers set the months and predicted the eclipses. The synodical month as defined by the designers of the quarter-remainder calendar of the 8th century B.C. was already accurate to within one day in 300 years. The synodical month was taken as the basic datum for calendar making until the 6th century. The day of the new moon was made the first day of every month. There were longer months with 30 days and shorter months with 29 days occurring alternately. The case in which two longer months came in succession would occur in a period of about every 17 months. This system suited well the real value of a synodical month—over 29.4 days.

The distinction between the synodical month and the sidereal month was known to Chinese astronomers from very ancient times. The book *Huai Nan Zi* compiled in the 2nd century B.C. stated that the moon travelled 13 7/19 degrees along its own path while the sun advanced one. That would mean that there were 27.3219 days in a sidereal month.

Like the earth's orbit around the sun the moon's orbit around the earth is eccentric, causing the periodic fluctuation in that satellite's velocity. The moon travels at a higher speed when close to the perigee where it is at the shortest distance from the earth, slowing down when it is near the apogee where it is farthest from the planet, Earth. The anomalistic month — time for the moon to return to its apogee — differs in length from the synodical month. The periodic changes in the moon's phase are thus of different durations. The so-called synodical month is therefore an average of such durations.

Shi Shen of the 4th century B.C. was aware of the fluctuations in the moon's motion but the writings he left on this subject were altogether too sparse. Liu Xiang who lived around 25 B.C. mentioned in his notes to the classics a diagram of the "nine roads of the moon", referring to the fluctuations in the satellite's motion. The same phenomenon was described by Jia Kui of A.D. 1st century who attributed it to the eccentricity of the moon's path and proceeded to point out that the apogee advanced 3° in every anomalistic month. This datum would mean that it would take 9.18 years for the apogee to make a whole cycle and that an anomalistic month would be 27.55081 days.

The technique of relying on the diagram of the "nine roads of the moon" won the support of the scientist Zhang Heng of the late 1st century. The technique, obviously very popular during that time, was the manifestation of rudimentary knowledge of the shifting of the apogee. Calendar makers who followed this technique allowed the successive occurrence of three longer months or that of two shorter months. Primitive as it was, the technique guaranteed an unprecedented accuracy.

By A.D. 206, Liu Hong, the astronomer who was responsible for the design of the Qianxiang Calendar and who pioneered in bringing the knowledge of the fluctuation of the lunar motion into calendar formulation, allotted 27.55336 days to an anomalistic month. The figure, near the 27.55445 days arrived at by modern observations and computation, was obtained through calculations made using the datum that the perigee would advance 3 1/19 degrees in one anomalistic month. Liu and his co-workers first observed the moon's daily advance along its path, worked out the difference between this value and its average velocity, and ascertained the total of these differences. The sum of the average plus the total of all the differences observed in the period starting from the last perigee to the day preceding a given day would be the figure for the predicted advance of the moon in that day. Liu's formula made it possible to ascertain the celestial longitude of the new moon or the full moon and, what is more, to predict the solar and lunar eclipses — an endeavour which this group of calendar experts took as their main task.

Ancient Chinese astronomers also studied the length of the nodical month for the prediction of the eclipses. In 462, Zu Chongzhi gave his figure for a nodical month — time for the moon to travel from one node between the ecliptic and its own path to another. Zu's figure of 27.21223 days differed only 0.00001 day from that obtained by modern astronomers. Basing themselves on these achievements, calendar makers of later generations worked and secured high precision in their researches on the value of the nodical month.

Zhang Zixin's discovery in the 6th century of the asymmetry in the sun's velocity further facilitated study on the duration of a lunation. Liu Zhuo and Zhang Zhouxuan of the 7th century initiated a significant reform in calendrical science by taking into account the fluctuations in the speed of the sun and of the moon when they tried to ascertain the real positions in order to determine the first day of the month.

It was He Chengtian who designed the Yuanjia Calendar published in 443 that proposed making the day of the sun's and the moon's occurrence on the same longitude the first day of every month instead of basing on the mean value of a synodical month. He's proposal was, however, neglected. Once adopted by the makers of the Wuyin Calendar of 619, this principle was not officially accepted until 664 when the Linde Calendar came into use, ending a two-century-long debate.

The 8th-century Buddhist astronomer Yi Xing improved on the formula initiated in 604 by Liu Zhuo for the fixation of the first day of the month. Later in the 13th century, Guo Shoujing challenged Liu's hypothesis that within a short period both the sun and the moon travelled at a symmetrically accelerated speed. He published his own hypothesis that the motion of the sun or the moon might be a function of the second degree instead of one of the first degree of time and that the gain which these two illuminators made in their velocity might be a third-degree function of time. This prominent scientist Guo Shoujing thus brought Chinese astronomical and calendrical science to an unprecedentedly high level by applying Chinese traditional mathematics to the formulation of calendars.

Researches on the Eclipses

Knowledge about the hows and whys of the solar and lunar eclipses is traced back to very ancient times. *Yi Jing* (*Book of Changes*), believed to have been compiled in the 2nd century B.C., states: "There is no lunar eclipse unless the moon is full." Two lines attributed to an anonymous 6th century B.C. poet read:

> As for the eclipse of the moon,
> There is always a law governing it.

Shi Shen, astronomer of the 4th century B.C., noted that a solar eclipse occurred only on the day of the new moon. In the 1st century B.C. Liu Xiang, in his notes to the classics, affirmed definitely: "It is the moon that covers the sun to make a solar eclipse." By the end of A.D. 1st century Zhang Heng the scientist had clearly explained the cause of the lunar eclipse in his book *Ling Xian* (*The Spiritual Constitution of the Universe*). He said that since the moon reflects the sunshine, it will be eclipsed when it travels into the shadow cast by the earth. Shen Kuo, author of *Meng Xi Bi Tan* (*Dream Stream Essays*), noted in the 11th century that the ecliptic and the lunar path are not in the same plane but intersect, and that no eclipse will occur unless the two are in the same longitude and approximately the same latitude, that is, unless they intersect. Shen further pointed out that whether it will be a full or partial eclipse depends on the precision of the intersection.

Chinese calendar makers were aware of a law governing the eclipses before the 3rd century B.C. Using their original methods in ascertaining the eclipse cycle, they obtained figures which led the world in accuracy. The figure for the eclipse cycle, the common multiple of the synodical month and the nodical year, was set at 135 synodical months by the makers of the Santong Calendar of 7 B.C. In 1199 Chinese astronomers obtained through their own research the figure which in the West was known as saros (223 synodical months). And the designers of the Wuji Calendar of the 8th century anticipated the Western astronomers in arriving at the more accurate value of 358 synodical months, or just twice the figure of the Newcomb period.

Ancient Chinese astronomers were not content with broad prediction of the eclipses but proceeded to develop a systematic method of computation to provide more accurate predictions. Scientists who made the Qianxiang Calendar of 206 obtained the value 6° for the intersection angle between the ecliptic and the lunar path, a value of considerable precision for that time. They also predicted that a solar eclipse could not occur unless the moon comes as near as 15° to the nodal point. This formula was accepted by later generations of astronomers and constituted the concept of the eclipse limit. As the researches on the solar and lunar motions advanced, the prediction of the eclipses became ever more accurate due to the improved techniques applied. The Jingchu Calendar formulated by Yang Wei in 237 was the first to include the degree of obscuration and the specific part of the disk where the first contact was to occur in every eclipse prediction. Liu Zhuo at the beginning of the 7th century pioneered in taking into account the parallax arising from the observer's distance from the centre of the earth. The 8th century saw Yi Xing's experiments on predicting an eclipse from different places. Guo Shoujing of the 13th century inherited the achievements made through hundreds of years to give more accurate predictions, and his technique was among the foremost in the world.

The Solar Terms and Intercalated Months

The solar terms and the intercalated months had very important places in China's traditional calendrical science. The ancient calendars were indigenously solar-lunar. For the resonance between the incommensurable tropical year, synodical month and the day, it was necessary to have intercalated months. In other words, this is because the period of 12 synodical months does not equal but is roughly 11 days shorter than a tropical year. *Shang Shu* (*Book of History*) mentions the intercalation of months in referring to the anecdotes of the Emperor Yao, believed to have reigned in the 21st century B.C. The system of the solar terms, an indispensable supplement to intercalation, was formulated later and in turn spurred the science of intercalation.

Even before the formulation of the simplest calendar, which was the quarter-remainder calendar, the ancient people told the seasons by observing the morning and evening stars and so readjusted the system of purely synodical months. They intercalated a month when obvious dissonance was observed. Books like *Yue Ling* (*Monthly Ordinances*) and *Xia Xiao Zheng* (*Lesser Annuary of the Xia Dynasty*) provide such information. The stars at the zenith and the handle of the Great Dipper were used as the chief indicators. No definite rule, however, was discovered, for inaccuracy was inevitable when the observers could but rely on the naked eye. Intercalation was done whenever it was felt necessary, but for convenience the intercalated month usually was put at the end of the year. Additional knowledge on the law of the resonance between the synodical month and the tropical year was acquired by experience. By the 7th century B.C. Chinese calendar makers started applying the principle of seven intercalations in every 19 years, a system known to the Western world only 200 years later.

The Taichu Calendar of the 2nd B.C. provided that when a month covering no *zhongqi* (one of the 12 among the 24 solar terms) was to occur, that month should be counted as an intercalated or extra month. This provision made it possible to make intercalations on a principle more rational scientifically and to keep the rate of dissonance within half a month. The system of seven intercalations for 19 tropical years facilitated the improvement of the quarter-remainder calendar, and this resonance period was accepted as an important factor in calendar making until the 6th century. The achievements in research during the period from the 1st to the 6th century led to greater accuracy in the calculations on the length of the tropical and sidereal years and the synodical month. They also made it possible and necessary to adopt new methods. In fact the practice of intercalating a month whenever the comparison between the system of synodical months and that of the 24 solar terms warranted rendered the resonance period unnecessary. Li Chunfeng of the 7th century took the lead in this transition.

The invention of a system of the solar terms must be attributed to the Chinese working people who made observations during long experience in production. Those terms, directly referring to the ecliptic position of the sun, are good indicators of the season, and that is why they were so valued by the industrious ancient Chinese, who were near the land. Among the total of 24, the two solstices were the first to be established and then the two equinoxes. Documentary evidence now available suggests that the solar terms gradually came into recognition by the 3rd century B.C. *Lü Shi Chun Qiu* (*Master Lü's Spring and Autumn Annals*) compiled in that century listed most of those terms, though it was the book *Huai Nan Zi* (*The Book of the Prince of Huai Nan*) of 120 B.C. that first mentioned them all. Another clue to the specific time of the inauguration of those terms is the setting by the Zhuanxu Calendar of 221 B.C. of the spring equinox, one of those terms, as the beginning of the year. We can thus assume that the entire system emerged before 221 B.C. — the year when China was unified into a single empire.

Below is a list of the 24 solar terms:

Chinese Name	Translation	Beginning
Lichun	Beginning of Spring	4 Feb.

Chinese Name	Translation	Beginning
Yushui	Rain Water	19 Feb.
Jingzhe	Waking of Insects	6 March
Chunfen	Spring Equinox	21 March
Qingming	Pure Brightness	5 April
Guyu	Grain Rain	20 April
Lixia	Beginning of Summer	6 May
Xiaoman	Forming of Grain	21 May
Mangzhong	Grain in Ear	6 June
Xiazhi	Summer Solstice	22 June
Xiaoshu	Slight Heat	7 July
Dashu	Great Heat	23 July
Liqiu	Beginning of Autumn	8 Aug.
Chushu	Limit of Heat	23 Aug.
Bailu	White Dew	8 Sept.
Qiufen	Autumn Equinox	23 Sept.
Hanlu	Cold Dew	9 Oct.
Shuangjiang	Frost's Descent	24 Oct.
Lidong	Beginning of Winter	8 Nov.
Xiaoxue	Slight Snow	23 Nov.
Daxue	Great Snow	7 Dec.
Dongzhi	Winter Solstice	22 Dec.
Xiaohan	Slight Cold	6 Jan.
Dahan	Great Cold	21 Jan.

Of the 24 terms, 12 are called *jieqi* while the rest are called *zhongqi*. Slight Cold, the first of the *jieqi* group, is followed by Beginning of Spring as the next in the same group. It comes 30 days later than Slight Cold; i.e., 30° longitude in terms of the sun's ecliptical position. The span between each of the other 10 in this series is identical. Winter Solstice leads the *zhongqi* series arranged in the same way. As has been noted earlier in this article, when a month is to occur that contains no term of the second or *zhongqi* series, that month will be considered as an intercalated month. Eight terms — the equinoxes, the solstices and the beginning of the four seasons

— are more important than the others.   There is an interval of roughly 46 days between the beginning of each of them.

The "nine nine-day periods" and the "three 10-day periods" have been significant supplements to the Chinese traditional calendar that have made it preferred for use in daily life.   They are, directly or indirectly, related to the sun's ecliptical position.   The "nine-day periods" start on Winter Solstice, Spring Equinox arriving soon after the total of 81 days are over.   The "10-day periods" begin with the third *geng* day in the 10-day series of the sexagenary cycle after Summer Solstice.   A popular saying to this day is: "The severest cold comes during the third nine-day period after Winter Solstice and the scorching heat occurs during the second 10-day period after Summer Solstice."   Zu Geng, working in the 6th century, looked into the cause of the coldest or warmest day not being the day of either solstices but occurring later.   He found that it was due to the "accumulation of cold and heat".

The system of the 24 solar terms, unique in the world, suggests the actual level of both production and science in ancient China. This has also been shown by data of high accuracy in Chinese calendar making and the appreciably adequate devices designed by the Chinese astronomers for the resonance between the synodical months and the tropical year.   Traditional Chinese calendrical science is, as we have seen, an important part of the world's wealth of knowledge in this branch of astronomy.   A challenge remains, however, to the scientists of China and other countries to carry out further research.

SOME IMPORTANT CALENDARS IN ANCIENT CHINA

Name	Designed by	Published in the Year	Tropical Year (days)	Synodical Month (days)	Used in the Period
Zhuanxu	—	221 B.C.	365.2503	29.53085	221-104 B.C.
Taichu	Deng Ping	104 B.C.	.2502	.53086	104 B.C.-A.D. 84
Sifen (Quarter = Remainder)	Bian Xin	A.D. 85	.2500	.53085	85-220, 221-263 in Shu Kingdom and 220-236 in Wei Kingdom
Qianxiang	Liu Hong	206	.2462	.53054	223-280 in Wu Kingdom
Jingchu	Yang Wei	237	.2469	.53060	237-265 in Wei Kingdom, 265-420 in Jin Dynasty, 420-444 in Liu Song Dynasty, 386-451 in Northern Wei Dynasty
Sanji	Jiang Ji	384	.2468	.53060	384-517 in Later Qin Kingdom
Yuanshi	Zhao Fei	412	.2443	.53060	412-439 in Northern Liang Kingdom, 452-522 in Northern Wei Dynasty
Yuanjia	He Chengtian	443	.2467	.53059	445-479 in Liu Song Dynasty; 479-503 in Qi Dynasty; 502-509 in Liang Dynasty
Daming	Zu Chongzhi	462	.2428	.53059	510-557; 557-589
Daye	Zhang Zhouxuan	597	.2430	.53059	597-618
Huangji	Liu Zhuo	604	.2445		
Wuyin	Fu Renjun	619	.2446	.53060	619-664
Linde	Li Chunfeng	665	.2448	.53060	665-728
Dayan	Yi Xing	728	.2444	.53059	729-761
Mingtian	Zhou Zong	1064	.2436	.53059	1065-1067
Jiyuan	Yao Shunfu	1106	.2436	.53059	1106-1127; 1133-1135
Revised Daming	Zhao Zhiwei	1182	.2436	.53059	1182-1234 in Kin Dynasty, 1215-1280
Tongtian	Yang Zhongfu	1199	.2425	.53067	1199-1207
Shoushi	Guo Shoujing	1281	.2425	.53059	1281-1383; 1383-1644
Shixian	Shcall von Bell	1645	.2422	.53059	1645-1723

Mathematical Classics

Du Shiran

Ancient China made great strides in mathematics, as in other sciences and technologies, as evidenced by the number of well-known mathematical classics handed down. The earliest, *Zhou Bi Suan Jing* (*The Arithmetical Classic of the Gnomon and the Circular Paths*) and *Jiu Zhang Suan Shu* (*Nine Chapters on the Mathematical Art*), were written near the beginning of the Christian era, about 2,000 years ago. Their existence today and continuous use and dissemination of the largely intact texts is itself no small accomplishment.

Mathematics was first studied and taught through the medium of hand-copied textbooks. The development of printing in the Northern Song period (960-1127) ushered in printed books on mathematics which are probably the earliest such works in the world. Today, some Southern Song (1127-1279) printed copies of five ancient mathematical classics including *Zhou Bi Suan Jing* and *Jiu Zhang Suan Shu* are preserved as rare cultural relics in the Beijing, Shanghai and Beijing University libraries.

Classical works on mathematics appeared successively in the Han, Tang, Song and Yuan (206 B.C. to A.D. 1368) dynasties. These works are either the traditional Chinese annotations on extant mathematical works, which added new approaches to the subjects of the annotators' own, or they were fresh works based on new theories or ideas on other topics.

Ten Major Mathematical Works

Suan Jing Shi Shu (*Ten Mathematical Manuals*) are the 10 major mathematical works produced in the millennium from the Han through the Tang Dynasty (206 B.C.-A.D. 907). They were used as textbooks in the faculty of mathematics of the Imperial College from the 6th to the 10th century. Their titles are: *Zhou Bi Suan Jing*, *Jiu Zhang Suan Shu*, *Hai Dao Suan Jing* (*Sea Island Mathematical Manual*) written in the 3rd century, *Wu Cao Suan Jing* (*Mathematical Manual of the Five Government Departments*) in the 6th century, *Sun Zi Suan Jing* (*Master Sun's Mathematical Manual*) in the 4th century, *Xia Hou Yang Suan Jing* (*Xiahou Yang's Mathematical Manual*) in the 8th century, *Zhang Qiu Jian Suan Jing* (*Zhang Qiujian's Mathematical Manual*) in the 5th century, *Wu Jing Suan Shu* (*Arithmetic in the Five Classics*) in the 6th century, *Ji Gu Suan Jing* (*Continuation of Ancient Mathematics*) in the 7th century, and *Zhui Shu* (*Art of Mending*) in the 5th century.

The oldest of the 10 manuals is *Zhou Bi Suan Jing* of unknown authorship. Researchers hold that the book could not have been written later than 100 B.C. It is not a book on mathematics in the strictest sense but an astronomical book of the "hemispherical dome" school, envisioning heaven as a hemisphere over the earth. As a mathematical work it describes the right triangle (*gougu*) as used in primitive astronomical calculation for measuring height and distance by proportion. Fairly complicated fractional calculations were

Page from *Jiu Zhang Suan Shu* (*Nine Chapters on the Mathematical Art*) printed in the Southern Song Dynasty.

sometimes involved, indicating that these methods of calculation must have been known before 100 B.C. Of historical sources available, however, *Zhou Bi* proves to be the oldest.

The most important of the 10 mathematical manuals is *Jiu Zhang Suan Shu* (*Nine Chapters on the Mathematical Art*), which is a comprehensive discourse on traditional Chinese mathematics. Its influence on subsequent developments in traditional Chinese mathematics is no less profound than that of Euclid on Western mathematics. *Jiu Zhang* was the main mathematical textbook studied for more than 1,000 years. It was also used as a textbook in other countries before the introduction of Western learning.

The author of *Jiu Zhang* is again unknown. We do know however that additions and deletions were made in it by such distinguished mathematicians as Zhang Cang and Geng Shouchang early in the Western Han (206 B.C.-A.D. 24) period. Yet there is no mention of *Jiu Zhang* in the bibliographical chapter of *Han Shu* (*History of the Han Dynasty*) compiled around A.D. 100, though the bibliographical chapter mentions *Suan Shu* (*Mathematical Art*) written by Xu Shang and Du Zhong of Western Han. Some scholars are therefore of the opinion that Xu and Du contributed to the writing of *Jiu Zhang*. In any case, *Jiu Zhang* certainly took its final form gradually through a long historical period of repeated revisions, and that some of the methods contained could have appeared before Western Han. The book presents a total of 246 problems and solutions which fall into nine groups, the nine groups constituting the nine chapters as indicated by the book title.

In appraising the mathematical value of *Jiu Zhang* we first see that it gives the rules of the four basic operations (addition, subtraction, multiplication and division) of fractions and the rules of proportion, in which respects it ranked high in the world at the time of writing. Besides elaborating the properties of the right triangle in calculations mentioned in *Zhou Bi*, *Jiu Zhang* shows how to calculate areas of various shapes. Still, its most important achievement is in algebraic inventions. Square and cubic roots methods are stated on the basis of which the numerical solutions of monadic quadratic equations are worked out, provided the coefficient of quadratic term is not negative. A whole chapter of *Jiu Zhang* is devoted to simultaneous linear equations, and this is 1,500 years earlier than their counterparts in Western mathematics. The same algebraic principle is still taught in secondary schools today. The rules of addition and subtraction of positive and negative numbers are also given in this chapter — the first time that negative quantities were dealt with in world mathematical history.

Some of the methods employed in *Jiu Zhang*, fractional and proportional calculations for example, could have been transmitted first to India, then to medieval Europe via the Arab world. The method of *yingbuzu*, or the rule of double false positions, is cited in traditional Arab and European mathematical works as "the Chinese method". A classical work of world fame, *Jiu Zhang* has been translated and published in many languages.

The third of the *Ten Mathematical Manuals* is *Hai Dao Suan Jing (Sea Island Mathematical Manual)*. Written by Liu Hui during the period of the Three Kingdoms (220-280), the book deals entirely with measuring height and distance using the surveyor's pole. For each unknown value determined, usually two, three, or even four poles at different positions are used in surveying. This surveyor's manual laid the mathematical foundation for ancient Chinese cartography, which was quite advanced.

Many mathematical achievements of world significance are found in the other seven of the *Ten Mathematical Manuals*, such as the problem of "The Unknown Number of Things" (involving linear congruences) in *Sun Zi Suan Jing*, and the problem of "The Hundred Fowls" (also involving indeterminate analysis) in *Zhang Qiu Jian Suan Jing*. The solution of cubic equations in *Ji Gu Suan Jing* and especially the geometrical problem from which the equations are derived are also known for their distinctive Chinese features.

Zhui Shu (Art of Mending) written by the celebrated mathematician Zu Chongzhi was unfortunately lost in about the 10th century. When the *Ten Mathematical Manuals* went into print in the Song Dynasty, *Zhui Shu* was already missing and had to be replaced with *Shu Shu Ji Yi (Memoir on Some Traditions of the Mathematical*

Art) which was of doubtful authorship.[1]

The mathematical terms used in the *Ten Mathematical Manuals*, such as *fenzi* (the numerator in a fraction), *fenmu* (the denominator), *kaipingfang* (the extraction of the square root), *kailifang* (the extraction of the cubic root), *zheng* (the positive), *fu* (the negative), *fangcheng* (simultaneous linear equations, later also applied to higher equations), etc., are still in current use. Some have a history of nearly 2,000 years.

Mathematical Works in the Song and Yuan Dynasties

By the 10th century, the traditional Chinese mathematics had become comprehensively systematic after more than 1,000 years of development. It was crowned with the highest accomplishments during the Song (960-1279) and Yuan (1271-1368) dynasties, the brightest period in China's history of mathematics with fastest development, largest number of works and highest academic level.

Four great mathematicians emerged within a few decades in the later half of the 13th century: Qin Jiushao, Li Ye, Yang Hui and Zhu Shijie. What people call the Song-Yuan mathematics is chiefly represented by the works of these four, which include:

Qin Jiushao's *Shu Shu Jiu Zhang* (*Mathematical Treatise in Nine Sections*) 1247,

Li Ye's *Ce Yuan Hai Jing* (*Sea Mirror of Circular Measurement*) 1248, and *Yi Gu Yan Duan* (*New Steps in Computation*) 1259,

Yang Hui's *Xiang Jie Jiu Zhang Suan Fa* (*Detailed Analysis of the Mathematical Rules in the "Nine Chapters"*) 1261, *Ri Yong Suan Fa* (*Method of Computation for Daily Use*) 1262, and *Yang Hui Suan Fa* (*Yang Hui's Method of Computation*) 1274-75,

Zhu Shijie's *Suan Xue Qi Meng* (*Introduction to Mathematics*)

---

[1] The late Professor Qian Baocong was of the opinion that *Shu Shu Ji Yi* might have been written by Zhen Luan of the 6th century in the name of a Han writer; see Qian Baocong, *A History of Chinese Mathematics*, Science Publishing House, Beijing, 1964, p. 93.

1299, and *Si Yuan Yu Jian* (*Precious Mirror of the Four Elements*) 1303.

Shu Shu Jiu Zhang by Qin Jiushao solves numerical higher equations and also problems involving linear congruences — both highly significant. The latter solution, called "the Chinese Remainder Theorem", was accomplished more than 500 years earlier than its counterpart in Western mathematics. A problem in Qin's book calls for solving equations of the 10th order; some problems yield 180 answers! Li Ye's *Ce Yuan Hai Jing* and *Yi Gu Yan Duan* introduce another major achievement of Song-Yuan mathematics: *tianyuanshu*, a type of matrix formed by rows of counting-rods, for laying out all the coefficients in a higher equation, and for solving the equation. Li Ye also explicates the relationship between the hypotenuse, the altitude, the base of the right triangle and the diameter of its inscribed circle. Li's method in mastering geometrical problems by the use of algebra is peculiar to traditional Chinese mathematics.

Yang Hui's works present another facet of Song-Yuan mathematics: applied mathematics and simplification of the counting-rod arithmetical operations by a number of mnemonic rhymes. These reflected the social developments of his time and helped pave the way for the invention of the modern abacus. Zhu Shijie's *Suan Xue Qi Meng* is an introductory textbook starting from the simple but leading to the complicated. *Si Yuan Yu Jian* embodies another two epoch-making achievements: the solution of multivariate higher simultaneous equations, and higher order arithmetic series and the method of higher power finite differences. These were much used in ancient Chinese calculations for the calendar year and other astronomical calculations.

Comparing Song-Yuan mathematical achievements with similar work in Western mathematics leaves one amazed at the extent to which Chinese mathematics led the West. The solution of higher numerical equations, of multivariate higher simultaneous equations, and of problems involving higher power finite differences appeared in China 400 to 500 years before Horner, Bèzout and Newton made their debut in the same fields. We may safely conclude that before

1500 China was considerably ahead of the West in science and technology.

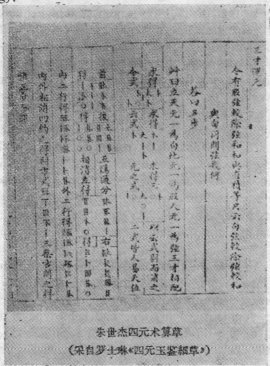

朱世杰四元术算草
(采自罗士琳《四元玉鉴细草》)

Page from *Si Yuan Yu Jian* (*Precious Mirror of the Four Elements*).

Many traditional mathematical works were also published in the Ming (1368-1644) and Qing (1644-1911) dynasties, the most famous being *Suan Fa Tong Zong* (*Systematic Treatise on Arithmetic*) 1592, by Cheng Dawei of the Ming Dynasty. A complete treatise on the abacus in its modern form, Cheng's book was very much in demand when it appeared in print. Although mathematical works published in the Qing Dynasty were numerous, they were no match for their Song and Yuan predecessors in creativeness. This reflected the decline in Chinese mathematical studies which occurred when China's progress was stopped by domestic feudalism and foreign imperialism in joint oppression. After 1,000 years of advance China lost the lead only in the past few hundred.

The Decimal Place-Value Numeration and the Rod and Bead Arithmetics

Mei Rongzhao

Chinese traditional mathematics has since ancient times laid emphasis on practical calculation and made remarkable achievement in this. Worth mentioning are the decimal place-value numeration and the rod and bead arithmetics, which are highly efficient and have done much in promoting mathematics.

Since the existence of written language, Chinese numeration has proved to be positional based on 10. In the oracle-bone inscriptions of the Yin time (c. 14th-11th century B.C.) and bronze inscriptions of the Western Zhou Dynasty (c.11th century-770 B.C.) all natural numbers within a hundred thousand could be expressed with the aid of a few symbols which stood for one, two, three, four, five, six, seven, eight, nine, ten, a hundred, a thousand, and ten thousand respectively plus a few compositions. For instance, the number two thousand six hundred and fifty-six was inscribed as 名 囪 文 ∩ on the bone; and the number six hundred and fifty-nine was 傘 予 帀 予 分 on the bronze. The two ancient forms were no doubt decimal place-value forms. They would be quite the same as the modern forms if the few place-value terms were deleted.

The Spring and Autumn (770-476 B.C.) and the Warring States (475-221 B.C.) periods in China witnessed the change from the slave to the feudal society. In agriculture the regular nine-square (in the pattern of the character 井, or well) farms broke down into small, irregular plots which made land mensuration essential. A more accurate calendar called for precise mathematical calculations. Technical innovation in both agriculture and handicrafts brought

57

about a new upsurge in social productivity which was naturally followed by commodity transactions and the issue of a currency. All these made new ways and means of calculation an acute problem. It was during this period that the counting-rods and rod arithmetic were developed. This is proved by the rod numerals cast on the coins then in circulation, and the appearance of such characters as *suan* and *chou*, or "to calculate" and "the counting-rod" in many literary classics written in that period. (No earlier bone or bronze inscriptions of either are known.) Much cited is the saying of Lao Zi in *Dao De Jing* (*Canon of the Dao and Its Virtue*) at the end of the Spring and Autumn Period: "Good mathematicians can do without the counting-rods."

The best and earliest account of the decimal place-value numeration is found in *Mo Jing* (*The Book of Master Mo*) written in about 330 B.C.: "One is less than two yet more than five. Explanation is given under 'establishing a position' " (the last 'one' is in a higher-than-unit position); and "There are ones in five and fives in one. The last 'one' is at the next higher place and therefore contains two 'fives'. "

According to "Memoir on the Calendar" in *Han Shu* (*History of the Han Dynasty*), the counting-rods were round bamboo sticks 0.23 centimetre in diameter and 13.86 centimetres long. 271 of these sticks bound together into a hexagonal bundle are conveniently held in the hand. By the 6th century, however, they had become shorter and square or rectangular in cross-section according to *Shu Shu Ji Yi* (*Memoir on Some Traditions of Mathematical Art*) of the 6th century, and also *Sui Shu* (*History of the Sui Dynasty*). The changes had been natural since shorter sticks took far smaller space in complicated calculations while square or rectangular shape prevented them from rolling about. Besides bamboo, the sticks could be made of wood, cast iron, jade or ivory. A carrying bag or container was usually attached. In the Tang Dynasty (618-907) all civil officials and military officers were obliged to carry a bagful of counting-rods with them wherever they went. In recent archaeological excavations a number of ancient counting-rods have been unearthed. More than 30 rods were found in August 1971 in

Qianyang County of Shaanxi Province. Dating back to the reign of
the Emperor Xuan Di (73-49 B.C.) of the Western Han Dynasty,
they are true to the descriptions in *History of the Han Dynasty*
except that they are made of bone. The bundle of rods unearthed

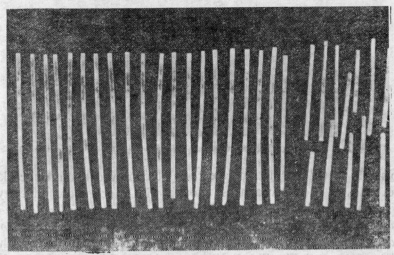

Counting-rods of the Western Han Dynasty unearthed in Qianyang
County of Shaanxi Province. They are made of animal bones.

in early 1975 in the No. 168 Han tomb at Fenghuangshan in Jiangling
County of Hubei Province are made of bamboo, but a bit longer
than the Shaanxi rods. They date from the reign of the Emperor
Wen Di (179-157 B.C.).

In rod arithmetic the rods could be placed either upright with
| || ||| |||| ||||| T TT TTT TTTT equivalent to modern 1, 2, 3, 4, 5, 6, 7, 8,
9 respectively, or horizontally with — = ≡ ≣ ≣ ⊥ ⊥ ⊥ ⊥ playing
the same roles. In fact, the rod digits were always placed alternately
upright or horizontally to avoid confusion in reading; the upright
for units and hundreds, the horizontal for tens and thousands, etc.
For instance, 6,708 would be ⊥ TT TTT, the blank space standing
for "zero". All arithmetical operations and even algebraic extrac-
tions of square and cubic roots could be thus carried out and traced
step by step with rods.

The same rod symbol had different absolute values depending on the position in which it occurred. For instance, = || = || was two thousand two hundred and twenty-two. The whole system was the same as the modern system except for the difference in sym-

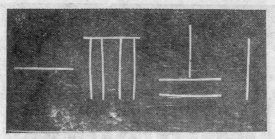

Counting-rods denoting the figure 1,971.

bols. In actual calculations the rods were moved about quite like the beads in the later abacus, though with much more difficulty. Systematic treatises on rod arithmetic were *Sun Zi Suan Jing* (*Master Sun's Mathematical Manual*) written in the 4th century; *Xia Hou Yang Suan Jing* (*Xiahou Yang's Mathematical Manual*) written in the 5th century, and *Shu Shu Ji Yi* written in the 6th century. In the 1st century, negative numbers appeared in calculations as black negative rods in contrast to the red positive rods. Later rod arithmetic was developed to solving algebraic linear, quadratic, cubic or even higher equations of nth order, having well entered the field of algebra. Some equations involved two, three, four, or even more unknowns. In short, the counting-rods were important mechanical aids of calculation which enabled ancient Chinese mathematicians to score many great achievements. To find the ratio of the circumference of a circle to its diameter, Zu Chongzhi gave values of π up to six significant figures after the decimal point. He had to calculate for the length of the 12,288 sides of a regular polygon inscribed in a circle and had to extract the square roots of nine-figured numbers for twenty-two times. Zu might not have succeeded had it not been for the aid of decimal positional numeration and rod arithmetic.

Ancient Babylonian numeration was also positional but to the power of sixty, hence very complicated in calculation. In ancient

Egypt, there were only two numerical symbols for the numbers from one to ten; only four symbols for the numbers from 100 to 10,000,000. Moreover, all ancient Egyptian numerical symbols were pictographs. The number 100,000 was signified for instance by the sketch of a bird. Despite the high level of civilization in ancient Greece the numeration there was backward due to emphasis being laid on logic or geometry and not on practical calculation. All numbers from one to ten thousand were written with the Greek alphabet. In case there were not enough letters to go round, the symbol "$_\iota$" was placed before them. One thousand was therefore "$_\iota\alpha$" and two thousand "$_\iota\beta$", etc. The modern standard numeration in use throughout the world was first developed long ago in India, where, however, before the 3rd century both the Greek and Roman numerations had been in use, and neither positional. In India a real decimal place-value system appeared only at the end of the 6th century, making China the oldest site of decimal place-value numeration. This system, together with the ancient Chinese rod arithmetic, has been prominent in world mathematical history.

Although the counting-rods were long in use in China and played an important part in the production, scientific experimentation and daily life of the Chinese people, and despite the inconvenience of some functions such as the solving of algebraic equations being assumed by the later abacus, their shortcomings were obvious. Inconvenient and for outdoor use, they took a large space when the calculations happened to be a little complicated. Errors occurred when the rods were moved too hastily. And, as time went by and society developed, these shortcomings became increasingly intolerable. Improvements on the rods became the order of the day. However, it took a good 700 years, or the latter third of the 2,000-year history of the rods, to even partially replace them with the modern Chinese abacus. (Algebraic calculations were later carried out with the aid of Chinese numerals on paper.) Improvements on rod operations were pioneered among commercial circles in the middle of the Tang Dynasty (roughly the 7th century). Throughout the dynasties of Song and Yuan from the 10th to the 14th century, computational mnemonic rhymes thrived urging the simplification of rod operations, but it was not until the Ming Dynasty (1368-1644) that the abacus became a household instrument. According to *Xin Tang Shu* (*New History of the Tang Dynasty*) written in 1060, and *Song Shi* (*History of the Song Dynasty*) written in 1343-45, a large number of mathematical works were produced during this 700-year period of transformation from rods to beads. Unfortunately most of the books fell victim to feudal obscurantism and have been lost. The fragmentary materials that are available tell us that the transformation was not initiated by the invention of new mechanical aids, but by shortcutting and quickening the rod operations with the aid of mnemonic rhymes which culminated in the birth of the modern Chinese abacus.

It is obvious that rod arithmetic was the predecessor of bead arithmetic. In the rod arithmetic, one upright rod above was equal to five rods lying transversely below it. An upper transverse rod was equal to five rods below and perpendicular to it. In the abacus, each of the beads above the long transverse bar dividing the abacus into the upper and lower parts is therefore equivalent to the five beads below it in the same vertical column. For multiplication and especially division in rod arithmetic it was sometimes necessary to indicate a number equal to or more than 10 at a single digital place, since carrying a 10 to the next higher digital place would be inconvenient. For instance, in 26,532÷8, the first step according to the dividing rhyme would be adding a 4 to the digit second from left in the dividend,[1] making the second from left digit 6+4=10. But it would be

---

[1] The rhyme shortcuts the steps in rod (and bead) division. Shen Kuo (1031-1095) in his *Meng Xi Bi Tan* (*Dream Stream Essays*) mentions the *zeng-cheng* method in division, which is the method of making up for the deficit (*zeng*) and depositing the surplus (*cheng*) in the dividing process. If the divisor is larger than the dividend at a certain digital place, the difference between 10 times the dividend and the product of the dividend and the divisor is called a "deficit", which should be made up by adding it to the next lower digit. In our particular case 10×2=20; 20-2×8=4; 4 therefore should be added to the 6 at the second from left digital place. Formerly in rod division the rods had to be placed in three rows: the quotient, the dividend, and the divisor. By the *zhengcheng* method the operations could be simplified to within a single row. This was important in leading to the birth of the modern Chinese abacus.

inconvenient to add a rod to the first from left digit (since so doing might cause confusion in the following steps), so 10 would have to be indicated in the second digital place; i.e. || ⊥ |||| ≡ || would have to be reshuffled into || 〇 |||| ≡ || . That is why in the modern Chinese abacus there are two upper beads in every vertical column.

A number of the computational mnemonic rhymes now applicable to the modern Chinese abacus were originally written for counting-rod operations, especially those written in the 13th and 14th centuries. Yang Hui in his *Cheng Chu Tong Bian Ben Mo* (*Origins and Details of Various Methods in Multiplication and Division*) of 1274 writes: "All calculations are carried out by placing the rods either upright or horizontally." Zhu Shijie writes in his *Suan Xue Qi Meng* (*Introduction to Mathematics*) of 1299: "Read the numbers accurately by the vertical or transverse counting-rods." Still later rhymes such as those in *Ding Ju Suan Fa* (*Ding Ju's Arithmetical Methods*) of 1355, He Pingzi's *Xiang Ming Suan Fa* (*Explanations of Arithmetic*) of 1373, and Jia Heng's *Suan Fa Quan Neng* (*All-Capable Mathematical Methods*) written in about 1373 are all rather comprehensive computational mnemonic rhymes for multiplication and division; yet none of them mention the abacus. On the contrary, draft calculations in the form of rod demonstrations are carried in He Pingzi's *Explanations of Arithmetic*. However, the growing popularization of the rod rhymes made the shortcomings of the counting-rods increasingly conspicuous. A more convenient mechanical aid became necessary.

A description of the abacus is given in *Lu Ban Mu Jing* (*Lu Ban's Manual of Carpentry*) written in the mid-15th century as follows: ". . . length of the abacus, one *chi*[1] and two *cun*[2]; width, four *cun* and two *fen*[3]; height, nine *fen*; thickness of the frame bars, six *fen* . . . . The upper and lower bars of the frame are connected by sticks forming a series of parallel columns. Each of the vertical columns

1 One *chi* was then equivalent to 0.321 metres.
2 1 *cun* = 1/10 *chi*.
3 1 *fen* = 1/10 *cun*.

has threaded on it seven slightly flattened wooden beads. A transverse wire is fixed one *cun* and one *fen* from the top bar. This wire divides the seven beads on each column into two unequal numbers, two above the wire and five below it."

Computational mnemonic rhymes written after the 15th century are all meant for the abacus. There are Xu Xinlu's *Pan Zhu Suan Fa* (*Arithmetical Methods of the Beads*) of 1573, Ke Shangqian's *Shu Xue Tong Gui* (*Rules of Mathematics*) of 1578, Zhu Zaiyu's *Suan Xue Xin Shuo* (*A New Treatise on the Science of Calculation*) of 1584, and Cheng Dawei's *Zhi Zhi Suan Fa Tong Zong* (*Systematic Treatise on Arithmetic*) of 1592. Cheng Dawei's book has been the most popular of all.

The abacus was mentioned in dramas or other literary works written in the latter half of the Yuan Dynasty (1271-1368). A special short poem about the abacus was written by Liu Yin in *Jing Xiu Xian Sheng Wen Ji* (*Collected Literary Works of Master Jingxiu*) in 1279. Tao Zongyi in his *Zhui Geng Lu* (*Talks in the Intervals of Ploughing*) in 1366 complains about his serving maids being "as lazy as the beads of an abacus, which never move unless one pushes them about". In *Yuan Qu Xuan* (*Selected Yuan Dramas*) appears "Master Pang's Blunder in Advancing to Others a Loan of Years in His Next Life" in which Master Pang asks his beneficiary to cancel some of the years of his next life on the abacus. Mentioning the abacus in literary works could have been motivated by the fact that it was still a novelty, or because the instrument was already fairly popular. In any case we may conclude that the abacus is a product of the 14th century, that towards the end of the Yuan Dynasty it had become fairly popular, and that from the Ming Dynasty it has been in wide and general use.

Some scholars outside China are of the opinion that the abacus and bead arithmetic first appeared in the Han Dynasty (i.e. before 220). These scholars base their view on a book entitled *Shu Shu Ji Yi* which we have mentioned above. The book was formerly believed to be written by a certain Xu Yue in the Han Dynasty, then annotated in the 6th century by Zhen Luan of the Northern Zhou

Dynasty (557-581). But the late Professor Qian Baocong[1] took a deeper look into the matter and found that *Shu Shu Ji Yi* was actually written by the annotator Zhen Luan himself under the pseudonym of a so-called Han Dynasty writer. In that sense the book is apocryphal. Numerical calculations in the Northern Zhou Dynasty were still carried out in rod operations. Multiplications and divisions were very complicated with the rods in three rows. No proofs can be found of any mnemonic rhymes then in existence, let alone the abacus in its modern form. So the bead arithmetic mentioned by Zhen Luan was probably a mechanical aid to memory or at most used for simple additions or subtractions. Neither the bead arithmetic as an instrument or its operation could possibly have been what they were 800 years later.

The abacus and bead arithmetic were early transmitted to Korea and Japan and played a certain role in the development of calculation techniques in those countries. In the middle of the 17th century the Japanese removed the second bead from the upper part of the abacus and made all the beads rhombic in form.

---

1 See Qian Baocong, *A History of Chinese Mathematics*, Science Publishing House, Beijing, p. 93.

The Out-In Complementary Principle

Wu Wenchun

Ancient Chinese geometry with its long history, rich content and many achievements forms a school of thought peculiar in style and systematically different from Euclidean geometry. Much of its history remains to be explored. However, the "out-in complementary principle" pervades it and is clearly defined in the following major classics handed down to date:

Zhou Bi Suan Jing (*The Arithmetical Classic of the Gnomon and the Circular Paths*), or *Zhou Bi* for short;

Jiu Zhang Suan Shu (*Nine Chapters on the Mathematical Art*), or *Jiu Zhang* for short;

Jiu Zhang Suan Shu Zhu (*Annotation on the Nine Chapters on the Mathematical Art*) by Liu Hui, or *Liu Zhu* for short;

Hai Dao Suan Jing (*Sea Island Mathematical Manual*), or *Hai Dao* for short;

Ri Gao Tu Shuo (*Theory with Diagrams of the Sun's Altitude*), or *Ri Gao Shuo* for short; and *Gou Gu Yuan Fang Tu Shuo* (*Theory with Diagrams of the Right Triangle Making Use of Circles or Squares*), or *Gou Gu Shuo* for short, both by Zhao Shuang.

As everywhere else, geometry in China arises from land mensuration and astronomical observation. These practices in ancient times gave rise to the calculation of planar areas and methods of surveying based on the properties of the right triangle. Later, solid figures were involved in earthwork, etc., leading to a theory of volumes. One of the characteristics of ancient Chinese geometry is its fairly high power of abstraction in formulating the seemingly most commonplace out-in complementary principle which arose from diverse experiences. It has, however, been applied successfully to solving problems of extreme diversity.

Simple Applications and
the Theory of Proportion

The essence of the so-called out-in complementary principle is the assumption of the following obvious facts: 1) The area of a planar figure remains the same when the figure is rigidly shifted to another place on the plane. 2) If a planar figure is cut into several sections, the sum of the areas of the sections is equal to the area of the original figure. It follows that the areas of the various sections involved before and after the out-in procedures possess simple arithmetic relations. The principle also applies to solid figures in space.

It is easy to apply this principle to obtaining the ordinary formula that the area of any triangle is equal to half the product of one side and the associated altitude. From this the area of any polygon can be calculated.

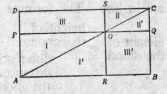

Another simple application is diagrammed as follows:

If △ACB is considered as △ACD shifted, and I'and II' as I and II shifted, then according to the out-in complementary principle III must be equal to III' in area, too.

Likewise, $\square$ PC= $\square$ RC, . . .

From this we know
OP × OS = OR × OQ, PQ × QC = RB×BC, . . .
Therefore AR:OQ = OR:CQ, AB:OQ = BC:QC, . . .
That is, the corresponding sides of the similar right triangles ARO and OQC and also of ABC and OQC are in proportion. From this we know that certain other corresponding parts are also in proportion.

Though these simple results are not explicitly stated in *Jiu Zhang*, they are time and again manifested in the solution of various practical problems (Ref. *Liu Zhu*).

Gnomon, Shadow and Double Differences

The method of using two gnomons to find the altitude of the sun is given in *Zhou Bi*. The formula appears below:

$$\text{Altitude of the sun} = \frac{\text{height of the gnomon} \times \text{distance between the gnomons}}{\text{difference between the lengths of shadows of the two gnomons}} + \text{height of the gnomon}$$

As shown in the following diagram:

A is the position of the sun, BI represents the ground level, ED and GF are the two gnomons, while DH and FI are the two shadows projected on the ground.

In *Hai Dao* the same method is used to measure the height of an island from the shore. In the same diagram above, AB is the height of the island, H and I are the observer's positions where the observer's eye, the tops of the gnomons and the top of the island are in line. The formula then becomes:

$$\text{Height of the island} = \frac{\text{height of the gnomon} \times \text{distance between the two gnomons}}{\text{difference between the distances of observer from the gnomons}} + \text{height of the gnomon}$$

Liu Hui's original proof and diagram have been lost. But we have pieced these together drawing inspiration from other sources as well as extant fragments of diagrams in *Ri Gao Shuo* to be roughly as follows:

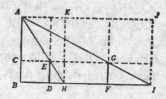

According to the out-in complementary principle, we know

$$\square \text{ JG} = \square \text{ GB} \tag{1}$$
$$\square \text{ KE} = \square \text{ EB} \tag{2}$$

(1)-(2) $\Box$ JG $-\Box$ KE $=\Box$ GD,

Therefore   (FI $-$ DH) $\times$ AC $=$ ED $\times$ DF,

That is

difference between
the distance of $\times \left(\begin{matrix} \text{height of} \\ \text{island} \end{matrix} - \begin{matrix} \text{height of} \\ \text{gnomon} \end{matrix}\right) =$
observer from the
two gnomons

$\begin{matrix} \text{height of} \\ \text{gnomon} \end{matrix} \times \begin{matrix} \text{distance between} \\ \text{the two gnomons} \end{matrix}$

From this we arrive at the formula for the sea island.

In *Hai Dao* altogether nine practical problems are listed, all having to do with the measurement of heights and distances. In all the nine formulae given, differences occurring from two observations are usually taken to be the denominator. Probably this is where the term "double differences" comes in. The other eight formulae can all be proved likewise on the out-in complementary principle.

Some of the problems carried in *Si Yuan Yu Jian* (*Precious Mirror of the Four Elements*), written by Zhu Shijie of the Yuan Dynasty 1,100 years later than *Hai Dao*, are essentially the same as the nine posed in *Hai Dao*. Zhu must have drawn heavily upon his predecessors' work. Careful analysis of Zhu's method as shown in the *tianyuanshu* brings us to the conclusion that Liu's proof of the

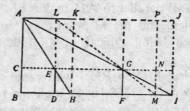

sea island formula is possibly somewhat more sophisticated than that given above. Accordingly, we suggest the following alternative proof to be considered as Liu's "original":

By the out-in complementary principle we have besides (1), (2) also

$\Box$ PG $=\Box$ GD in the diagram above.   (3)

From (1), (2) and (3) we get

□ JN = □ EB = □ KE,

Therefore IM = DH,   (4)

FM = FI − IM = FI − DH = difference between the distances
from observer to the two gnomons

From (3) we arrive at the formula for the sea island.

If done in the usual manner according to Euclidean geometry, an auxiliary line GM' should naturally be drawn parallel to AH to make the proving plain, as shown in the diagram on the right. The rest can then be proved by making use of the similar triangles and the theory of proportion.   In fact the proving of the formula

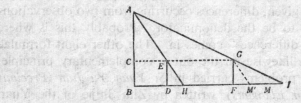

has been so traced by historians of mathematics in China and elsewhere in recent times, including Li Huang of the Qing Dynasty (1644-1911).   But this is surely not the original method of Liu Hui; it is in fact totally out of accord with the spirit of ancient Chinese geometry.   Note GM' parallel to AH makes FM'=DH.   The constructed point M' here and the M point taken for equation (4) are quite different, each being typical of an independent school of geometry.

The Italian priest Matteo Ricci who came to China near the end of the Ming Dynasty (1368-1644) took the teaching of Euclidean geometry as one of his academic missions.   In the book *Method and Theory of Surveying* dictated by him there appears a problem almost identical with the sea island problem.   However, instead of proving it according to the Euclidean method he takes without reason a point M on FI to meet the requirement of (4) above, then goes on to prove the formula by proportions.   This runs counter to Euclidean geometry but coincides with the Chinese tradition. Why Matteo Ricci should have done so is quite puzzling.

The *Gougu* Theorem

The Pythagorean theorem is called the *gougu* theorem in traditional Chinese geometry, and in both *Zhou Bi* and *Jiu Zhang* it is
clearly prescribed in the written texts: Multiply the shorter and
longer arms enclosing the right angle by their own values respectively
and add up the squares; the sum is equal to the hypotenuse multi-
plied by its own value; i.e., $gou^2 + gu^2 = xuan^2$.   Though the original proof has long been lost, we can still trace it from the texts of
Gou Gu Shuo, *Liu Zhu*, and especially from the few diagrams left
from Zhao Shuang.   It is clearly stated that the proof is based on the
out-in complementary principle; therefore it can be something like this:

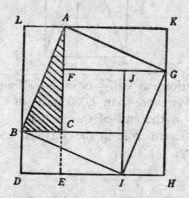

In the diagram on the left, ABC is the right triangle.   BCDE
is the square on the *gou* (the shorter arm), while EFGH is equal to
the square on the *gu* (the longer arm).   In the planar shape DBCFGH,
cut off the triangle △BDI and shift it to the position of △ABC; cut
off △ GHI and move it to the position of △ AFG. We then have
ABIG equal to the square of the hypotenuse AB, and hence the
gougu theorem.

In Euclid's *Elements of Geometry* the Pythagorean theorem
is proved as illustrated in the diagram below:

It is clear that before the Pythagorean theorem is tackled, a
lot of preparatory work must be done.   First, a few theorems with

regard to identical triangles and triangular areas must be established. That is why the Pythagorean theorem does not appear in the first volume of *Elements of Geometry* until near the end of the book. Euclid's book gives practically no applications of the theorem, but in ancient China the *gougu* theorem was widely employed

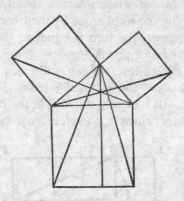

in diverse applications as early as in *Jiu Zhang*. It was a source of development over more than 2,000 years of Chinese mathematics (cf. the diagrams at the end of this article). The same theorem played quite a different role in the Eastern and Western systems of ancient geometry.

Gou, Gu, Xuan, Their Sums and Differences and Methods of Finding One from the Others

Gou, gu and *xuan*, the sum of and the difference between any two of the three, give out nine values. One can find the unknown from two knowns. Any one of the three sides can be found provided the other two are given. This is mainly a problem of extracting a square root. But the sum of or the difference between two sides is more often employed in solving practical problems such as those listed in the *gougu* chapter of *Jiu Zhang*:

1. Given the difference between *xuan* (the hypotenuse) and

gu (the longer arm), and *gou* (the shorter arm); find *xuan* and *gou*. Five problems are listed.

2. Given the difference between *gou* and *gu*, and *xuan*; find *gou* and *gu*. One problem.

3. Given the difference between *xuan* and *gou*, and *gu* respectively; find *gou*, *gu* and *xuan*. One problem.

4. Given the sum of *xuan* and *gu*, and *gou*; find *gu* and *xuan*. One problem.

Formulae are given for the problems in *Jiu Zhang*. The propositions in *Gou Gu Shuo* are of the same nature. In *Liu Zhu* proofs of the formulae are worked out, making use of the out-in complementary principle; sometimes also the theory of proportion. Take Problem No. 13 in the *gougu* chapter, the problem of the "broken bamboo", for example:

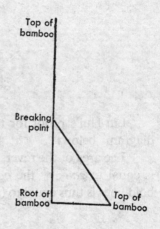

The height of the bamboo (*gu* plus *xuan*) is known. When bent the top touches the ground at a known distance from the stem (*gou*). Find the height of the break (*gu*).

The formula is given as follows:

$$xuan - gu = \frac{gou^2}{xuan + gu};$$

$$xuan, gu = \frac{(\text{sum of } xuan \text{ and } gu \pm \text{ difference between } xuan \text{ and } gu)}{2}$$

Liu Zhu provides another formula: $gu = \dfrac{(\text{sum of } xuan \text{ and } gu)^2 - gou^2}{2 \times \text{sum of } xuan \text{ and } gu}$.

To prove the former formula, see in the diagram below:

The side of the squares ABCD or AEFG is equal respectively to the *xuan* or *gu* of the right triangle. According to the *gou gu* theorem the area of EBCDGF is equal to *gou*². Shift □ FD to the position of □ CH, then according to the out-in complementary principle, the area of □ BH is equal to *gou*², while the longer and shorter sides of this rectangle are equal to the sum of *xuan* and *gu* and the difference between them respectively. From this we get the former formula.

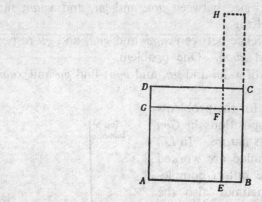

Liu Hui's proof for the other formula is done likewise. In the diagram below:

The area of the reversed L-shaped figure in the lower right corner is equal to gou^2 by the *gougu* theorem. The area bordered by the bold lines is thus equal to $(xuan + gu)^2 - gou^2$. Shift I to the position

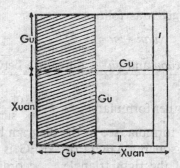

of II and we see according to the out-in complementary principle that this area is two times the shaded area; i.e., $2 \times gu \times (xuan + gu)$. The formula is therefore proved.

Qin Jiushao's Formula[1]

In Qin Jiushao's *Shu Shu Jiu Zhang*[2] (*Mathematical Treatise in Nine Sections*) 1247, there is a problem of finding the area of a scalene triangular plot. Given the three unequal sides of the triangle *da*, *zhong*, *xiao*, (the longest side, the medium side and the shortest side). Qin Jiushao's solution can be formulated as follows:

$$\text{Area}^2 = \frac{1}{4}\left[xiao^2 \cdot da^2 - \left(\frac{da^2 + xiao^2 - zhong^2}{2}\right)\right]^2.$$

Qin says nothing about the source of this formula. The proof of the formula has also been lost. Making use of the results and methods in *Liu Zhu*, we may infer the lost proof to be somewhat as follows:

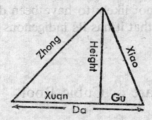

Draw an altitude of the triangle perpendicular to *da*, dividing *da* into two parts. Let the longer and the shorter parts be the *xuan* and *gu* of a right triangle. From *Jiu Zhang* we know the area of a triangle to be $^1/_2 \times$ altitude $\times$ *da*, therefore our problem becomes one of finding the altitude, then further boils down to finding the *gu* of that right triangle. Since

$$xuan + gu = da,$$
$$gou^2 = xuan^2 - gu^2 = zhong^2 - xiao^2,$$

our problem is the same as that of finding *gu*, given *gou* and the sum

---

1 Qin Jiushao was one of the greatest Chinese mathematicians of the 13th century.

2 A very important mathematical classic written by Qin, known especially for its treatments of numerical equations of higher degree and indeterminate analysis.

of *xuan* and *gu*. From Liu Hui's formula we have:

$$gu = \frac{(xuan + gu)^2 - gou^2}{2 \times (xuan + gu)} = \frac{da^2 - (zhong^2 - xiao^2)}{2 \times da}.$$

$$\text{Altitude}^2 = xiao^2 - gu^2 = xiao^2 - \frac{(da^2 + xiao^2 - zhong^2)^2}{2 \times da}.$$

From this we get Qin's formula.

Qin's formula looks rather odd. But the proof traced above is quite natural and perfectly in line with ancient Chinese mathematical tradition. We may even regard it as the original proof.

Heron's formula in Western geometry, however, is neat in form and good-looking:

Area of a triangle $= \frac{1}{4}\sqrt{(a+b+c)(b+c-a)(c+a-b)(a+b-c)}$,

where a, b, c are the three sides of the triangle.

Qin's formula is not likely to have been derived from Heron's, and we may conclude that it has its indigenous origin independent of Heron's influence.

Extracting the Square or Cubic Root

To find the hypotenuse from the two arms enclosing the right angle in a right triangle, we add up the two squares on the arms and extract the square of the sum. Thus the application of the *gougu* theorem inevitably leads to the extraction of the square root. In fact, in the ancient mathematical classic *Zhou Bi* the square roots of many concrete numbers are provided. Detailed steps in extracting square roots are stated in *Jiu Zhang*. The method is geometric, based on the out-in complementary principle. Suppose the task is to find the square root of the number 55,225. In geometry this is to find the side of a square the area of which is 55,225. Note the decimal system has long been in use in China. First we must decide on how many digits the root is going to have. The square root of a five-digit number has three digits. So our task is to ascertain the first, second and third digits successively. Since our number 55,225 lies between 40,000 and 90,000, its square root must lie between 200 and 300. Our first digit is therefore a 2. (In *Jiu Zhang* this process of ascertainment is called *yi*,[1] or "to suggest".) In the diagram let ABCD be the square the area of which is 55,225. On one side AB we take a point E and let AE be equal to 200. Draw the square AEFG. Cut off AEFG from ABCD. The area of the

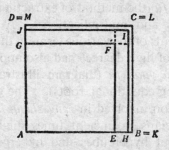

remaining inverted L-shaped figure is therefore $55,225 - 200^2 = 15,225$. We then suggest that the second digit be a 3. On EB we again take a point H making EH equal to 30. Draw the square AHIJ. Cut the inverted L-shaped figure into three parts: ☐ FH, ☐ FJ, ☐ FI. Their areas are respectively $30 \times$ EF, $30 \times$ FG, 30^2. But EF = FG = 200, so the area of the remaining inverted L-shaped figure is equal to

$$15,225 - (2 \times 30 \times 200 + 30^2) = 2,325.$$

Let us then suggest that the third digit be a 5, and on HB we take a point K making HK equal to 5. Draw the square AKLM. The area of the remaining inverted L-shaped figure, if any, must be

$$2,325 - (2 \times 5 \times 230 + 5^2) = 0$$

In that case K and B must coincide, and the square root of 55,225 is 235.

The same method is used in extracting the cubic root. It will of course be more complicated to dissect a cube but the principle is still geometric and still that of out-in complementation. The method is described in detail in *Jiu Zhang*.

These methods of extracting the square and cubic roots date back to very ancient times in China. They are clearly geometric

---

[1] Some say the character means "to discuss".

and display a superiority in the decimal place-value system of numeration employed.

By the middle of the 11th century Chinese mathematicians had already improved the methods of extracting the square and cubic roots to the solution of equations of higher degree. This is called *zeng zheng kai fang fa* (the method of extracting equational roots by successive additions and multiplications). A diagram illustrating the different coefficients of the various terms in the expansion formulae of binomial powers of high degrees had also appeared and was called *kai fang zuo fa ben yuan tu* (diagram illustrating the origin and method of extracting equational roots). The geometric nature and the high degree notion involved in *zeng zheng kai fang fa* show that Chinese mathematicians in ancient times might already have had primitive ideas about hypercubes and hypergeometry.

Quadratic Equations

In extracting the square root, we make use of the diagram on p.74. 2 × EF in the diagram is called the *dingfa*. Having obtained AE, we come to find EB from the known area of the inverted L-shaped figure EBCDGF. Shift □ DF to the position of □ CH, the area of □ BH is the same as that of the inverted L-shaped figure according to the out-in complementary principle. Note that the difference between the longer and shorter sides of □ BH is equal to 2 × EF (*dingfa*), which is also known. The problem of finding EB is therefore a problem of

(A) finding the longer and shorter sides of a rectangle, given its area and the difference between the two sides.

Conversely, the solution of problem (A) can be reduced to one as from the second step onwards in the method of square-root extraction, which in *Jiu Zhang* is called *kai dai cong ping fang fa*. The solution of (A) in *Jiu Zhang* is stated in the following words:

(B) "Take [the area of the rectangle] as *shi* and [the difference between the length and width] as *congfa*, then *kai fang chu zhi* (literally "to extract the square root" which means here *kai dai cong ping fang*) and the root is the [width]."

The term *congfa* comes from *dingfa* in extracting the square root. The term *kaifang* (root extraction) shows its origin.

The following problem is taken from *Jiu Zhang*. In the diagram on the right, ABCD is a square walled city. At point G there is a big tree of known distance in terms of human steps northward from the north gate (north steps for short). A man takes a definite number of steps southward out of the south gate (south steps for short), then turns west and also counts his steps till he is just able to see the tree (west steps for short). Find the length of each side of the square city. The answer given in *Liu Zhu* is obtained on the out-in complementary principle as follows: ☐ EJ = 2 ☐ EG = 2 ☐ KG = 2 × north steps × west steps. In ☐ EJ the difference between the length and the width is equal to the sum of north steps and south steps. The problem is thus reduced to one in the form of (A) above. According to *Jiu Zhang* its solution is as follows: Take 2 × north steps × west steps as *shi*, and the sum of north steps and south steps as *congfa*, *kai ping fang chu zhi* and we find the length of one side of the city as represented by EI in the diagram.

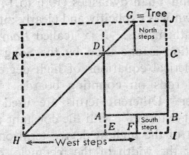

Not only the numerical value of problem (A) can be found by means of the *kai dai cong ping fang fa* method, but also a precise expression of the solution of (A) may be obtained on the out-in complementary principle. In fact, if in the rectangle we take the width as the *gou* and the length as the *gu* of a right triangle, then problem (A) becomes the following:

(C) Given the product of *gou* and *gu*, and the difference between them in a right triangle, find *gou* and *gu*.

Let us examine a diagram left by Zhao Shuang in which there are two squares the sides of which are equal to the sum of, and the

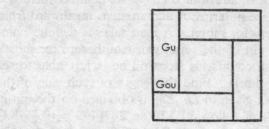

difference between, *gou* and *gu* of the right triangle respectively. We therefore have

$$(gou + gu)^2 = 4\,(gou \times gu) - (gu - gou)^2.$$

From this we get the sum of *gou* and *gu*, and *gou* and *gu* consequently. Similarly, *gou* and *gu* can be found given their sum and their product. Reference can be made to the last proposition in *Gou Gu Shuo*.

In the Song and Yuan dynasties (10th to 14th century) the notion of the unknown was explicitly and clearly introduced into traditional Chinese mathematics. If x (called *tianyuanyi*[1] then; while the *tianyuan* notation is one used by the Song algebraists for the expression of numerical equations of high degree. It is a way of arraying counting rods on counting boards. The array is of a "matrix" character. Different terms are used for distinguishing figures on different "storeys", with the constant term on the lowest, and the coefficient of the highest degree term on the highest storey above;) stands for the width of the rectangle, our problem (A) is equivalent to solving a quadratic equation of the form

$$x^2 + bx = c, \text{ with } b \text{ as } congfa \text{ and } c \text{ as } shi.$$

Ancient Chinese mathematicians furnished both numerical and accurate solutions to quadratic equations of the above type (with b

---

[1] *Tianyuanyi* has different meanings in the works of Song and Yuan dynasty mathematicians.

and c positive).   During the Song and Yuan dynasties the *kaifangshu* (method of root extraction) was extended to solving numerical equations of high degree.   As for the method of accurate solution of equations of higher degree, historical traces have long been lost. Judging from what Wang Xiaotong wrote in the early years of the Tang Dynasty (618-907) and from historical comments on Zu Chongzhi (429-500), we cannot totally rule out the possibility that geometrical approaches have been attempted with some success in accurate solution of cubic equations.

In other countries, the Arab mathematician Al-Khowārizmi in his well-known classic on algebra (A.D. 829) gives accurate solutions for quadratic equations of various types.   His method was geometrical in spirit, similar to ours on the out-in complementary principle.   Later, Italian mathematicians in the 16th century worked out solutions for cubic equations.   Their methods were also geometrical.

Theory of Volumes and Liu Hui's Principle

Since the area of a rectangle is the product of its length and width, it is easy to infer on the out-in complementary principle that

(1) the area of a triangle = $^1/_2$ × its height × its base.

It is also easy to derive further the formulae for areas of polygons.   All these fall within the category of plane geometry.

In solid geometry, however, although we know that the volume of a rectangular parallelepiped must be equal to its length × its width × its height, it is by no means definite whether we can on the out-in complementary principle reason that

(2) the volume of a tetrahedron = $^1/_3$ × its altitude × the area of its base surface, and hence form a theory for volumes of polyhedra. In fact this constitutes a most difficult problem in geometry which was presented as one of the 23 unsolved problems at the International Congress of Mathematicians in 1900 by the celebrated David Hilbert. This problem has been solved by Max Dehn who proved that besides being of equal volumes certain conditions must further be satis-

fied before two polyhedra can be cut into a number of mutually congruent smaller ones. These conditions have since been called Dehn's conditions. In 1965 the Swiss mathematician Sydler proved that Dehn's conditions are also sufficient. Even so, it appears that the problem may still be regarded as not yet satisfactorily settled. Dehn's conditions are too complicated to be accepted as final.

A probe into how the problem was dealt with by ancient Chinese mathematicians would probably provide us with some food for thought.

In both *Jiu Zhang* and *Liu Zhu* the starting point from which problems of polyhedra volumes are solved is to cut some regular

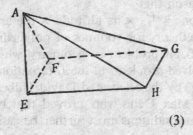

(1)

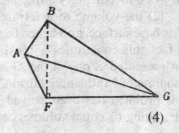

(2)

polyhedra into several basic solid figures which will be helpful in analysis. A rectangular parallelepiped can be cut diagonally (passing through two diagonally opposite edges) into two *qiandu* (right triangular prisms), as shown in diagrams (1) and (2). A *qiandu* in turn can be cut into a *yangma* (pyramid) and a *bienao* (tetrahedral wedge) as shown in (3) and (4). The basic features of a *bienao* are that it has AB perpendicular to the plane BFG, and FG perpendicular to the

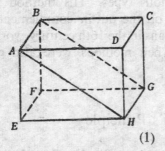

(3)

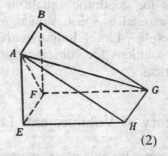

(4)

plane ABF as shown in the diagram. Since any polyhedron can be cut into tetrahedra and any tetrahedron can be cut into six *bienao* as shown in the diagram below, the whole problem boils down to finding the volumes of the *bienao* (and the *yangma*) so produced.

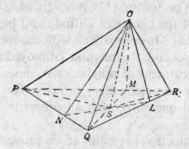

In Liu Hui's own words *yangma* and *bienao* are the "basic figures for the whole theory and practice involving volumes of polyhedra".

We then come to the problem of finding the volumes of *yangma* and *bienao*. If our parallelepiped is simplified into a cube, it will be easy to see that the volume of the pyramid cut from the prism is twice that of the tetrahedral wedge. Liu Hui proved in a long dissertation that this is the case not only in the *qiandu* from a cube, but in all *qiandu* alike. In Liu Hui's words, "In a *qiandu* the volume of the *yangma* is always twice that of the *bienao*." We may well call this statement Liu Hui's principle. In modern language, "If any rectangular parallelepiped is cut diagonally into two prisms, and the prisms are further cut into pyramids and tetrahedra, the ratio between the volumes of the pyramid and tetrahedron so produced is always 2:1."

From this principle it will be easy to arrive at the formulae for volumes of *yangma* and *bienao*. It is then no problem to prove formula (2) above. The whole theory for volumes of polyhedra may then be based on the principles of Liu Hui and of out-in complementation.

Liu Hui's long and detailed dissertation is proof of his principle, proof based on some limit considerations. What has been made clear by Hilbert and his followers can be construed as that volumes are different from planar areas in that the mere out-in complementary principle is insufficient for a satisfactory theory. In fact, it must be supplemented by some axiom or principle of continuity. Though in 1903 Shatunovsky argued that the principle of continuity could be omitted and that the foundation of the theory of polyhedra volumes could be built on formula (2), it nevertheless requires a proof of the independence of the choice of altitude and base which is neither plain nor trivial at all. In comparison with the method of exhaustion of the ancient Greeks and the method employed in Legendre's *Eléments*, Liu Hui's treatment of polyhedra volumes based on his principle and the out-in complementary principle can be safely regarded as the most natural one surpassing all others in simplicity and elegance.

It seems that much yet remains to be proved in the field of the polyhedra. It might be an aid if the conceptions and methods in ancient Chinese geometrical approaches were duly taken into account.

The *Xianchu* Theorem

The term *xianchu* (a wedge with trapezoid base and both sides sloping, see the diagram below) as well as other strange terms for polyhedra have come down from ancient Chinese architecture and earthwork.

In *Jiu Zhang*, volumes of polyhedra are calculated on the out-in complementary principle and by the *yangma* and *bienao* formulae.

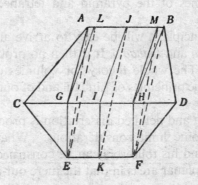

Take the *xianchu* in the diagram for instance.   ABCD form a trapezoid on the ground surface.   CDEF is another trapezoid in a plane
perpendicular to the ground.   ABEF is a slope.   The whole solid
ABCDEF in the form of a tunnel is *xianchu*.   Plane IJK is perpendicular both to the ground and plane CDEF.   It bisects *xianchu*
into two symmetrical parts.   EG, FH and KI show the depth of
xianchu.   IJ is the length of *xianchu* on the ground.   CD, EF and
AB are called the upper width, the lower width, and the hind width
of *xianchu*.   The formula for the volume of *xianchu* given in *Jiu
Zhang* is as follows:

$$\text{Volume of } xianchu = \frac{1}{6} \left(\frac{\text{upper}}{\text{width}} + \frac{\text{lower}}{\text{width}} + \frac{\text{hind}}{\text{width}} \right) \times \text{depth} \times \text{length}.$$

To prove this, Liu Hui in his book *Liu Zhu* cuts *xianchu* into several
parts, and supposes CD> AB> EF as in the diagram above.   *Xianchu* is therefore regarded as composed of a *qiandu* EFGHLM, two
small *bienao* AGEL and BFHM, and two big irregular *bienao* ACEG
and BDFH.   From formula (2) above and the formulae for *qiandu*
and *bienao*, the formula for the volume of *xianchu* is therefore obtained.   The same method is employed in *Jiu Zhang* in calculating
the volumes of *chumeng* (wedge with rectangular base and both sides
sloping), *chutong*, *panchi*, *minggu* (three variations of a frustum of a
pyramid with rectangular base of unequal sides), and other polyhedra.

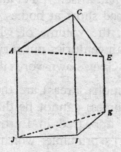

The formula of the *xianchu* volume is of special importance in
that half of the *xianchu* standing erect on the right triangular base
IJK will be equal to a right-angled prism cut slantwise at the upper
end.   Its volume will simply be the product of the average height

and the right triangular (*gougu* form) base. Now a pillar bounded at the top by any curved surface can be regarded as composed of such slant-topped prisms approximately. Therefore the integral approximate formula of a function f(x, y) can be obtained analogous to Simpson's integral approximate formula in the case of an area under a curve. This shows the particular significance of the *xianchu* formula.

In Western mathematics, the earliest formula for the volume of a pillar cut slantwise at the top appeared in 1794 in Legendre's *Elèments de géométrie*, and has since been called Legendre's formula. Legendre's book is the earliest work to take the place of Euclid's *Elements*. Legendre's proof of his own formula is also based on the volume of the tetrahedron but with different method of dissection from that in *Liu Zhu*. Reference can be made to both for comparison.

Volume of the Sphere and the Principle of Zu Geng

Within the 300 years or so between the writing of *Jiu Zhang* and that of *Liu Zhu* a fairly complete theoretical system with regard to volumes of polyhedra had arisen. Yet ancient Chinese mathematicians at that time stopped short at bodies bounded by curved surfaces, especially spheres, the volume of which remained unsolved till Zu Geng of the 5th-6th centuries put forward a famous principle named after him. In Zu Geng's own words the principle is as follows:

"If the *mi* (cross-sections, areas) are the same on the same *shi* (level), the *ji* (whole volumes) cannot be different."

The same principle appeared in Europe in the 16th century by the name of Cavalieri's principle, which was an important step towards the invention of calculus.

Zu Geng's proof for his formula of spherical volumes is described in detail in an annotation by Li Chunfeng (in about 656) to *Jiu Zhang*. The arguments are very clear in three successive steps:

1. Within a cube draw two inscribed cylinders at cross direc-

tions. That part in common of the two cylinders is called *mou he fang gai* (literally "the common square cover"). Cut a small cube $1/8$ of the original cube. According to Zu Geng's principle, the following proportion is obtained:

$$1/8 \text{ volume of sphere} : 1/8 \text{ } fanggai = \pi:4.$$

2. That part of *fanggai* within the $1/8$ cube is the inner *qi*, and those three parts within the small cube but left out of *fanggai* are the outer *qi*.

From the small cube cut an inverted *yangma*. Prove by the *gougu* theorem that if we cut the *yangma* horizontally at a certain level from the base, the cross-section of the *yangma* is equal to the total cross-sections of the outer *qi* cut at the same level in area.

3. Prove by Zu Geng's principle that the total volume of the outer *qi* is equal to that of the *yangma*.

From these the formula for the volume of the sphere is immediate.

The idea of *mou he fang gai* was first introduced by Liu Hui. The first step of Zu Geng had actually been worked out by Liu also. In fact, in *Liu Zhu* he had time and again made use of what was later called Zu Geng's principle to find the volumes of solid bodies bounded by curved surfaces, such as the volumes of the cylinder from the polygonal pillar, of the cone from the pyramid, of the frustum of cone from the frustum of pyramid, etc. Zu Geng's merits not only consist in actual solution of volumes of *mou he fang gai* and the sphere, but also in his summing up of practical experiences and objective facts in the form of a general principle. Whether the principle should be called the Liu-Zu principle to give Liu Hui his due is a matter that deserves discussion.

Other Applications

Jiu Zhang is so comprehensive that, leaving other topics aside, the out-in complementary principle is by no means applied merely to the various problems above. The problem of the inscribed circle in a right triangle in *Jiu Zhang* treated on this principle has since been further developed. It is fully treated in *Ce Yuan Hai Jing*

(*Sea Mirror of Circle Measurement*, 1248) by Li Ye. In the works of Qin Jiushao and Li Ye, the problem of the "square city" above has been replaced by a problem of "circular walled city" which was beyond the masters of older times. The invention of such methods as *tianyuanshu* in the Song and Yuan dynasties not only solves heretofore unsolvable problems but also largely simplifies old problems. Compared with the older methods, the new methods give results with far less effort. The essence of the new methods and new theories lies in the algebraization of geometry, which blazed the trail for both analytical geometry and modern algebra.

Conclusion

The out-in complementary principle together with the principles of Liu Hui and Zu Geng demonstrated the considerable abilities of ancient Chinese masters in scientific abstraction. Drawing intrinsic conclusions from objective facts, they summed up the conclusions into succinct principles. These principles, plain in reasoning and extensive in application, form a unique character of ancient Chinese mathematics. The emphasis has always been on the tackling of concrete problems and on simple, seemingly plausible principles and general methods. The same spirit permeates even such outstanding achievements as the algebraization of geometry and the place-value decimal system of numbers. Western mathematics, in contrast, lays emphasis on conceptions and the logical relationships between them.

The majority of the ancient Chinese mathematical classics have sunk into oblivion because of feudal obscurantism — a most deplorable loss in human society. Zu Geng's contributions would also have been lost had it not been for the rather casual entry by Li Chunfeng in his annotation to *Jiu Zhang*. However, judging from what is still available, the historical facts that ancient Chinese mathematics had its origin in human productive activities and had thrived in its own, independent way before the 15th century are still clear, as pithily shown in the following two diagrams:

Diagram I

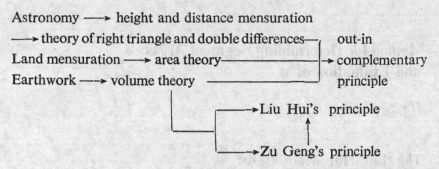

Astronomy ⟶ height and distance mensuration
⟶ theory of right triangle and double differences⟶, out-in
Land mensuration ⟶ area theory⟶ complementary
Earthwork ⟶ volume theory ⟶ principle

⟶Liu Hui's principle

⟶Zu Geng's principle

Diagram II

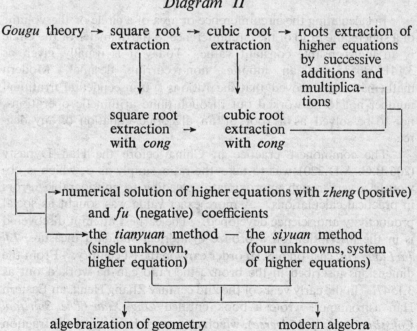

Gougu theory ⟶ square root ⟶ cubic root ⟶ roots extraction of
extraction extraction higher equations
by successive
additions and
multiplica-
tions

square root cubic root
extraction ⟶ extraction
with *cong* with *cong*

⟶numerical solution of higher equations with *zheng* (positive)
and *fu* (negative) coefficients

⟶the *tianyuan* method — the *siyuan* method
(single unknown, (four unknowns, system
higher equation) of higher equations)

algebraization of geometry   modern algebra

Method for Determining Segment Areas and Evaluation of π

He Shaogeng

The Liu Hui Method for Determining Segment Areas

In calculating the circumference or area of a circle or the volume of a sphere, the ratio of the circumference to the diameter of the circle is an indispensable constant value. Today it is usually given as 3.1415926535. . . , an infinite, non-recurring decimal. Modern mathematics has proved that the ratio is a transcendental irrational number not to be worked out through finite arithmetic operations, nor to be solved as a root of an algebraic equation of any degree.

The commonest practice in China before the Han Dynasty (206 B.C.-A.D.220) was to take the ratio simply as 3, which is obviously a very rough approximation leading to considerable errors in practical calculations. A more exact value was sought as social productivity and science developed. The earliest attempt discovered is in the inscription on a bronze cylindrical standard measure, *Lü Jia Liang Hu*, cast on official order early in the 1st century. From the dimensions inscribed in the bronze the ratio can be worked out as 3.1547. In the early years of the 2nd century Zhang Heng, an Eastern Han astronomer, wrote a book entitled *Ling Xian (The Spiritual Constitution of the Universe)*, which expresses the ratio by the fraction $\frac{730}{232}$, or 3.1466. However, Zhang's formula for the volume of a sphere gives the ratio as $\sqrt{10}$, or 3.1622. In the Three Kingdoms period (220-280), Wang Fan of the state of Wu gave $\frac{142}{45}$, or 3.1556, as the ratio, as recorded in *Hun Tian Xiang Shuo* (*Discourse on Uranographic Models*). All these values are of course much closer than 3. All were given empirically, however. A theory for evaluating the ratio was still pending.

A little later Liu Hui did excellent work in that field. In A.D. 263, in his annotation of the ancient mathematical classic *Jiu Zhang Suan Shu* (*Nine Chapters on the Mathematical Art*), he points out that in a circle 3 is not the ratio of the circumference, but that of the perimeter of an inscribed regular hexagon, to the diameter. The hitherto supposed area of a circle, he says, is in fact the area of an inscribed regular 12-sided polygon. Liu Hui finds that the more sides an inscribed regular polygon has, the nearer its perimeter approaches the circumference of the circle. He seeks better values of the ratio by inscribing polygons of as many sides as possible in the circle in an attempt to "exhaust" the residual areas of the segments so formed. His approach is fairly accurate in theory.

According to Liu Hui:

1. The length of each side of an inscribed regular hexagon is equal to that of the radius of the circle.

2. On the basis of the properties of the right triangle (the *gougu* properties), the length of each side of an inscribed 2n-sided

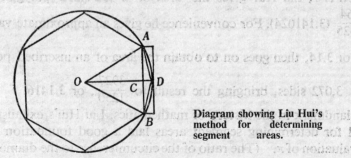

Diagram showing Liu Hui's method for determining segment areas.

regular polygon can be calculated from the length of each side of an inscribed n-sided regular polygon.

3. The area of an inscribed 2n-sided regular polygon is given

directly by multiplying the length of the radius by the length of one side of an inscribed n-sided regular polygon and dividing the product by 2.

4. The area S of the circle satisfies the inequalities:

$$S_{2n} < S < S_{2n} - (S_{2n} - S_n) .$$

In the diagram on p.91, if the areas of the two right triangles taking AD and DB as their hypotenuses are added to the area of the triangle $\triangle$ OAB, the total area will be that of the quadrilateral OADB. But if an area equivalent to that of the aforesaid two right triangles is again added to that of the quadrilateral OADB, the total area will then exceed the area of the sector OAB.

5. "The finer we cut the segments," says Liu Hui, "the less will be the loss in our calculation of the area of the circle. The exact area of the circle is obtained when such segments so cut off come to be infinitesimals." In other words, the area of a circle is considered a limit to the increasing areas of the inscribed polygons; as the number of sides of the polygons is increased, their areas approach increasingly closer to the area of the circle. Likewise, when the number of sides of the inscribed polygon increases to infinity, the perimeter of the polygon will become the circumference of the circle.

Starting from a regular hexagon inscribed in a circle the radius of which is 1, Liu Hui gives the area of a 192-sided polygon as $3.14 \frac{64}{625}$ (3.141024). For convenience he gives an approximate value $\frac{157}{50}$, or 3.14, then goes on to obtain the area of an inscribed polygon of 3,072 sides, bringing the result to $\frac{3927}{1250}$, or 3.1416.

A landmark in the history of mathematics, Liu Hui's exhaustive method for determining segment areas laid a good foundation for later evaluation of π. (The ratio of the circumference to the diameter. In a circle in which the radius is 1, π is equal to the area of the circle in numerals.) Liu Hui was quite in advance of his contemporaries. He took into consideration only the inscribed polygons, leaving the circumscribed ones alone, and that stood him in good stead, making his calculations much simpler than that of Archimedes. Liu Hui's method of approximation and his primitive idea of the dialectical unity of arc and chord were indeed great contributions 1,500 years ago.

π Value Given by Zu Chongzhi

In the Southern and Northern Dynasties (420-589), Zu Chongzhi (429-500), a mathematical star of the first magnitude in China's history, calculated the value of π to $\frac{355}{113}$, a truly remarkable feat in ancient mathematics. According to the Calendrical and Astronomical Chapters of *Sui Shu* (*History of the Sui Dynasty*), Zu gave a "deficit value" of 3.1415926 and an "excessive value" of 3.1415927

Zu Chongzhi's calculation of the value of π is recorded in *Sui Shu* (*History of the Sui Dynasty*).

for π. The true value, as pointed out by Zu, must lie between the two. That is

$$3.1415926 < \pi < 3.1415927.$$

For practical use, Zu gave two fractional values for π; an "inaccurate value" (*yuelü*) $\frac{22}{7}$, and an "accurate value" (*milü*) $\frac{355}{113}$.

It was not until 1,000 years later that Zu's record was bettered by the Arab mathematician al-Kāshī and the French mathematician Vièta.

Before decimal numbers became popular in China, Chinese mathematicians were accustomed to using fractions for the approximate values of certain constants. $\frac{355}{113}$ is the best fractional value for π with both the numerator and the denominator less than 1,000. In Europe $\frac{355}{113}$ was obtained 1,000 years later by the German Otto and the Dutch Antonisz. It is therefore known as the "Antonisz value" in Western mathematics.

π has very wide practical use. In ancient times, however, when science was at best primitive, it was very complicated and difficult to evaluate π. The values given by Zu Chongzhi testify to the high level of Chinese mathematics in Zu's days. People therefore suggest that the ratio $\frac{355}{113}$ be named after Zu to commemorate the great Chinese mathematician.

Zu's evaluation of π and other important achievements were recorded in his book *Zhui Shu* (*The Art of Mending*). Unfortunately the book has long been lost. Had Zu employed Liu Hui's method, he would have laboured over polygons of 12,288 and 24,576 sides to obtain his "deficit" and "excessive" values respectively.

Determination of Segment Areas and Use of Infinite Series for Expressing π

In the early years of the Qing Dynasty (1644-1911), Ming Antu (?-1765), a Mongolian mathematician, wrote *Ge Yuan Mi Lü Jie Fa*

(*Quick Method for Determining Segment Areas*).   By that method he proved independently or derived nine formulae of infinite series, including some expressions of trigonometric functions by power series and one expression of π.   Ming was a pioneer in tackling these formulae by a traditional Chinese method.

By Ming Antu's time China had already fallen behind in science and technology.   Hard-working Chinese mathematicians like Ming felt, through practice and studies, an irresistible impact of vital Western ideas upon the stagnant Chinese mind.   They aspired after such Western achievements as trigonometric functions and loga-rithms.   But owing to incomplete or incompetent media of transmis-sion, often what they were trying to arrive at was but a rough idea of a certain formula, leaving them to their own resources for its proof and derivation.   The Promethean fire flickered but had to be rekindl-ed by themselves.   One of such formulae was given by Gregory in 1667:

$$\sin x = x - \frac{1}{3!}x^3 + \frac{1}{5!}x^5 - \cdots;$$

and another by Newton in 1676:

$$\frac{\pi}{3} = 1 + \frac{1^2}{4 \cdot 3!} + \frac{1^2 \cdot 3^2}{4^2 \cdot 5!} - \cdots$$

But when these two formulae were introduced to China, no methods of deriving and proving accompanied them.   Ming Antu had to grope in the dark for nearly all the latter half of his life before he finally succeeded not only in giving them proofs but also in develop-ing new trigonometric functions and inverse trigonometric functions in power series.   Ming Antu's method is illustrated by a diagram in his own book:

1.   Arc $\overset{\frown}{ACD}$ has a chord AD; the length of AD is L.   Divide $\overset{\frown}{ACD}$ into *m* equal parts.   Every part of the arc has a subtending chord $L_{1/m}$.

2.   When m = 2 (or any even number), L can be expressed in the form of an infinite power series having L_1/m as the variable.   In proving this, Ming Antu takes the following steps: Since the ratio of corresponding sides in two similar triangles are the same, Ming

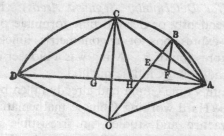

Diagram showing Ming Antu's method for determining segment areas.

formulates the relation between the chord of an arc and the chord of a part of that arc as shown in the diagram

$$AD = 2AC - \frac{AC \cdot BE}{OA}.$$

3. When m=5 (or any odd number), L can be a polynomial of $L_{1/m}$. Ming Antu learns this from Western mathematics.

4. From the second and the third steps Ming finds expressions for L in the forms of power series of variables $L_{1/10}$, $L_{1/100}$, $L_{1/10,000}$ . . . (by ways of $1/2 \times 1/5 = 1/10$, $1/10 \times 1/10 = 1/100$, $1/100 \times 1/100 = 1/10,000$, etc.).

Ming Antu then proceeds to secure a succession of results approaching a desired result. He "approximates" the length of an arc by trying to sum up the lengths of infinitesimal chords subtending infinitesimal parts of that arc. In modern mathematical parlance, what Ming expects is this: When m approaches infinity

$$\overset{\frown}{ACD} = \lim_{m \to \infty} mL_{1/m}.$$

Since r (radius of the circle) = OA, $\frac{L}{2} = r \sin \alpha$ $(\angle \alpha = \angle AOC)$ in the diagram, Ming substitutes the former for the latter in his series and gets the proofs of Gregory's and Newton's formulae.

Ming makes further advances on the method for determining segment areas. Liu Hui sought the true value of π by inscribing polygons within circles in an attempt to exhaust the residual areas so formed. Ming Antu applies the method to deriving and proving formulae of infinite series. To Ming the idea of summing up infinitesimals to approximate a certain value is not limited to finding the circumference of a circle. He takes a step forward to develop trigonometric functions in power series. Ming sees a dialectical unity not only of arc and chord within a circle, but of curve and straight line in general. This in the days that followed paved the way for Xiang Mingda in Xiang's *tuo yuan qiu zhou shu* (method for finding the circumference of an ellipse), which appeared early in the 19th century.

Xiang Mingda's Method for Finding the Circumference of an Ellipse

Xiang Mingda (1789-1850), another mathematician of the Qing Dynasty, also did fruitful research into the development of trigonometric functions in power series. In his book *Xiang Shu Yi Yuan* (*Common Origin of Shape and Number*) Xiang derives a formula for the circumference of an ellipse:

$$P = 2\pi a \ (1 - \frac{1}{2^2}e^2 - \frac{1^2 \cdot 3}{2^2 \cdot 4^2}e^4 - \ldots);$$

in which $e^2 = \dfrac{a^2 - b^2}{a^2}$; e being the eccentricity of the ellipse; a and b being the lengths of the semimajor and semiminor axes respectively.

Xiang also derives a formula for $\dfrac{1}{\pi}$:

$$\frac{1}{\pi} = \frac{1}{2} \cdot \ (1 - \frac{1}{2^2} - \frac{1^2 \cdot 3}{2^2 \cdot 4^2} - \ldots).$$

Xiang did not finish writing his "Method for Finding the Circumference of an Ellipse" (which forms an appendix to Vol. 6 of *Xiang Shu Yi Yuan*) because of illness and consequent death. Later Xiang's friends drew diagrams illustrating his method:

Divide the circle on the major axis of the ellipse into n equal parts, then draw lines from the points of division on the circle parallel to the minor axis of the ellipse. The lines cut the ellipse also into n parts but of unequal lengths. By joining together the points where the parallel lines cross the ellipse, we get a number of chords subtending arcs which form the circumference of the ellipse as a whole. By adding the lengths of the chords we get a result approaching the circumference of the ellipse. When the chords become few in number but each infinitesimal in length, their sum-total will be the circumference of the ellipse.

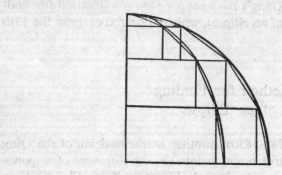

Diagram illustrating Xiang Mingda's method for finding the circumference of an ellipse.

Ming Antu's and Xiang Mingda's methods smack of calculus. They tended in that direction but fell short of it. The reasons could have been fortuitous. In any case, they helped prepare Chinese mathematicians for a new era of Descartes, Newton and Leibniz; for a leap forward from the constant to the variable which is the essence of modern mathematics.

The Chinese Remainder Theorem

Li Wenlin and Yuan Xiangdong

In China, many mathematical games have been handed down from ancient times in the form of folk rhymes. Among the most popular of these are *Ge Qiang Suan* (*Computations Behind the Wall*), *Jian Guan Shu* (*Method of Cutting the Tube Lengths*) and *Qin Wang An Dian Bing* (*The Prince of Qin's Method of Counting Soldiers by Heart*). A folk song entitled *Sun Zi Ge* (*The Song of Master Sun*),[1] which is known also in Japan, reads as follows:

> *Not in every three persons is there one aged three score and ten,*
> *On five plum trees only twenty-one boughs remain,*
> *Every fifteen days rendezvous the seven learned men,*
> *We get our answer by subtracting one hundred and five over and again.*

The translation of the original song seems to make little sense, and its meaning cannot be construed literally from the text. In fact the lines are but mnemonic rhymes hinting at the procedure of solving the famous Sun Zi problem which first appeared in *Sun Zi Suan Jing* (*Master Sun's Mathematical Manual*) of the 4th century:

> There is an unknown number of things. When counted by threes, they leave a remainder of two; when counted by fives,

---

[1] *Sun Zi Ge*, also known as *Han Xin Dian Bing* (*General Han Xin's Method of Counting Soldiers*), Han Xin being a well-known general 2,200 years ago. The song appears in *Suan Fa Tong Zong* (*Systematic Treatise on Arithmetic*) (1592) by Cheng Dawei of the Ming Dynasty; actually it was known far and wide among the common people long before. Although the text might have been a little different, the numbers hinted in it had remained the same.

they leave a remainder of three, and when counted by sevens they leave a remainder of two. Find the number of things.

The modern theory of numbers views the problem as one of linear congruences, as equivalent to finding solutions (positive integers) to the following simultaneous indeterminate equations of the first degree:

$$N = 3x + 2, \quad N = 5y + 3, \quad N = 7z + 2$$

which can be expressed in the modern form

$$N \equiv 2 \ (\text{mod } 3) \equiv 3 \ (\text{mod } 5) \equiv 2 \ (\text{mod } 7)[1]$$

where the least integer for N is required.

Sun Zi Suan Jing gives the answer as 23, which obviously can be obtained by inspection (trial and error) since the problem is simple. But this is not how the answer is obtained in *Sun Zi Suan Jing*, which suggests the following procedure:

If we count by threes and have the remainder 2, then take the number 70 and multiply it by 2: $70 \times 2 = 140$.

If we count by fives and have the remainder 3, then take the number 21 and multiply it by 3: $21 \times 3 = 63$.

If we count by sevens and have the remainder 2, then take the number 15 and multiply it by 2: $15 \times 2 = 30$.

Add the three products, and we get 233. From this we subtract 105 twice, and we get the result:

$$N = 70 \times 2 + 21 \times 3 + 15 \times 2 - 2 \times 105.$$

Here 105 is the least common multiple of the moduli 3, 5 and 7. Obviously the 23 given in *Sun Zi Suan Jing* is the smallest positive integer qualified for our answer. For remainders other than 2, 3 and 2, *Sun Zi Suan Jing* points out that we should only substitute them for the 2, 3 and 2 in the above procedure. Let R_1, R_2, R_3 be the other remainders respectively, then according to *Sun Zi Suan Jing* the formula is

$$N = 70 \times R_1 + 21 \times R_2 + 15 \times R_3 - p \times 105 \quad (p \text{ being an integer}).$$

---

[1] When each of the two integers A and R is divided by a given integer m (called the modulus) and gives the same remainder, we say A is congruent to R, modulus m. The whole process is written in the form: $A \equiv R \ (\text{mod } m)$.

The crux of the Sun Zi method lies in finding out the three key numbers 70, 21 and 15, which are precisely those contained in the first three lines of *The Song of Master Sun*. *Sun Zi Suan Jing* gives no explanation of how the numbers were obtained. However, we find something unusual in them which can be expressed as follows:

$$70 = 2 \times \frac{3 \times 5 \times 7}{3} \equiv 1 \ (\text{mod } 3);$$

$$21 = 1 \times \frac{3 \times 5 \times 7}{5} \equiv 1 \ (\text{mod } 5);$$

$$15 = 1 \times \frac{3 \times 5 \times 7}{7} \equiv 1 \ (\text{mod } 7).$$

This is equivalent to saying that the three numbers 70, 21 and 15 can be obtained by dividing the least common multiple 105 by the three moduli 3, 5 and 7 respectively, and multiplying the quotients by the three integers 2, 1 and 1. Let $k_1 = 2$, $k_2 = 1$, and $k_3 = 1$. Then, by selecting a set of integers k_i ($i = 1, 2, 3$), we may obtain three numbers so that, when divided by their respective modulus, the remainder will always be 1.[1] From the above we know that when the remainders are R_1, R_2, and R_3 respectively, the following can be deduced:

$$R_1 \times k_1 \times \frac{M}{3} = R_1 \times 2 \times \frac{3 \times 5 \times 7}{3} \equiv R_1 \ (\text{mod } 3);$$

$$R_2 \times k_2 \times \frac{M}{5} = R_2 \times 1 \times \frac{3 \times 5 \times 7}{5} \equiv R_2 \ (\text{mod } 5);$$

$$R_3 \times k_3 \times \frac{M}{7} = R_3 \times 1 \times \frac{3 \times 5 \times 7}{7} \equiv R_3 \ (\text{mod } 7).$$

When added together, they give

---

[1] That is, when we are counting by threes, then among those numbers which can leave a remainder 1, we select 70, as it can be divided by 5 and 7 with no remainder. When we are counting by fives, then among those numbers which can leave a remainder 1 we select 21, as it can be divided by 3 and 7 with no remainder. When we are counting by sevens, among those numbers which can leave a remainder 1 we select 15, as it can be divided by 3 and 5 with no remainder.

$$R_1 \times 2 \times \frac{3 \times 5 \times 7}{3} + R_2 \times 1 \times \frac{3 \times 5 \times 7}{5} + R_3 \times 1$$

$$\times \frac{3 \times 5 \times 7}{7} \equiv R_1 \ (mod \ 3)$$

$$\equiv R_2 \ (mod \ 5)$$

$$\equiv R_3 \ (mod \ 7).$$

Since $M = 3 \times 5 \times 7$ which can be divided by any of its three factors with no remainder, we get

$$(R_1 \times 2 \times \frac{3 \times 5 \times 7}{3} + R_2 \times 1 \times \frac{3 \times 5 \times 7}{5} + R_3 \times 1$$

$$\times \frac{3 \times 5 \times 7}{7}) - pM \equiv R_1 \ (mod \ 3)$$

$$\equiv R_2 \ (mod \ 5)$$

$$\equiv R_3 \ (mod \ 7).$$

Here p is an integer.   This proves the method in *Sun Zi Suan Jing*.

The Sun Zi method can be generalized if the above reasoning is followed.   Let a number N be divided by n divisors $a_1, a_2 \ldots a_n$ (where the divisors are prime in pairs which have no common factors except 1), and let the remainders be $R_1, R_2 \ldots R_n$ respectively, i.e.

$$N \equiv R_i \ (mod \ a_i) \quad (i = 1, 2, \ldots n);$$

then all we have to do is to find a group of numbers k_i that satisfy

$$k_i \ \frac{M}{a_i} \equiv 1 \ (mod \ a_i) \ (i = 1, 2, \ldots n);$$

then the smallest positive integral solution of any given problem of linear congruences is

$$N = (R_1 k_1 \frac{M}{a_1} + R_2 k_2 \ \frac{M}{a_2} + R_3 k_3 \ \frac{M}{a_3} + \ldots \ldots$$

$$+ R_n k_n \ \frac{M}{a_n}) - pM$$

$$(p \text{ is an integer}, M = a_1 \times a_2 \times \ldots \times a_n); \text{ i.e.}$$

$$N \equiv \sum_1^n \ R_i k_i \ \frac{M}{a_i} - pM.$$

This is known as the remainder theorem in the modern theory of numbers.   Its primitive procedure, though not clearly explained and generalized, lies in what is suggested in solving the problem of the "Unknown Number of Things" in *Sun Zi Suan Jing*.

The emergence of the Sun Zi problem in a mathematical classic early in the 4th century was not accidental. The study of problems concerning linear congruences had obviously long been urged by the necessary practice of mathematical calculations in ancient astronomy and calendars, especially those related to the number of years elapsed since *liyuan* (the beginning of the calendar) or *shangyuan* (the Grand Beginning) called *shang yuan ji nian* in ancient China. Proofs in this respect can be found in ancient Chinese astronomical and calendrical documents. All calendars need a certain point of time in the past as a beginning for their calculations. For instance the Jing Chu calendar initiated in the 3rd century during the dynasty of Wei took as the starting point the last time the winter solstice fell exactly at the beginning of a lunar month, which also happened to be the midnight of the first day of a 60-day cycle, named Jia Zi in ancient China. Let a denote the number of days in a tropical year, b the number of days in the synodic month, R_1 the cyclical-day number of the winter solstice (i.e., the number of days in the 60-day cycle between the winter solstice and the last Jia Zi preceding it), R_2 the number of days between the winter solstice and the beginning of the lunar month, then N, the number of years since the Grand Beginning, can be calculated from the linear congruences

$$aN \equiv R_1 \pmod{60} \equiv R_2 \pmod{b}.$$ [1]

During the Southern and Northern Dynasties (420-589) the Da Ming calendar compiled by the celebrated mathematician-astronomer Zu Chongzhi (initiated in the year 462) demanded that the Grand Beginning should start with the beginning of the Jia Zi year (60-year cycle), and when the sun, the moon, and the five planets

---

[1] From the definition of *liyuan* or *shangyuan* we know exactly N number of tropical years, a total of $a \times N$ days, had elapsed up to the winter solstice of a certain year in the past (generally the year in which a new calendar was initiated). The 60-day cycle took the first day of the tropical year as its starting point. To divide aN by 60, the remainder should be the number of days since the last Jia Zi preceding it. From this $aN \equiv R_1 \pmod{60}$ was obtained. The ancient calendar experts arrived at the second congruence on the same principle.

should be in the same direction. Moreover, the moon should be at the perigee and the ascending node. To calculate the years elapsed since such a Grand Beginning, the calendar experts would have to solve a set of ten linear congruences. Astronomical and calendrical data are complicated. From the 3rd to the 6th century, when

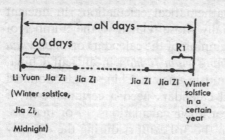

Sun Zi Suan Jing appeared, Chinese mathematicians had certainly long been able to solve problems concerning indeterminate analysis much more complicated than those involved in the "Unknown Number of Things". They could have mastered a general procedure in so doing,[2] as reflected in the practical problems and their solutions provided in Sun Zi Suan Jing. But certainly the inner logic of the procedure had never been clearly explained or even truly mastered. However, further application of the procedure in calculating shang yuan ji nian called for more intensive studies in the procedure. Finally in the 13th century the great mathematician Qin Jiushao worked out the first general mathematical formulation for solving linear congruences, moving mathematical discourse onto a new level of abstraction and crowning others' previous attempts with success.

Qin Jiushao, who lived in the 13th century, liked mathematics since childhood. After long years of hard study he completed the

---

[2] The calculation of shang yuan ji nian was undertaken in the Han Dynasty. The Han calendar experts, making use of extant astronomical data, found that the number of shang yuan ji nian could be obtained by solving one or two fairly simple linear congruences. For instance, we can prove that when the Santong calendar was initiated in the 1st century B.C., the value obtained merely satisfies the form $145 \times 4617 \times p \equiv 135 \pmod{1728}$, p being an integer and shang yuan ji nian $x = 4617 \times p$. Such a solution could have been obtained by trial and error. After the 3rd century astronomical surveying technique became more sophisticated, which called for more complicated linear congruences in calculating shang yuan ji nian. These had urged the calendar experts to work out a general rule in indeterminate analysis.

writing of his *Shu Shu Jiu Zhang* (*Mathematical Treatise in Nine Sections*) in 1247. A masterpiece of mathematical work, *Shu Shu Jiu Zhang* records creative works in many fields, including *Da Yan Qiu Yi Shu* (*The Great Extension of Seeking Unity*) in indeterminate analysis and *Zheng Fu Kai Fang Shu* (*Method of Numerical Extraction of Positive Roots of Algebraic Equations*) which are of world significance. Here, however, we shall discuss only Qin Jiushao's great contribution to linear congruences.

In *Shu Shu Jiu Zhang*, Qin clearly and systematically explores the procedure in solving problems of linear congruences. Qin's method is what we now call the remainder theorem, which reduces the problem of linear congruences to the finding out of a set of numbers k_i which satisfy the formula $k_i \dfrac{M}{a_i} \equiv 1 \pmod{a_i}$. Qin calls k_i *chenglü* (multiplying terms). In Volume 1 of *Shu Shu Jiu Zhang*, entitled *Da Yan Zong Shu* (*General Method of the Great Extension*), Qin expounds the method of finding k_i, and names the method *da yan qiu yi shu*, or the *dayan* method.

To elaborate, let us show how k_i is found out. If $G_i = \dfrac{M}{a_i}$ $> a_i$; Qin first divides G_i by a_i, and obtains the remainder $g_i < a_i$; then $G_i \equiv g_i \pmod{a_i}$, and $k_i G_i \equiv k_i g_i \pmod{a_i}$. Since $k_i G_i \equiv 1 \pmod{a_i}$, the method boils down to finding k_i so that $k_i g_i \equiv 1 \pmod{a_i}$. Qin calls a_i *dingshu* (fixed number), and g_i *qishu* (odd number). Qin's *da yan qiu yi shu*, in modern mathematical language, is to apply the division algorithm (Euclidean algorithm) to g_i and a_i so that the quotients $q_1, q_2 \ldots q_n$, and the remainders $r_1, r_2, \ldots r_n$ are obtained. In doing so, Qin calculated the values of c as in the right column below:

	QUOTIENTS	REMAINDERS	C VALUES
a_i/g_i	q_1	r_1	$c_1 = q_1;$
g_i/r_1	q_2	r_2	$c_2 = q_2 c_1 + 1;$
r_1/r_2	q_3	r_3	$c_3 = q_3 c_2 + c_1;$
$\vdots$	$\vdots$	$\vdots$	$\vdots$
r_{n-2}/r_{n-1}	q_n	r_n	$c_n = q_n c_{n-1} + c_{n-2}$

Qin points out that when $r_n = 1$ while n is an even number, the c_n obtained in the final step is the k_i required. If n is an odd number, we may apply the division algorithm to r_{n-1} and r_n, in form let $q_{n+1} = r_{n-1} - 1$, so that the remainder r_{n+1} will still be 1. Then find $c_{n+1} = q_{n+1} c_n + c_{n-1}$. This time $n+1$ is an even number, and c_{n+1} is therefore the k_i required. In either case the remainder will be 1 in the last step, and the whole procedure ends there. Qin calls this the "Method of Seeking Unity". As for the meaning of "Great Extension", he says in his preface to *Shu Shu Jiu Zhang* that the term comes from *Yi Jing* (*Book of Changes*), and adds that what he means by the term in *Shu Shu Jiu Zhang* is exactly the same as in *Yi Jing*. Here Qin seems to wander a bit from the point. However, the remainder theorem Qin elucidates in his book proves to be perfectly correct and logical.[1]

In Qin Jiushao's days, all calculations were carried out with counting-rods. On a counting board Qin places the odd number g in the upper right corner, and the fixed number a in the lower right corner. In the upper left corner he places *1* (which he calls *tianyuan* 1 (Celestial Monad One). Then, within the right column, Qin divides the larger number by the smaller number, multiplies the quotient by the upper (or lower) left number and adds the product to the lower (or upper) left number, till *1* appears in the upper right corner. Following is a demonstration in symbols of Qin's counting-rods calculation. On the right is an example in figures ($g = 20$, $a = 27$, $k = c_4 = 23$):

---

[1] In fact, if $l_2 = q_2$, $l_3 = q_3 l_2 + 1$, $l_4 = q_4 l_3 + l_2 \ldots l_n = q_n l_{n-1} + l_{n-2}$, then $r_1 = a_1 - g_1 q_1 = a_1 - c_1 g_1$,

$r_2 = g_1 - r_1 q_2 = g_1 - (a_1 - c_1 g_1) q_2 = c_2 g_1 - l_2 a_1$,

$r_3 = r_1 - r_2 q_3 = (a_1 - c_1 g_1) - (c_2 g_1 - l_2 a_1) q_3 = l_3 a_1 - c_3 g_1$,

. . . .

When $r_n = 1$ while n is an even number (if n is an odd number, we can always transform the situation to meet our requirement of making the remainder a "unity", as shown in the above); obviously we have $r_{n-1} = l_{n-1} a_1 - c_{n-1} g_1$; $r_n = c_n g_1 - 1_n a_1 = l$, that is $c_n g_1 \equiv 1 \pmod{a_1}$. This proves that c_n is the k_i required.

Tianyuan	1	odd	g_i	1,	20
		fixed	a_i		27

1		g_i	1, 20
$c_1 = q_1$,		r_1	1, 7
(q_1)			

$c_2 = c_1 q_2 + 1$,	(q_2)	3, 6
	r_2	
c_1,	r_1	1, 7

$\vdots$

c_{n-2},	r_{n-2}	3, 6
$c_{n-1} = c_{n-2} q_{n-1} + c_{n-3}$,	r_{n-1}	4, 1
	(q_{n-1})	

	(q_n)	23, 1
$c_n = c_{n-1} q_n + c_{n-2}$,	1	
c_{n-1},	r_{n-1}	4, 1

In *Shu Shu Jiu Zhang*, Qin Jiushao teaches by demonstration how he has solved a number of practical problems in calendrical calculations, engineering works, state taxes and levies of service, and even military matters such as those of the number of years, months and days of the conjunction of the solar year, the lunar month and the sexagenary cycle according to obsolete calendars; the different amounts of earthwork done by different groups of labourers from different prefectures with different productivities; the distance covered and days spent by different messengers in sending messages from an army in field action to the capital; the amounts of bricks of various sizes needed in construction, etc. In calculation Qin differentiates between what he calls *yuanshu* (a_i as integers), *shoushu* (a_i as decimals), and *tongshu* (a_i as fractions), and works out a method in dealing with each. By the *dayan* method he converts decimals and fractions into integers. In practical problems the moduli in the congruences involved are sometimes integers not prime in pairs. In such cases, Qin appropriately selects the *yuanshu* factors to be *dingshu* and reduces the problems to those providing integral moduli prime in pairs.[1] Qin's formulations in these respects, rigorous and systematic in theory and resourceful and profound in thinking, are marvellous achievements even by modern mathematical standards.

The *dayan* method was most probably Qin's summation of the method of calculation for *shang yuan ji nian*. However, his work was so far above the heads of his contemporaries that his book was almost forgotten and lost after the middle of the Ming Dynasty (after about 1500). When it was rediscovered during the Qing Dynasty it aroused the interest of many mathematicians, who did a lot of work in elucidating, simplifying and improving on the *dayan* method. One of them, Huang Zongxian, in his *Qiu Yi Shu Tong Jie* (*A Thorough Explanation of the Method of Seeking Unity*), gives a much simpler method in dealing with moduli which are integers not prime in pairs. But Huang's book appeared only towards the end of the 19th century.

Ancient Chinese mathematicians' contributions to linear congruences, from the solution of the problem of the "Unknown Number of

---

[1] Qin's method is to cancel the common factor among the *yuanshu* a_i by a definite process, so that in each *yuanshu* a factor t_i is obtained, and $m = t_1 \times t_2 \times \ldots \times t_n$. Note that m is the least common multiple of a_i, t_i being prime in pairs. Qin then takes t_i as *dingshu* and finds k_i according to the method of *da yan qiu yi shu*.

Things" in *Sun Zi Suan Jing* to Qin Jiushao's *dayan* method were brilliant achievements in mankind's history of mathematics. The earliest European attempt at linear congruences was made in Italy by Qin's contemporary Leonardo Pisano (Fibonacci), who presents two problems in his *Liber Abbaci* (1202) about the same in nature as the problem of the "Unknown Number of Things" in *Sun Zi Suan Jing* and scarcely more brilliant than the latter. It was not until the 18th and 19th centuries that two most learned European mathematicians, L. Euler (1743) and C. F. Gauss (1801), succeeded in profound research in linear congruences and produced a theorem the same as that worked out by Qin Jiushao. Euler and Gauss arrived at a strong proof for the character of moduli prime in pairs. Obviously Euler and Gauss knew nothing of Qin Jiushao's work. In 1852 the English missionary A. Wylie published his *Jottings on the Science of the Chinese: Arithmetic*, in which he discusses the problem of the "Unknown Number of Things" in *Sun Zi Suan Jing* and Qin Jiushao's method, which immediately attracted the attention of European mathematicians. In 1876 a German scholar, L. Matthiessen, pointed out the identity of Qin's method with Gauss' formula. M. Cantor, a great German historian of mathematics, read Matthiessen's article and spoke highly of the *dayan* method as having been created by the "most fortunate ingenuity" of the Chinese mathematicians. Even today the *dayan* method in solving congruences remains a most interesting subject for further research among Western historians of mathematics. Ulrich Libbrecht, a Belgian mathematician and sinologist, writes in *Chinese Mathematics in the Thirteenth Century* which was published in 1973: "If we take into account the early date of Ch'in's[1] work within the field of indeterminate analysis, we can see that Sarton[2] did not exaggerate when he called Ch'in Chiu-shao 'one of the greatest mathematicians of his race, of his time, and indeed of all times'."

---

[1] Qin Jiushao was spelt Ch'in Chiu-shao in all publications in English before the introduction of the Chinese Phonetic Alphabet.

[2] G. Sarton is an American scholar specialized in the history of science. Libbrecht quotes from Sarton's *Introduction to the History of Science* (1927).

Ancient mathematicians of India also made important contributions to linear congruences. Between the 6th and 12th centuries, several centuries later than *Sun Zi Suan Jing*, they developed their own kuṭṭaka method in solving simultaneous indeterminate equations of the first degree[1]. Problems similar to that of the "Unknown Number of Things" were written in the works of Brahmagupta (7th century) and Máhavîra (9th century). We do not claim that the kuṭṭaka method was inspired by the Chinese. But the L. Van Hee's allegation that the Chinese *dayan* method came from the kuṭṭaka method of India is at the same time groundless. L. Van Hee argued that Indian influence on the Chinese could be proved by the writing of numerical figures from left to right horizontally — a practice seemingly contrary to the ancient Chinese writing habit which was upright and from right to left. But what Van Hee did not know was that counting rods had appeared in China long before the Warring States Period (475-211 B.C.) and all counting-rod figures were arranged on a decimal place-value basis from left to right horizontally as borne out by the figures cast on ancient coins of the 3rd century B.C. The profound studies done by ancient Chinese mathematicians in linear congruences bore the brilliance of the ancient Chinese traditional creativeness and continuity. The prestige enjoyed by the *dayan* method is fully justified. Western works on the history of mathematics are quite right in referring to the theorem in solving linear congruences as the Chinese remainder theorem.

---

[1] Indeterminate equations of the first degree are closely related to linear congruences. For example, to find the integral solutions of
$$by - ax = c$$
we obtain $by = ax + c$
This is equivalent to $by \equiv c \pmod{a}$.

The Numerical Solution of Higher Equations and the *Tianyuan* Method

Guo Shuchun

In ancient and medieval China the method of solving algebraic equations was called *kaifangshu*[1], which in the Song Dynasty (960-1279) was developed into *zeng cheng kai fang fa*, i.e. the method of extracting equational roots by successive additions and multiplications. Further, *tianyuanshu* and *siyuanshu* were invented. *Tianyuanshu* was a method of building up a higher equation by supposing the *tianyuan* to be the unknown; *siyuanshu* was a method of solving simultaneous higher equations. These constituted medieval China's original contributions to mathematics far in advance of any other contemporary civilized country.

Jia Xian Triangle — Numerical Cues for *Kaifangshu*

An important achievement in algebra in the Song Dynasty was the *kai fang zuo fa ben yuan tu* (basic diagram for solving equations) introduced by Jia Xian, a methematician who lived in the first half of the 11th century, in his book *Huang Di Jiu Zhang Suan Fa Xi Cao* (*Detailed Solutions of the Problems in the Nine Mathematics*

---

1 In ancient and medieval China *kaifangshu* was not limited to extracting the roots of binomial equations in the form of $x^n = N$ as the Chinese term now denotes. It had the much broader application of finding the positive roots of numerical equations. The term *fangcheng* then meant simultaneous linear equations, not just an equation with a single variable.

111

Chapters of Huang Di). The diagram tabulates the binomial coefficients up to the 6th power, the exponents being positive integers. The same diagram is known in Europe as the Pascal Triangle, drawn by Blaise Pascal in 1654, more than 600 years after Jia Xian's. In 1527 the same triangle appeared on the title page of *Arithmetic*, written by the German mathematician Petrus Apianus. But even Apianus was more than 500 years later than Jia, giving further good reason to name the diagram the Jia Xian Triangle.

Jia Xian's "basic diagram for solving equations" reproduced from *Yong Le Da Dian* (*Great Encyclopaedia of the Yongle Reign*).

The first three sentences of an annotation under the Jia Xian Triangle explain the structure of the Triangle and the roles the numbers play in solving equations. The numbers on the $n+1$st row show the coefficients of the expanded terms of the binomial equation $(a+b)^n$, n being positive integers. The numbers in the left and right outer slanting lines, called the *ji* (a^n) and the *yusuan* (b^n), are coefficients of the first and the last terms respectively. The inner numbers "2", "3, 3", "4, 6, 4" . . . on the 3rd, 4th, 5th . . . rows, called the *lian*, are coefficients of the inner (except the first and the last) terms when binomial equations of the 2nd, 3rd, 4th . . . degrees are expanded. The term *ji*, *yu* and *lian* have their origins in the ancient *kaifangshu*[1]. Appended to the diagram is a list of "Methods of Calculating the Coefficients by Additions and Multiplications" (*zeng cheng kai fang qiu lian fa cao*), which Jia Xian called "the basic method for finding the coefficients"(*qiu lian ben yuan*).

The Jia Xian Triangle was at first meant solely for root-extraction, and that is why it is called *kai fang zuo fa ben yuan tu* (literally, the basic method for "root-extraction"). The last two sentences

---

[1] The extraction of the square or cubic root in ancient China was illustrated by means of geometrical figures. The square or cubic figure of the *chushang* ("primary deliberation") for the root is a large square (a^2) or a large cube (a^3), called the *ji* or the *fang*. The square or cubic figure of the *cishang*

in the annotation sum up the method of extracting the root of the nth degree with the help of the coefficients. That is: to multiply the coefficients on the $n + 1$st row by a suggested value for the root; then subtract the nth power of the suggested root from *shi* (the constant the root of which is to be extracted). Divide the difference by the product of the suggested value and the coefficient and an additional value for the root is obtained. For instance, to extract the square root of N, we learn

$$(x_1 + x_2)^2 = x_1^2 + 2x_1 x_2 + x_2^2 = x_1^2 + (2x_1 + x_2)x_2$$

from the 3rd row of the Jia Xian Triangle. A primary value x_1 for the root is suggested by trial and error. An additional value x_2 is suggested by dividing $N - x_1^2$ by $2x_1$. Add x_2 to $2x_1$ and multiply the sum by x_2. Finally subtract the product from $N - x_1^2$. Similarly, the cubic root of N is extracted with the help of the coefficients 1, 3, 3, 1 on the 4th row of the Triangle.

Obviously the same procedure can be applied to extracting the root of nth degree of N by means of the coefficients on the $n + 1$st row in the Jia Xian Triangle. The method of square or cubic root extraction, which had been in use for more than a thousand years before Jia, was extended to extracting arbitrary root of higher degree. This was a great improvement on the old root-extraction method.

The Jia Xian Triangle paved the way for mathematicians of the Song and Yuan dynasties (10th-14th century) in China to scale academic heights of their times. The Triangle was extended to the 9th row (the 8th power) by Zhu Shijie in the early years of the Yuan Dynasty (1271-1368). It was a great achievement which facilitat-

---

("additional deliberation") for the root is a small square (b^2) or a small cube (b^3), which occupies a small corner in the whole figure $(a+b)^2$ or $(a + b)^3$, and was thus called the *yu* (literally, "corner"). The product of the primary and the additional deliberations is a rectangle (ab) or a rectangular parallelepiped or flat square parallelepiped (a^2b or ab^2), which occupies a place along one or two of the sides of the whole figure, and was thus called the *lian* ("side"). To extract a square root, in every step we have to calculate for two rectangles, as shown on the 3rd row in the Jia Xian Triangle. To extract a cubic root, in every step we have to calculate for three rectangular parallelepipeds and three flat square parallelepipeds, as shown on the 4th row in the Jia Xian Triangle.

ed the development of the limited difference method and the summation of arithmetical series of higher order. The basic procedure of additions and multiplications in finding the coefficients in the Triangle ushered in a new method in solving numerical higher equations — *zeng cheng kai fang fa*. The Jia Xian Triangle was still widely referred to in the Ming (1368-1644) and Qing (1644-1911) dynasties.

Zeng Cheng Kai Fang Fa

In modern computational mathematics, the very simple Qin Jiushao procedure is popular in solving higher algebraic equations. Actually the procedure was introduced by Jia Xian in the Northern Song Dynasty (960-1127), only to be brought to maturity by Qin Jiushao in the earlier half of the 13th century during the Southern Song Dynasty (1129-1279). A similar method was originated by William George Horner in 1819, nearly 600 years later than Qin and more than 700 years later than Jia. The Qin Jiushao procedure marked a new epoch in the algebra of numerical higher equations.

The *zeng cheng kai fang fa* introduced by Jia Xian was not a single operation using the coefficients in the Triangle. Instead, a method of successive multiplications and additions was employed for the same result. Let us follow the method and try to find x in $x^4 =$ N. First, the root extraction form Diagram 1 (formerly with counting-rods) is set up:

1. *shang* (root) 2. *shi* (constant) 3. *fang* (coefficient of x)
4. *shanglian* (coefficient of x^2) 5. *xialian* (coefficient of x^3)
6. *yu* (coefficient of x^4, the lowest divisor)

	(1)	(2)	(3)
1.		x_1	x_1
2.	N	$N - x_1 x_1^2 = N - x_1^4$	$N - x_1^4$
3.	0	$x_1 \cdot x_1^2 + 0 = x_1^3$	$x_1 \cdot 3x_1^2 + x_1^3 = 4x_1^3$
4.	0	$x_1 \cdot x_1 + 0 = x_1^2$	$x_1 \cdot 2x_1 + x_1^2 = 3x_1^2$
5.	0	$x_1 \cdot 1 + 0 = x_1$	$x_1 \cdot 1 + x_1 = 2x_1$
6.	1	1	1

Chushang (primary value for the root) is suggested by trial and error to be x_1. Multiply *yu* by x_1 and add the product to *xialian*. Multiply *xialian* by x_1 and add the product to *shanglian*. Multiply *shanglian* by x_1 and add the product to *fang*. Finally multiply *fang* by x_1 and subtract the product from *shi*, and we obtain Diagram 2 as shown above. If there is no remainder, x_1 is the root. If there is a remainder, start the operation all over again in Diagram 2 by multiplying the lower row term by x_1 and add the product to the term in the next row above, starting from *yu* up and stopping at the row of *fang*. We then obtain Diagram 3. Repeat the operations in Diagram 3 and later in Diagrams 4 and 5, but notice that in every round we stop at the row just next to the row where the preceding round of multiplications and additions stopped. Diagram 5 is a reduced equation from

$$x^4 + 4x_1 x^3 + 6x_1^2 x^2 + 4x_1 x = N - x_1^4.$$

In Diagram 5 we suggest a *cishang* (additional value for the root) and start the operations all over again. If the subtraction from *shi* leaves no remainder, $x_1 + x_2$ is the root. If there is a remainder, repeat the operations in Diagram 3 and subsequently Diagrams 4 and 5; find a solution for the second reduced equation and so on till the required result is obtained.

x_1	x_1
x_1^4	$N - x_1^4$
$4x_1^3$	$4x_1^3$
$x_1 \cdot 3x_1 + 3x_1^2 = 6x_1^2$	$6x_1^2$
$x_1 \cdot 1 + 2x_1 \;= 3x_1$	$x_1 \cdot 1 + 3x_1 = 4x_1$
1	1
(4)	(5)

Compared with earlier methods of root extraction, the above method was neat and simple. It could be applied to both root-extraction and solution of any numerical higher equation. In fact the reduced equation shown in Diagram 5 contains not only the highest power of x but also lower powers of x as well. There is an illustration of the extraction of the 4th degree root in Jia Xian's works, according to Yang Hui of the 13th century.

The root-extraction method by successive additions and multiplications initiated by Jia Xian performed an immortal feat in the history of mathematics. But Jia's own applications of the method were limited to solving binomial equations in the forms of $x^2 = N$, $x^3 = N$, and $x^4 = N$. The coefficients of the unknowns and the constants in ancient Chinese algebraic equations had always been positive numbers down to Jia Xian. The first person who attempted a breakthrough, according to Yang Hui, was Liu Yi of the 12th century. In Liu's book *Yi Gu Gen Yuan* (*Discussion on the Old Sources*) he mentions the negative coefficients respectively of x and x^2 (*fufang* and *yiyu*) in quadratic equations, i.e. equations in the forms of $x^2 - ax = b$ and $-x^2 + ax = b$, a and b being greater than zero. Liu gives several methods for solving equations like these. Worthy of mention among them are *yijishu* (method of augmenting the constant) and *jiancongshu* (method of diminishing the coefficient of x). *Yijishu* is so called because after suggesting *chushang* x_1, it augments ax_1 with the *fufang* equation or x_1^2 with *yiyu* and subtracts x_1^2 or ax_1 from it respectively. *Jiancongshu* is so called because after suggesting *chushang* x_1, it acquires *jiancong* $x_1 - a$ with *fufang* or $a - x_1$ with *yiyu* and subtracts $(x_1 - a) x_1$ or $(a - x_1) x_1$ from the constant respectively. The two methods were not derived from *zeng cheng kai fang fa*. But of the two the method of "diminishing the coefficient of the x term or eliminating it" was nearer to the *zengcheng* method. Liu Yi's two methods can be regarded as the first two links in the chain of thoughts which led to the solution of higher equations of the nth degree with a negative coefficient for either x^n or a lower power of x by the *zengcheng* method. In *Yi Gu Gen Yuan* there is a quartic equation as follows:

$$-ax^4 + bx^3 + cx^2 = N \ (a, b, c, N > 0)[1]$$

which is solved by *zeng cheng kai fang fa*. It paved the way for using *zeng cheng kai fang fa* in finding positive roots of numerical equations of all degrees with negative coefficients — a method brought to per-

---

[1] The equation is actually $-5x^4 + 52x^3 + 128x^2 = 4096$.

fection by Qin Jiushao as stated in *Shu Shu Jiu Zhang* (*Mathematical Treatise in Nine Sections*) 100 years later.

, Since all algebraic equations in ancient and medieval China arose from practical problems, mathematicians before Qin Jiushao had always regarded *shi* as a practical known quantity. *Shi* had always been the positive constant term to be placed on the right side of the equation. However, Qin Jiushao thought it better to place *shi* together with the unknown terms so that a negative quantity could be added to the positive and be mutually cancelled, equalizing the whole equation to zero. In a root-extraction form so disposed, the multiplications and additions in the *zengcheng* method could be carried through to the end. In his works Qin therefore always makes the constant term negative. His root-extraction form becomes

$$f(x) = a_0x^n + a_1x^{n-1} + a_2x^{n-2} + \ldots + a_{n-1}x + a_n = 0,$$

while $a_n < 0$, $a_0 \neq 0$. The coefficients in such an equation can either be positive or negative, integers or decimals. That is to say, Qin was capable of finding the positive roots of all numerical equations. Qin calls his method *zheng fu kai fang shu*, i.e. the method of solving equations with "positive-negative" coefficients. When the first coefficient $a_0 \neq 1$, Qin calls his method "the extraction of the joined branch nth root" (*kai lian zhi mou cheng fang*). When an equation has only even powers of x, Qin calls his method "the extraction of harmonious alternating nth root" (*kai ling long mou cheng fang*). During the extraction operations, the constant term of the reduced equation often approaches steadily nearer to zero. Sometimes the constant term changes from negative to positive, the changed term being called by Qin *huangu* (literally "changed bones"). Sometimes a negative constant term remains negative but its absolute value increases, the term being called by him *toutai* (literally "brought to birth again"). When an irrational root is obtained, Qin further develops the idea of *weishu* (minute numbers) initiated by Liu Hui of the 3rd century and goes on with the extraction for the decimal part of the root to obtain an approximate value of the root. Liu Hui's idea of *weishu* was that of decimal numbers in the bud. Qin Jiushao appears to have been the first to grasp this.

Tianyuan Notation

The *zengcheng* method of finding positive roots of higher equations matured with the growth of *tianyuanshu* — setting up of an equation on the counting board and disposing the rods in solving the equation. To suppose an unknown and to write an algebraic equation for this unknown is easy today for anyone with a little knowledge of mathematics. But it was a tall order before the invention of the *tianyuan* notation. To give expression to a cubic equation in mind, Wang Xiaotong, great mathematician early in the Tang Dynasty (618-907), had to resort to literal paraphrasing. That made his style necessarily cumbersome and his reasoning very difficult for others to follow. As practical problems in life demanded more and more complicated forms of higher equations, their expressions became more and more difficult. A simple and speedy formulation of the equations was urgently sought after. It was under such circumstances that the *tianyuan* notation came into being. History tells us that books on the *tianyuan* notation were legion in North China in the dynasties of Kin and Yuan (1115-1368). Unfortunately nearly all, including *Huang Di Jiu Zhang Suan Fa Xi Cao* and *Yi Gu Gen Yuan* are lost. Among the few extant works on the subject are *Ce Yuan Hai Jing (Sea Mirror of Circle Measurement)* 1248, *Yi Gu Yan Duan (New Steps in Computation)* 1259 by Li Ye, *Suan Xue Qi Meng (Introduction to Mathematics)* 1299 and *Si Yuan Yu Jian (Precious Mirror of the Four Elements)* 1303 by Zhu Shijie.

The *tianyuan* method started with putting down on the counting board as *tianyuanyi* (a unity symbol serving as a place-indicator) an unknown. Then two equal polynomials in terms of the unknown were set up to meet the demands of a particular problem. One polynomial was subtracted from the other to produce an equation which was equal to zero. Finally the positive root of the equation was extracted by the *zengcheng* method. It is obvious that there is not the slightest difference between the *tianyuan* notation and the way a modern algebraic equation is set up. But the *tianyuan* notation had grown up independently from the ancient Chinese *kaifangshu*, while in Europe the modern algebraic equation and its solution appeared only in the 16th century.

However, the evolution from the ancient Chinese *kaifangshu* to the *tianyuan* notation was not smooth. At first, nine characters from *tian* ("heaven") to *xian* ("god") and nine more from *di* ("earth") to *gui* ("spirit") were used to indicate the positive and negative exponents of the unknown respectively, while *ren* ("man") was employed to denote the constant term. Later the symbols were reduced to *tianyuan* and *diyuan* for the positive and negative exponents respectively, while the constant term was indicated by *tai* ("the absolute"). It was Li Ye who further simplified the notation and used only the *tianyuan* for the powers of the unknown. In *Ce Yuan Hai Jing* Li observed the ancient rule of placing the positive exponents above the negative and the constant term. Later in *Yi Gu Yen Duan* he reversed the rule by placing the negative exponents above the positive and the constant term in conformity with the traditional array of *kaifangshu*. Sometimes the unknown was labelled *yuan* ("the primary"), sometimes the constant term as *tai*, because a single place-indicator was enough. For negative numbers there was a diagonal stroke across the unit digit. Li, unlike Qin Jiushao, did not make it a rule that the constant term must be negative. His constant term could be either positive or negative. Li's arrangements in these respects were observed by later mathematicians such as Zhu Shijie and Guo Shoujing. For instance, the equation

$$25x^2 + 280x - 6905 = 0$$

was represented on the counting board as either

Li Ye's works show that the four basic arithmetical operations of addition, subtraction, multiplication and division were skilfully applied to polynomials in the first half of the 13th century. Only in division the divisor was the single term for unknown. All the *tianyuan* equational terms except the constant term had to be composed of unknowns the exponents of which were positive integers. Radicals and fractions were first eliminated by raising them or multiplying them by a common multiple to the rational integral expressions before solution of the equation was attempted.

Siyuanshu

Early in the Western and Eastern Han dynasties (206 B.C.-A.D. 220) Chinese mathematicians were able to solve simultaneous linear equations, the method being called *fangchengshu*. The application of the *tianyuan* notation to the simultaneous equations brought forth successively *eryuanshu* (solving higher simultaneous equations with two variables), *sanyuanshu* (solving higher simultaneous equations with three variables), and finally *siyuanshu*. This was another brilliant achievement scored by Chinese mathematicians in the latter half of the 13th century and the beginning of the 14th. *Si Yuan Yu Jian* written by Zhu Shijie was a masterpiece on *siyuanshu*.

Siyuanshu was an array of four higher simultaneous equations with four variables represented by *tian* (heaven), *di* (earth), *ren* (man) and *wu* (matter) respectively. The arrangements are shown in the

$\vdots$	$\vdots$	$\vdots$	$\vdots$	$\vdots$	$\vdots$	$\vdots$
$\cdots y^3 u^3$	$y^2 u^3$	$y u^3$	u^3	$z u^3$	$z^2 u^3$	$z^3 u^3 \cdots$
$\cdots y^3 u^2$	$y^2 u^2$	$y u^2$	u^2	$z u^2$	$z^2 u^2$	$z^3 u^2 \cdots$
$\cdots y^3 u$	$y^2 u$	$y u$	u	$z u$	$z^2 u$	$z^3 u \cdots$
			$\boxed{yz}$			
$\cdots y^3$	y^2	y	*Tai*	z	z^2	$z^3 \cdots$
		$\boxed{xu}$		$\boxed{xyz}$		
$\cdots xy^3$	xy^2	xy	x	xz	xz^2	$xz^3 \cdots$
$\cdots x^2 y^3$	$x^2 y^2$	$x^2 y$	x^2	$x^2 z$	$x^2 z^2$	$x^2 z^3 \cdots$
$\cdots x^3 y^3$	$x^3 y^2$	$x^3 y$	x^3	$x^3 z$	$x^3 z^2$	$x^3 z^3 \cdots$
$\vdots$	$\vdots$	$\vdots$	$\vdots$	$\vdots$	$\vdots$	$\vdots$

following diagram with x, y, z and u representing the four variables respectively.

"Heaven" (x) is placed below the constant term which is denoted by *tai*, so that the power of x increases as it moves downwards; "earth" (y) is placed to the left of the constant so that the power of y increases as it moves towards the left; "man" (z) is placed to the right of the constant so that the power of z increases as it moves towards the right; and "matter" (u) is placed above the constant so that the power of u increases as it moves upward. The products of linear terms, squares, cubes, and higher powers of two neighbouring unknowns such as xy, x^2z, y^3u^3 are shown at points where vertical and horizontal lines representing the respective powers meet. The products of unknowns not neighbouring such as xu, yz, xyz are shown at places between the lines. For instance, the simultaneous equations are shown

$$-x - y - xy^2 - z + xyz = 0, \tag{1}$$
$$x - x^2 - y + z + xz = 0, \tag{2}$$
$$x^2 + y^2 - z^2 = 0, \tag{3}$$

in the following forms:

  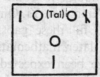

In fact the forms are expressions of the detached coefficients of a set of multivariate simultaneous higher equations.

The first steps of *siyuanshu* were eliminations. Four equations with four variables were reduced to three equations with three variables, then to two equations with two variables, and finally to one equation with one variable suitable for root-extraction by the *zengcheng* method. The whole procedure was quite similar to that employed today in solving simultaneous equations. Yet systematic treatment of the rule of elimination in solving simultaneous higher equations was given only in 1775 by a French mathematician —

Etienne Bézout.

Time and again the Chinese attained the highest academic summits in mathematics. Time and again they contributed to the civilization of mankind in a way unparalleled in the contemporary world. But ancient and medieval Chinese algebra laid emphasis only on practical problems, neglecting the study of theories, especially those related to the properties of equation. Underestimation of theory nipped many future marvels in the bud. For instance, the negative number has its longest history and earliest applications in China, but in almost 2,000 years the solution of an equation was limited to the positive root. The negative root was simply ignored. No research was done in either the relationship between the order of an equation and the number of its roots, or that between the roots and the coefficients. The answer to two neighbouring problems in *Yi Gu Gen Yuan* happen to be the two roots of the same quadratic equation. But neither Liu Yi nor Yang Hui noticed this. Moreover, *siyuanshu* at its apogee was limited to the possibilities of the two-dimensional counting board. No attempt was made to tackle problems involving more than four variables. Though it would be too harsh to blame such masters of old as Jia Xian, Liu Yi, Qin Jiushao, Li Ye and Zhu Shijie for not being able to fill these gaps, the shortcomings should have been overcome by later mathematicians had things developed in China as might have been expected. However, feudal obscurantism not only killed the hope of carrying forward the Song and Yuan creative spirit; it almost entirely obliterated the trial blazed by the masters. Later, imperialist aggression joined in delaying the development of capitalism and modern science in China. Not until the end of the 18th century or the beginning of the 19th were the aforesaid shortcomings tackled by such Chinese mathematicians as Li Rui, Wang Lai, Jiao Xun and Luo Shilin. Li Rui found that an equation may have negative roots and multiple roots. He introduced a method for judging the roots and coefficients of an equation. When a polynomial has one variation of sign, it may have one positive root. When a polynomial has two variations of sign, it may have two positive roots. When a polynomial has three variations of sign, it has

three or one positive root.   When a polynomial has four variations
of sign, it has four or two positive roots.   This method of judgement
is the same as the rule of signs initiated by Rene du Perron Descartes
in 1637.   But Li lagged far behind Descartes.

SOME MECHANICAL SECRETS OF THE WATER POT OF...

the centre of gravity shifted. When the pot hangs full, its centre
of gravity has further force position requiring balance arched of equipment
with same as the body of some unearthed above balance for earth
in 1973, but chipped the behind locators.

Mechanics

Dai Nianzu

All objects in nature are constantly in mechanical motion, and
mechanics, as the science of this mechanical motion, is necessarily
wide in scope.   Ancient China had ample knowledge of mechanics,
as summed up in the following points.

Centre of Gravity and Moment of Force

The centre of gravity of things has found practical use in China
since ancient times.   A water pot unearthed among the Yangshao
cultural relics at Banpo Village near Xi'an topples on water when
empty, allowing water to enter, but it rights itself again when full.

A water pot unearthed at Banpo Village near Xi'an.

Designed to draw water from wells, it demonstrates a profound principle of mechanics. From time immemorial man has learned from practice to use the centre of gravity of objects, and the relationship between the metacentre and the stability of a floating body. By the Qin Dynasty (221-207 B.C.) this water pot was developed into an inclining vessel. "It leans over on one side when empty, stands up straight when the desired amount of water has entered it, and topples when that amount is exceeded," according to an essay in *Xun Zi* (*The Book of Master Xun*), dating from the 3rd century B.C. Obviously functioning here is the centre of gravity, which is slightly high in the empty vessel so that it is unbalanced. When a certain amount of water has entered the vessel, the lowered centre of gravity causes the vessel to straighten up like a tumbler or roly-poly. More water, however, again raised the centre of gravity and toppled the vessel. This type of inclining vessel was produced as late as in the Tang Dynasty (618-907).

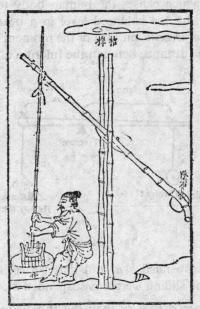

Illustration in *Wang Zhen Nong Shu* (*Agricultural Treatise of Wang Zhen*) showing a well sweep.

The concept of force and moment of force is developed from man's practical experience over a long period of time. The ancient Chinese grasped the concept through their use of simple devices such as the steelyard, pulley, wheel and axle, well sweep (counterweighted bailer bucket) and windlass. A good account of these is given in *Mo Jing* (*Mohist Canon*) of the 5th-4th centuries B.C. Mo and his disciples were known for their asceticism, hard work, serious experimentation, great valour and martial skill. Many of their works deal with natural science.

In *Mo Jing* the definition of force is derived from the strength of the human body, which is called in the book *xing*, or "shape"; while lifting, holding, casting, striking, etc. of an object by the human body is called *fen*, or "exertion". "Force", according to *Mo Jing*, "is that which causes the 'shape' to 'exert' ". In other words, force is that which causes the human body to move an object, or in other words that force is equivalent to weight. It gives an example of force being that which causes the human body to lift a weight.

When discussing the balanced level in a beam scale such as a steelyard, *Mo Jing* presents an idea of the moment of force in Mohist terms. It calls the distance between the fulcrum and the point where

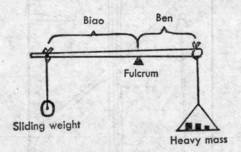

The steelyard.

a heavy mass is suspended *ben*, and that between the fulcrum and the point where the sliding weight is suspended *biao*.

If the mass is heavier than the sliding weight yet the level is horizontally balanced, this is because *ben* is shorter than *biao*. If now at both points of suspension the same weight is added, the *biao* side must go down. To sum up, the heavier a mass and the farther it is from the fulcrum, the more it tends to go down; the lighter a mass and the nearer it is to the fulcrum, the more it tends to go up.

What is important in the Mohist view of a balanced level is: Not only the different magnitudes of the force and the weight, but also the different distances of the two from the fulcrum are considered. The distances, called *biao* and *ben* by the Mohists, are the level arms of effort and load respectively. Although *Mo Jing* falls short of a numerical summary of the relationship between the magnitudes in a balanced level, it shows that the Mohists, before Archimedes (c. 287-212 B.C.), had noticed that a balanced level has much to do with its two arm lengths.

In the early Han Dynasty (206 B.C.-A.D. 220) the importance of the fulcrum in a level was common knowledge in China. *Huai Nan Zi* (*The Book of the Prince of Huai Nan*), a compendium of natural philosophy written by the group of scholars gathered by Prince Liu An in about 120 B.C., makes the observation that

> a wooden column with a circumference which takes 20 times the span between the tips of the thumb and the index finger to cover may prop up a roof 1,000 *jun*[1] [15,000 kg.] in weight; a door bolt five *cun*[2] [11.5 cm.] long may bolt a door. These do not mean that the column and the bolt would in any case be equal to the stresses that might put on them by the roof and the door, but it is because of the pivotal positions they are in where the stresses are the least.

Indeed, if a door is bolted at the top or the bottom instead of being bolted at an appropriate point in the middle, it naturally takes much less effort to force it open.

---

[1] 1 *jun* = 15 kilogrammes.
[2] 1 (Han Dynasty) *cun* = 2.3 centimetres.

Stress and Deformation of Solids

Man learns about the properties of various materials through manufacturing goods and construction over a long period of time. *Xun Zi* says: "Strong and rigid materials are for making columns while pliable and tough materials are for binding things up." This conclusion was inevitably drawn from man's extensive practical experience.

Since olden times man has learned that materials may deform in use or during processing. Analysing the deformation of wooden beams, *Mo Jing* states:

> When a cross-beam remains straight under a load, the beam is strong enough to bear the load.
>
> When another load is laid upon the beam yet it shows no signs of bending, that means the beam is excellent in bearing loads. But when a rope is horizontally held, it bends under its own weight. Ropes in that position are very poor or simply useless in withstanding a perpendicular load.

Modern material mechanics tells us that rigid beams are bend-resisting while ropes can only withstand pulling. The different properties of wooden beams and ropes in this respect are noted in *Mo Jing*. Though a wooden cross-beam also deforms under a load, within allowable stress the deformation is not visible to the naked eye.

Xun Zi states that some material retains its deformed shape even when the external force that has deformed it no longer exists.

> A wooden staff straight as an arrow can be baked and bent into a ring which keeps its shape even when wizened in use. Such is the effect of baking and bending [on a wooden staff].

This property of materials is called in modern mechanics their plastic property. *Kao Gong Ji* (*Artificers' Record*) written in about the 5th century B.C. tells about the twist and strength of leather. It says that leather should be strained before use to see whether it is flat and straight. If not, a disequilibrium of stress must exist within the leather. The part where greater signs of strain are shown tears easily in later use.

Perhaps the best discourse on stress in ancient China is that on the suspension of a weight on a hair, the discourse probably having arisen from the use of hair braid as rope. Some hairs snap easily while others do not, and *Mo Jing* states that this depends on whether the cohesive substance in a hair is homogeneously distributed along the whole length, and whether the load is evenly borne by the whole length without a weak link when the hair is taut. The same explanation is given in an essay entitled "Tang Wen" ("Emperor Tang's Query") in *Lie Zi* (*The Book of Master Lie*) written sometime between the 5th and 1st century B.C. Zhang Zhan of the Jin Dynasty (265-420) says in an annotation on the text of *Lie Zi*:

> A hair, though slender and delicate, bears a load without being torn apart since its tensile strength comes from a homogeneous structure. Yet some hairs snap easily because they are not evenly formed and there are weak links in them.

Lie Zi cites a number of writers on this subject. Gongsun Long, one of the Logicians in the later part of the Warring States Period (475-221 B.C.), based himself on the Mohist explanation of the tensile strength of a hair and boldly imagined: "A hair may withstand a pulling load of a thousand *jun*" ("Essay on Confucius", *Lie Zi*). Mou, son of a feudal prince, elaborated on the view soon after: "Let a hair be perfect in its homogeneous structure so that any strain could be borne evenly to perfection, then the hair would be able to withstand a pulling load of 1,000 *jun*" ("Emperor Tang's Query", *Lie Zi*). The last two statements are obviously extreme even in theory, since the strength of any material has a limit beyond which it will break. It is certainly impossible for a single hair to withstand a pulling load of 15,000 kilogrammes. Yet the view "the more homogeneous a material is in structure, the greater the load it can bear" is correct. In it the modern theory of stress in mechanics is conceived.

Buoyancy and Specific Gravity

Some properties of fluid have been known to man since very ancient times. "Water has no steady form," says Sun Wu, the great strategist, in an essay "The False and the True" in *Sun Zi Bing Fa* (*Master Sun's Art of War*) written probably in 345 B.C. Zhuang Zhou, a great philosopher in the Warring States Period, says in *Zhuang Zi* (*The Book of Master Zhuang*) c. 290 B.C.: "Water levels when calm." These are the earliest extant statements on the properties of fluid.

In *Mo Jing* there appears this account on buoyancy:

> When a very large body floats on water with only a very small part of it submerged, that means the constant equilibrium between the submerged part and the whole body has already been established. This can be likened to the exchange of goods on a market. Some articles are exchangeable only for five other articles.

This Mohist statement errs seriously in failing to recognize that the submerged part of the body is exactly the volume of the water displaced, and that the weight of that water is exactly equal to the upward pressure keeping the body afloat. What has been established is an equilibrium between the buoyancy and the weight of the floating body, and not that between the body and its submerged part. However, the statement senses that some sort of relationship exists between the submerged part and the whole floating body. However, the submerged part is yet to be construed as a certain volume of water displaced, while the equilibrium is yet to be calculated in weight and not in volume. The Mohist formulation is close to the Archimedian.

Buoyancy was ingeniously made use of in ancient China. There is an amusing anecdote about Cao Chong of Wei in the period of the Three Kingdoms (A.D. 220-280) weighing an elephant by means of water buoyancy in the absence of other weighing devices. On Cao's suggestion the elephant was led onto a sufficiently large boat and the water level on the boat was marked. The elephant was then led ashore and rocks loaded in the boat till it sank to the level

Cao Chong weighing an elephant.

marked. The rocks were weighed individually and their total weight reckoned, correctly, as the weight of the elephant.

Surface Tension and Its Applications

Surface tension is a property peculiar to liquids caused by unequal molecular cohesive forces near the surface of the liquid. In China, surface tension of liquids has from time immemorial been noted and used in production as well as in daily life.

A method of testing the quality of tung oil is mentioned in *You Huan Ji Wen* (*Things Seen and Heard on My Official Travels*) written by Zhang Shinan in 1233 in the Song Dynasty (960-1279). A ring made of a thin strip of bamboo is dipped into the oil. If a very thin layer of oil is found across the ring, the oil is pure. Adulterated oil forms no oil film. This test is possibly much older than the record.

The formation of a thin layer of a liquid across such a ring depends mainly on the particular surface tension of the liquid. If it is not pure, the impurities will lessen the surface tension.

Water has a rather weak surface tension. But when a tiny object such as a sewing needle is carefully laid on its surface, especially when there are many minute bubbles in the water, the needle will not sink because of the water's surface tension. Young Chinese women have had the tradition since ancient times of doing this as a festival game or for divination on the seventh day of the seventh moon according to the lunar calendar, the practice being called *diuzhen* (casting needles). In *Di Jing Jing Wu Lu* (*Descriptions of Things and Customs at the Imperial Capital*) written by Liu Tong and others around 1638, the phenomenon of *diuzhen* is explained: "Since there is an invisible membrane on water, it is this membrane that prevents the needles from going down."   All of these facts and references indicate that the physical effect of surface tension of liquids has long been observed.

Siphons and Atmospheric Pressure

Siphons in ancient China included the *zhuzi* or wine pipette, *pianti* or sideways lifter, *kewu* or "thirsty sun god", and *guoshanlong* or "water-dragon lying across a mountain". The *kewu* for irrigation first appeared towards the end of the Eastern Han Dynasty (A.D. 25-220). Later, around A.D. 450, it was used in the balanced clepsydras devised by the Taoist Li Lan of the Northern Wei Dynasty (386-534). The national minorities in southwestern China sometimes still drink wine from long bamboo tubes, which are also siphons. In *Wu Jing Zong Yao* (*Collection of the Most Important Military Techniques*) compiled by Zeng Gongliang in 1044, a siphon in the form of a long bamboo pipe is recommended to draw rivulet water over a hill which otherwise prevents its flow.

Pumps have been widely used in daily life and production since very ancient times. Various military writings mention water pumps as important fire extinguishers in war. The fourth chapter of *Dong Po Zhi Lin* (*Journal and Miscellany of Dongpo*) by Su Shi (1037-1101) contains a detailed account of the bamboo tube-buckets used to draw brine from the bottom of salt-wells in Sichuan Province.

> The bucket has no fixed bottom, and possesses an orifice in the top. A piece of leather several *cun* in size is attached to the bottom, forming a valve. As the bucket enters and is brought

out of the brine, the liquid opens the valve as it enters the bucket. Each bucket brings up several *dou*[1] of brine.

Pumps for water spraying are mentioned in *Zhong Shu Shu* (*Book of Afforestation*) written in the Ming Dynasty (1368—1644).

Atmospheric pressure is what makes siphons and pumps work, and the wide use of these tools in ancient China naturally caused their users to seek the principle involved.

In an essay entitled "Jiu Yao" ("Nine Drugs") contained in *Guan Yin Zi* (*The Book of Master Guan Yin*) written in the Southern and Northern Dynasties (A.D. 420-589), there is this interesting description of "a bottle with two holes":

> Fill the bottle with water and turn it upside down, the water will come out of one hole. But if one of the holes is covered with a finger, then water will not leave the other hole. This is because if something does not go up, something else will not come down.

The reason for this is because air could enter one hole to let the water flow out of the other when both holes were left open. Atmospheric pressure prevented the water from flowing out the other hole when one was closed. Although Master Guan Yin failed to attribute this phenomenon to atmospheric pressure, his dialectical approach is inspiring.

In an annotation on the text of *Su Wen* (*Questions and Answers*), Wang Bing of the Tang Dynasty more clearly observes the physical phenomenon of atmospheric pressure thus:

> Dip an empty tube into water and let it fill with the water. Close the upper opening with a finger and pull the tube out. The water inside the tube will not flow out since the tube is too thin to let air go in and up through the lower opening. It is difficult to pour water into a bottle through its single orifice because the air inside the bottle cannot easily escape.

---

[1] 1 *dou* = 10 litres.

In the Song Dynasty a scholar named Yu Yan recorded in a book entitled *Xi Shang Fu Tan* (*Hackneyed Talk on a Mat*):

> When I was young, a Taoist priest showed me a trick. A piece of paper was set ablaze inside an empty bottle which was then quickly turned over with its mouth thrust into a thin layer of water in a silver basin. The water was immediately sucked into the bottle with a gurgling sound. This was because of the role played by the fire and the air. The same process was repeated with the bottle made to stand upside down on a man's belly. It sucked at the man's belly so tight as not to be easily removed.

Another such account is given in "Nei Pian" ("Esoteric Chapter") of *Shu Qu Zi* (*The Book of the Hemp-Seed Master*) written by Zhuang Yuanchen in the 16th century: "When an empty bottle gourd is turned upside down in water, the water will not be able to enter because of the resistance of the air inside."

The Wonder of the Fish Basin

Though this fish basin may also be of potter's clay, the one described here is of bronze. Shaped like an ordinary wash basin, it has a number of fish or dragons cast on the inside bottom, and is therefore called a fish or dragon basin. Its first appearance was in the Tang or Song Dynasty at the latest. The wonder of the basin is that when its two handles are rubbed slowly and rhythmically with one's hands, the basin vibrates like a struck bell. Fountains of spray shoot up near the sides of the basin as high as the handles, as if from the mouths of the fish. With skill, the spray may be made to shoot a metre into the air. At full vibration the water surface is covered with a very complex pattern of waves, the waves, spray and shimmering fish images presenting quite a life-like sight. The basin is obviously built with some knowledge of the transformation of metallic vibrations into liquid waves, which then wax or wane with propagation or interference.

Knowledge of Crystals

The study of crystals constitutes the main field of solid-state physics. The most salient features in the appearances and geometrical forms of certain crystals, notably those of snowflakes, were very early observed in China. At about the beginning of the 1st century snowflakes were found to be composed of tiny hexagonal crystals. The same was discovered in the West more than 1,500 years later, by Johann Kepler in 1611.

But description of forms of crystals in ancient Chinese classics goes beyond that of snow and ice. The crystals and forms of crystals of more than 100 different substances are outlined in various Chinese pharmaceutical and alchemists' writings. Even the processes and conditions of the formation of certain crystals are given in some books. This is remarkable, as the books were written long before the birth of modern science.

A type of white quartz was described by Tao Hongjing in the period of the Southern and Northern Dynasties (420-589). In Tao's words: "It is the breadth of a man's finger and about two or three *cun* long. Its six facets are as smooth as if neatly cut. Limpid, it shines with a white lustre." The crystalline form of another mineral, selenite, a variety of calcium sulphate $CaSO_4 \cdot 2H_2O$, is described in *Meng Xi Bi Tan* (*Dream Stream Essays*) by Shen Kuo in 1086:

Selenite is formed in the brine of the salt marshes at Xiezhou [near Yuncheng, Shanxi Province], and can be extracted from the earth dug from the bottom of the ditches. The crystals range in size from that of an apricot-tree leaf to that of fish scales. All the crystals are in the shape of a regular hexagon, like the figures seen on the shells of turtles. A closer examination shows that the crystals have tiny flanges which cock up at the front and droop down at the rear, so that the crystals overlap and hang on each other as the scales on a pangolin do, presenting a whole appearance just like the shell of a turtle. The colour is transparent emerald, and when knocked the 'shell' breaks into longitudinal fragments with a shining lustre. The rifts are zigzag lines composed of the partial sides of tiny hexagons. When burnt, the 'shell' disintegrates into tiny, thin flakes, snow-white and lovely.

In *Ben Cao Gang Mu* (*Compendium of Materia Medica*) compiled by Li Shizhen in 1578, the various facets, angles and prisms of all the mineral crystals listed are minutely described.

Huang Di Jiu Ding Shen Dan Jing Jue (*The Yellow Emperor's Canon of the Nine-Vessel Spiritual Elixir*) compiled at the beginning of the Tang Dynasty describes a method of making crystals of what is now called potassium sulphate from *puxiao* (sodium sulphate) and *xiaoshi* (potassium nitrate) as follows:

> The two ingredients in their coarse mineral state are first pulverized, mixed and rinsed with hot water. The clear solution is then decanted and simmered over a slow fire for some time. It is then removed from the fire and the solution allowed to cool in the air till it is lukewarm, when it is poured into a small basin set in cold water to cool. The following morning there will be white and angular potassium sulphate crystals in the basin.

This very early work describes not only the form of the crystals but the process and conditions in making them as well.

Some classics tell even the optical and mechanical properties of certain crystals.

Motion

Different kinds of motion and their relationships, problems of time and space, the motion of a spheric body and its indifferent equilibrium, the loads and efforts on the wheel and axle and the inclined plane etc. are given in *Mo Jing*, which was written in the 4th century B.C. From that early time to the end of the Ming Dynasty (1368-1644), a great number of tools and machines based on the principles of simple mechanical motions or combinations of motions were invented in China, keeping abreast of the times.

The inertia of moving bodies was noticed and recorded early in the latter Spring and Autumn Period (770-476 B.C.). *Kao Gong Ji* tells about the inertia of a moving horse-drawn cart in these words: "When a cart is moving forward drawn by a horse but the horse is suddenly halted, there is a tendency for the cart itself to keep moving forward for a certain distance." Inertia was made use of when Zhang Heng (A.D. 78-139) of the Han Dynasty invented a seismograph in A.D. 132.

A water-lifting device which demonstrates the principle of the parallelogram of forces, called the *hudou*, first appeared in about the Yin time (c. 14th to 11th century B.C.). A water-bailing wicker bucket, the *hudou*, is fastened between two ropes which are pulled jointly by two persons standing apart ashore. Water from a river or pond below is scooped up and emptied into the field above.

The *hudou* at work—an illustration reproduced from *Wang Zhen Nong Shu* (*Agricultural Treatise of Wang Zhen*).

Other applications of the component and composite forces are found in the ship sail and rudder. The chapter on land and water vehicles in *Tian Gong Kai Wu* (*Exploitation of the Works of Nature*) written by Song Yingxing in 1637 is a pithy discourse on this subject. The chapter analyses in detail how the sail and helm of a ship should be handled in order to sail down, haul upon, or advance against the wind; or how to attain a desired speed.

Although no theoretical summary of the projectories of missiles and the velocity of free falling bodies can be found in ancient Chinese classics, the accuracy of missiles has always been much emphasized in all ancient and medieval writings concerning the making of projectiles. *Kao Gong Ji* written about 2,500 years ago and *Tian Gong Kai Wu* written on the eve of the birth of modern science are cases in point. *Kao Gong Ji* specifies standards for the making of the head, shaft and feathered part of an arrow so as to guarantee a good shot even in a strong gale. "A shaft which is uneven in weight, or too light or heavy as a whole, will seriously hamper a straight flight," says this book. "The amount of feather used on an arrow affects the speed and accuracy of the shot." Such statements concerning the relationship between the structure of an arrow and its flight made in the 2nd century B.C. show a meticulous observation and analysis that are advanced as compared with those of the Aristotelean school of physics in medieval Europe, in which the flight of all missiles was thought to be in a straight line.

Acoustics

Dai Nianzu

Ancient China made many achievements in acoustics, from the concept of vibrational waves to practical application, and from the fabrication of musical instruments to the study of musical scales.

Concepts of Wave and Vibration

People living on the shores of rivers and lakes in prehistoric times witnessed the waves formed on water surfaces by dropping stones, and when weaving nets they often saw rope waves. As early as the Neolithic Age such wave phenomena were artistically drawn on various ceramic artifacts. The figure on p. 140 shows a design on a pottery vase made during the Yangshao cultural period of the Neolithic Age depicting water waves produced by dropping a stone.

After long historic periods, men often explained the propagation of sound in terms of water waves.

Wang Chong, who flourished in the early Eastern Han Dynasty (A.D. 25-220), said in *Lun Heng* (*Discourses Weighed in the Balance*), Chapter Ephemeral Changes:

> A man sitting on a balcony looking at the ground cannot discern ants, even less the sound of ants. Why? The body of an ant is very small compared to that of man, and the sound emitted from its organ hole will not reach the ear of man. . . . A fish one *chi*[1] in length moving in water will cause the water on

---

[1]   1 (Eastern Han Dynasty) *chi* = 0.237 metre.

Pottery jar of the neolithic Yangshao cultural period, with water wave relief.

either side to vibrate. The central area of vibration would be only a few *chi* in diameter or at the most as big as that caused by a person. The extent of the vibration would reach no farther than a hundred steps, and at a distance of one *li*[1] all would be quiet and still on the water surface, because the distance is too great. A man producing sound by manipulating air is like a fish, the change of air is like that of water.

The above statement of Wang Chong says that sound is produced by the vibration of the throat and tongue and that sound is propagated by the air medium. It further explains the relationship between volume of sound and distance of propagation. We now know that due to the difference of the energy of sound vibration and

---

[1] 1 *li* = 1.800 *chi*.

to the damping of vibration a sound of specified strength can be heard only within a specified distance. The description and example given by Wang Chong contains rudiments of science.

Lun Qi (*Discourse on Air*), Chapter on Air and Sound, written by Song Yingxing of the Ming Dynasty (1368-1644), describes water waves more explicitly as an analogy for the propagation of sound in air. He wrote:

> Air (*qi*) has substance. . . . When an arrow flies through it, sound is produced by striking it; when the string of a musical instrument is plucked, sound is produced by vibration. . . . When a body strikes air, it is like a body striking water. . . . When one throws a stone into water, the place where the stone drops is no larger than a fist, but waves will spread outwards circularly. The vibration of air is the same.

Of course a sound wave is longitudinal, while a water wave is transverse. It is understandable however that the ancients were unable to distinguish between them.

Production and Transmission of Sound

Many Chinese books describe the various physical phenomena during the emission and transmission of sound, such as intensity, tone and resonance, etc.

The book *Kao Gong Ji* (*Artificers' Record*) which appeared during the later Spring and Autumn Period (770-476 B.C.) says: "The sound of a large squat bell is sharp and can be heard only a short distance away. The sound of a small and tall bell is mellow and can be heard a long distance away." This is a statement about the relationship between the structure of the bell and its intensity (volume) and distance of transmission. Some books before the 2nd century B.C. carry accounts of the tonal quality of various materials.

It was known very early in China how to control the pitch of stone chimes (*qing*). *Artificers' Record* says: "The chime makers when making chimes . . . will grind a small portion from the face if the pitch is too high; if the pitch is too low they will grind a small portion off the ends."

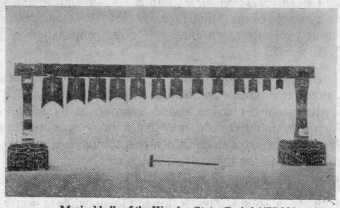

Musical bells of the Warring States Period (475-221 B. C.) unearthed in Xinyang, Henan Province.

Ancient Chinese musical bells were moulded in the form of two tiles put together, the reason being to prevent the sound waves from interfering with each other. The Song Dynasty scholar Shen Kuo (1031-1095) in his book *Meng Xi Bi Tan* (*Dream Stream Essays*) explained this theoretically, saying,

> All the ancient musical bells are flat like tiles; flat bells give out short sounds whereas round bells give out long sounds. Short sounds are abrupt, long sounds are undulating. Short and abrupt sounds will interfere with each other producing noise, and will not form musical notes. Later generations did not know this and made all bells round. When these bells are struck, they give out tremulous sounds, the clear sound no longer distinguishable from the dull sound.

Here the words abrupt (*jie*) and undulating (*qu*) represent two opposite qualities of sound. The abrupt refers to a short, clipped sound; an undulating sound is comparatively longer. "Abrupt and short" refers to the short rhythms in a fast melody and is not the same as in "short sounds are abrupt". It is the prolonged sounds plus the short rhythmic striking that results in interference with each other that produces confused noise. Shen Kuo's analysis of the sounds produced by ancient musical bells is quite scientific. The vibration of bells is similar to the vibration of boards. Round boards vibrate more continuously than any other material and shape. When a round bell is struck, the air inside the bell is whirled into turbulence, and the compression and rarefaction of the air at the orifice of the bell takes a longer time, prolonging the sound. In a fast-tempo melody the sound waves overlap and collide, preventing rhythmical music from being formed. When musical bells are moulded like two tiles combined, the sound waves are prevented from interfering with each other.

The pitch of string instruments is determined by three factors: length, density and tension of the strings. In the development of musical scales, the quantitative relationship between the length of string and its pitch was known to the Chinese: the longer the string, the lower the pitch. They also knew the variation of pitch with the density of string. Many books which appeared during the period of the Warring States (475-221 B.C.) contain accounts to this effect: "A thick string gives a low sound, whereas a thin string gives a high sound." The above is a qualitative description of such relations in *Han Fei Zi* (*The Book of Master Han Fei*). They also knew the effect of tension on pitch. When a thick string is too tautly adjusted so that the pitch is too high, there is the danger that a thin string on the same musical instrument will break when a' pitch is adjusted. Moreover, they also knew that atmospheric humidity affects the tension of strings. When the weather is humid, taut strings will loosen, causing a change in musical notes. (See *Huai Nan Zi* and *Lun Heng*.) The illustration on p. 144 shows the tuning experiment of ancient craftsman making a *qin* (Chinese lute).

Descriptions of resonance abound in ancient Chinese literature. When a body vibrates with the vibration of another body, this is called resonance. The two bodies in resonance have the same intrinsic frequency or frequencies in the ratio of simple integers. *Zhuang Zi* (*The Book of Master Zhuang*), which first appeared in

Detail of a painting by Gu Kaizhi of the Jin
Dynasty (265-420) showing a tuning experiment.

the 4th or 3rd century B.C., contains a description of resonance
encountered during the tuning of a *se* (zither):

> When one puts the *se* in a quiet room and tunes it, if one
> plucks the *gong* string [corresponding to the C string] then the
> other *gong* strings will also vibrate; if one plucks the *jiao* string
> [corresponding to the E string], the other *jiao* strings will also
> vibrate, because the notes are the same. If one tunes another
> string which corresponds to none of the other notes, then all
> twenty-five strings will vibrate.

This illustrates the resonance of prime notes with its harmonics.
This was quite a remarkable discovery in the history of acoustics.

An interesting story appears in *Liu Bin Ke Jia Hua Lu* (*Discourses of Liu, the Prince's Companion*) of the Tang Dynasty (618-907)
about a monk in the city of Luoyang who had a musical chime suspended in his room. The chime often sounded spontaneously,
and this made the monk so nervous that he became ill. Cao Shaokui, a close friend of the monk, heard about this and went to see
him. During his visit the monastery bell began to sound, and the

chime in the room sounded with it. Cao told the monk he could cure him of his illness. He visited the monk again the next day, bringing a steel file with which he filed off a small portion of the chime. After that it no longer sounded spontaneously. But the monk was perplexed and asked Cao why this was. Cao told him that his chime originally had the same note as the monastery bell, and when the bell was struck the chime would also sound. The monk was delighted to hear this and soon recovered from his illness. What this story shows is that the ancient Chinese people not only understood resonance; they also knew how to eliminate it. Filing off a portion of the chime changed its intrinsic frequency so that it no longer resonated with the bell.

In the 11th century, Shen Kuo of the Song Dynasty experimented on resonance with paper dolls. He cut out a small paper doll and put it on the string of a musical instrument. When the string in resonance with it was struck, the paper doll jumped up and down, whereas if other strings were struck, the paper doll did not move. This experiment was several centuries earlier than a similar experiment performed in Europe. In the 15th century, Leonardo da Vinci of Italy began to experiment with resonance, and it was not until the 17th century that William Noble and Thomas Pigott[1] of Oxford experimented with paper sliding marks to prove the resonance relation of prime note and harmonics.

Acoustics Applied in Architecture, Manufacturing Technique and Detection in War

In ancient wars, resonance was often used for detecting enemies, as frequently recorded in Chinese military books of different ages. The earliest account appeared in *Mo Zi* (*The Book of Master Mo*) which was written in the period of the Warring States. The story is as follows: Below the base of the city wall deep shafts were dug a few paces apart. Then within each shaft was placed a pottery jar

---

[1] Cf. Florian Cajori: *A History of Physics*, MacMillan Co., 1928, p. 104.

whose volume was about 80 litres. The orifice of the jar was closed by a leather membrane — actually an underground resonance box. Men with good hearing were posted to listen. If there were enemies planting mines to attack the city, the men would be alerted by the sound and could judge the direction and position of the enemies. Then countermines were planted within the city for counterattack. Another way of doing this was to bury two jars slightly apart in a deep shaft at the base of the city wall. Then according to the difference of the sound volume the direction of the enemy could be determined. In *Wu Jing Zong Yao* (*Collection of the Most Important Military Techniques*) written in the Song Dynasty (960-1279), such jars were called "listening jars". Shen Kuo wrote about a portable sound detector that was used when encamping in military march, this was an arrow quiver made of ox skin which served as a pillow for detecting sounds of enemy horses and troops several miles away.

Reflection and resonance of sound were often applied in building such special architectural structures as temples and palaces. The sound of the ancient *qin* (lute) was weak and low, and ancient Chinese built *qin* rooms where pottery jars were buried underground to serve as resonating chambers and enhance the sound effect. Some buildings had walls of pottery jars, their orifices facing inward so that the sound could not be transmitted outside the room. Each jar served as a sound absorber.

The Temple of Heaven in Beijing, which was built at the beginning of the 15th century, is a world-famous gem of architecture. Within the Temple grounds are three structures with good acoustical properties: Echo Wall, Triple Sounds Stone and Circular Mound.

Echo Wall is circular and about six metres high, with a radius of about 32.5 metres. There are three buildings inside it. The northern one called Huang Qiong Yu (Imperial Heavenly Vault), is the nearest to the wall, its closest approach being 2.5 metres. The entire circular wall is smooth and regular and is an efficient sound reflector. Two persons standing beside the wall at any two points can converse in low voices. When A speaks in *sotto* voice close to the wall towards the north, B will hear him distinctly. The sound that B hears does not come directly from A, but is bounced back

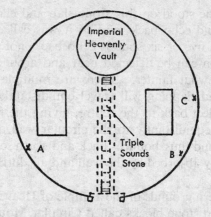

**Plane view of Triple Sounds Stone and Echo
Wall inside the Temple of Heaven in Beijing.**

from C.   So long as the sound at A comes out at an incidence angle
(relative to the tangent at A) smaller than 22°, the sound waves
will be continuously reflected by the circular wall and will not be
dispersed by the Imperial Heavenly Vault.

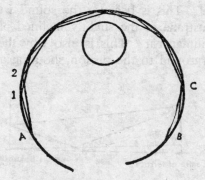

**Illustration of sound reflection of
Echo Wall in the Temple of Heaven.**

A stone path leads to the south of the Imperial Heavenly Vault,
the third stone of which is the centre of the circular wall. It is

said that when one stood on this stone slab and clapped his hands, three sounds would echo back, and so it was called Triple Sounds Stone. Actually, five or six sounds echo back, not just three. The same phenomenon can be detected near and around Triple Sounds Stone, only somewhat fainter. There are multiple echoes because the sound of hand-clapping will travel equidistantly to the circular wall and then reflect back to the centre, giving the first echo, which will again travel equidistantly to the circular wall and be again re-echoed. The sound thus bounces back and forth several times until its energy is totally absorbed by the wall and air during reflection and propagation.

In the southern grounds of the Temple of Heaven is a round terrace built of green stone blocks called Circular Mound. Its highest level is about five metres above the ground, its radius about 11.5 metres. Except for the four entrances on the west, east, north and south, the round terrace is surrounded by a green stone balustrade. The round platform is actually not flat, but slightly inclined from the centre outward. The whole Circular Mound is built of green stone and marble which have good acoustical properties. A person standing at the centre of the terrace and shouting will hear his own voice louder than usual. This is because the sound waves are reflected by the stone balustrade to the slightly inclined platform and then reflected to the human ear. This is also why the sound seems to come from underground to the person shouting at the centre.

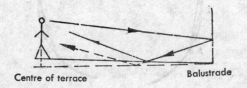

Centre of terrace   Balustrade

Illustration of sound reflection at Circular Mound.

Ancient architectural structures with acoustic qualities of such ingenuity are quite rare in the world.

Among other remarkable acoustic achievements of ancient China were musical instruments shaped like a bowl or plate into which different amounts of water were poured. When struck, these harmonized with other string instruments in an orchestra — an application of controlling the frequency of a vibrating body by varying amounts of water. This method came into use at least as far back as the 4th century. In *Shui Jing Zhu* (*Commentary on the "Waterways Classic"*), written by Li Daoyuan (466?-527) during the Northern Wei Dynasty (386-534), there appears an account of the architect Chen Zun who invented an approximate method of measuring distances by the speed of sound, similar to the method used today. The book *Chao Ye Jian Zai* (*Brief Record of Contemporary News*), which was written in the Tang Dynasty, tells about a craftsman who made a wooden mechanical monk that could beg in the streets and pronounce the simple words: *"Bushi* (Give donations)."

The Development of the Study of Musical Scales

The study of musical scales goes far back into Chinese history. By the beginning of the Western Zhou Dynasty (c. 11th century-770 B.C.) China had already established her own musical system: In one scale 12 pitches were determined and from the 12 pitches five or seven notes were selected to form a gamut. In the Spring and Autumn Period, the method of trisection was already formulated and used to determine the relationship between the length of pipes or strings and the pitch of the notes.

According to the rule of trisection, one starts from a length of string or pipe that is given as the prime note, divides it into three equal parts, then adds or subtracts one part from it to determine the length of another note. Mathematically speaking, that is to multiply the length of the string giving the prime note by 2/3 or 4/3 and follow this procedure until one arrives at the string or pipe giving out a note whose frequency is twice or 1/2 of the frequency of the prime note. Here the calculation of the 12 notes in one music scale ends. The law governing the variation of pitch with the length of string or pipe was first recorded in *Guan Zi* (*The Book of Master Guan*) about the 6th century B.C.

The differences of lengths or frequencies of two consecutive pitches in the 12 notes calculated by the rule of trisection are not all the same. This rule is therefore also called the rule of 12 unequal intervals (temperament).

After calculating all the notes in a music scale by the rule of trisection, the frequency that is one octave higher or lower than the prime note will only approximate 12 or 1/2 that of the prime note. Suppose the relative frequency of the prime note *do* is 1, then the relative frequency of *do* one octave higher, which is calculated by the rule of trisection, will not be exactly 2 but (let's say) 2.003. The difference then of 0.003 crops up. It took almost 2,000 years in the long history of musical scales to find how to eliminate this difference and return to 2.

Jing Fang (1st century B.C.) of the Han Dynasty, Qian Lezhi (5th century) and Shen Zhong (6th century) of the Southern and Northern Dynasties (420-589) had all tried to add more pitches in a gamut in order to eliminate or minimize this difference. Jing Fang used the trisection rule to increase the number of notes in a scale to 56 (nominally 60), Qian Lezhi and Shen Zhong had increased the number to 360. Even so they did not eliminate this difference, but only minimized it. When increasing the number of notes failed to solve the problem, Cai Yuanding (11th century) of the Song Dynasty turned back to a gamut of 18 notes.

He Chengtian of the Southern and Northern Dynasties made a daring innovation. He divided the difference in length into 12 equal parts then added this to the notes after the prime note in sequence. This made the ratio of the prime note and the note one octave higher exactly 1:2 in length. This method distributes the difference according to length, not according to frequency, and therefore falls short of the ideal. However it paved the way for the final completion of the rule of 12 equal intervals (temperament).

Zhu Zaiyu (16th century) of the Ming Dynasty invented the rule of equal temperament (rule of 12 equal intervals). In 1584 he used the geometric progression with ratio $12\sqrt{2}$ to complete the calculation of equal temperament, entirely eliminating the difference in the trisection rule and establishing the theoretical foundation for the production of modern keyboard musical instruments. His invention preceded similar work in Europe by 50 years. In the second half of the 19th century the German physicist Ferdinand von Helmholtz (1821-1894) appraised it highly.

Magnetism and the Compass

Lin Wenzhao

The magnetic compass was invented in China more than 2,000 years ago, and is often mentioned in ancient classics. Following is a short account of the invention of the compass, of ancient China's knowledge of magnetism and its impact on humanity.

The Magnetic Compass

A magnetic compass is an instrument showing direction based on the phenomenon that a magnetic bar or needle swinging freely in the earth's magnetic field will direct itself to lie in a magnetic south-north position. It differs from the south-pointing chariot in that the latter has differential gearing. The term "compass" in this article includes all the various models in different historical periods, among them the *sinan* (south-pointing ladle), the *zhinanyu* (south-pointing fish) and the *zhinanzhen* (south-pointing needle). The exact invention dates remain unknown, but it is clear that a primitive magnetic south-pointing instrument appeared very early in China as a result of the knowledge people gained over long years of productive labour.

Stages of Development

Perhaps the first primitive magnetic compass was *sinan*, which came into use in about the Warring States Period (475-221 B.C.).

A lodestone in the shape of a ladle swinging freely on a very smooth "earth-board", its "handle" naturally points south as the bulk of its body points to the magnetic north. *Sinan* is mentioned in many classics written between the Warring States Period and the Tang Dynasty (i.e. between 475 B.C. and A.D. 907). To quote from *Han Fei Zi* (*The Book of Master Han Fei*) written in the 3rd century B.C.: "The great emperors of the past dynasties created *sinan* to determine the four cardinal directions." An essay in *Gui Gu Zi* (*Book of the Devil Valley Master*) written in the 4th century B.C. says that when the people of the state of Zheng went far away for quarrying jade they took *sinan* along so as not to lose their bearings.

Magnetic ladles in ancient China were carved of lodestone as jade was carved. Jade-carving had already attained a fairly high level by the Shang and Zhou dynasties (c. 16th to 8th century B.C.). In the Spring and Autumn Period (770-476 B.C.) many jade articles were carved of materials of 5-7 degrees hardness on the Mohs scale. Lodestone with a hardness of 5.5-6.5 degrees Mohs was easily carved into the simple shape of a ladle.

The difficult part was making the "handle" of the ladle point exactly south. Moreover, repeated concussion during cleaving, chiselling and grinding inevitably caused the lodestone to lose a part or all of its magnetism, so the production rate was very low. There was also considerable friction between the ladle and the "earthboard" which placed a further limitation on the popularity of this magnetic compass, and eventually it was superseded by improved models.

The demand for a better direction-finding instrument became greater with the development of social productivity and navigation. Artificial magnetization was discovered in the course of man's productive activities, paving the way for more advanced models of magnetic compass; these were the "south-pointing fish" and "south-pointing needle".

The south-pointing fish and the south-pointing needle are mentioned in *Wu Jing Zong Yao* (*Collection of the Most Important Military Techniques*) compiled by Zeng Gongliang in 1044, and in *Meng Xi Bi Tan* (*Dream Stream Essays*) written by Shen Kuo in 1086.

A thin leaf of iron is cut into the shape of a fish and magnetized in the earth's magnetic field. When needed to direct troops in a march, the fish is made to float on a bowl of water. The fish's head will point south. The south-pointing needle is made by rubbing a needle with a lodestone, magnetizing it, after which it tends to point south. In Shen Kuo's words: "Magicians rub the point of a needle with a lodestone, making it capable of pointing to the south."

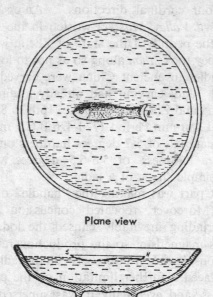

Plane view

Cross section

The "south-pointing fish" as described in *Wu Jing Zong Yao* (*Collection of the Most Important Military Techniques*).

Note that almost all compasses made before the 19th century (when electro-magnets appeared in industrial production) were made by this kind of artificial magnetization. The south-pointing needle was a great improvement upon both the lodestone ladle and the floating fish. Even the modern magnetic compass has an artificially magnetized needle as the principal part, though magnetization is now done in a electro-magnetic field and the needle is more sophisticatedly mounted to suit different purposes.

To install the magnetized needle, Shen Kuo made four experiments. He thrust the needle into a rush and floated it on water,

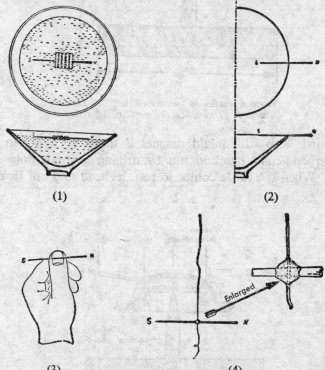

(1)   (2)

(3)   (4)

Shen Kuo's experiments illustrating different ways of installing a magnetic needle: (1) needle thrust through a floating rush, (2) needle pivoting on the rim of a bowl, (3) needle pivoting on a fingernail, (4) needle suspended by a thread.

balanced it on a fingernail or the rim of a bowl, or suspended the needle in air by a thread.

Two other methods often used for installing the magnet are given in *Shi Lin Guang Ji* (*Through the Forest of Affairs*) written by Chen Yuanjing in the Southern Song Dynasty (1127-1279): A wooden fish with a magnet inlaid is made to float on water. A

Wooden turtle as described in *Shi Lin Guang Ji* (*Through the Forest of Affairs*).

wooden turtle also with inlaid magnet is uniquely made to pivot atop a perpendicular bamboo pin by drilling a small hole under its body. When the turtle comes to rest, its head and tail lie north-south.

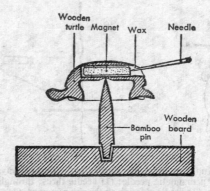

Cross section of the wooden turtle.

The fish, needle and turtle were apparently not at first attached to a disc marked with the 24 points of direction as no such disc is mentioned in Shen Kuo's book. Very soon, however, something like the "earth-board" used together with the lodestone ladle was found to be convenient with the fish and the needle as well, only the square board was replaced with a round disc. The number of points of direction remained at 24. Yet this was another major improvement on the previous models. A magnetic needle with a disc indicating the directions beneath it was already a prototype of the modern mariner's magnetic compass, though in the Song Dynasty such a compass was chiefly used by necromancers or geomancers. The oldest record of such discs, called *luojingpan* or *diluo*, is found in *Yin Hua Lu* (*Discourse on the Cause of Things*) by Zeng Sanyi of the Southern Song Dynasty, not much later than Shen Kuo. In the book Zeng says: As for the use of the *diluo*, there is the magnetic north-south system, but there is also a system which uses a needle to indicate the geographical north-south determined by the shadow of the sun. From Zeng Sanyi's words we know Zeng and the people of his time had already noticed that there is an angle between the magnetic north-south and the geographical north-south. The angle is now called the magnetic deviation or declination.

Diluo was first used with the floating needle (thrust in a rush), and the whole instrument, called *shuiluopan*, was equivalent to a primitive liquid magnetic compass. Xu Jing of the Northern Song Dynasty (960-1127) writes in *Xuan He Feng Shi Gao Li Tu Jing* (*Illustrated Record of an Embassy to Korea in the Xuanhe Reign*) A.D. 1124 that when bad weather prevented the use of heavenly bodies, the floating south-pointing needle was used by mariners at sea. Zhu Jifang, a Southern Song scholar, wrote in a poem about a voyage at sea:

Killicks were sunk in ascertaining a lonely islet,
A needle was set afloat to determine the four directions.

The first dry-disc magnetic compass, called *hanluopan*, appeared in the Jiajing reign (1522-1566) of the Ming Dynasty. What was new in a *hanluopan* was that the magnet was made to pivot atop a pin so that the friction was reduced to a minimum. This new model naturally prevailed over the floating needle since the pivot was fixed. When at sea, a pivoted axis gave better results than a floating needle, which was subject to water disturbance.

China was anticipated by the West in using the dry-disc magnetic compass, though the idea of the fixed pivot did originate in China.

Sinan had in fact a fairly fixed pivoting point. In Shen Kuo's experiments the magnetic needle was made to rotate on a fingernail or the rim of a bowl. The south-pointing magnetic turtle described by Chen Yuanjing and the "earth-disc" beneath it were actually a prototype dry-disc magnetic compass, only the turtle was to be replaced by a needle. In the 12th and 13th centuries, all models of the Chinese primitive compass could well have been introduced via the sea route to the Arab countries, and further via the Arab world to Europe, since navigation and trade by sea were already flourishing by the end of the Song Dynasty (1279), with Quanzhou and Guangzhou as two of the largest seaports in the contemporary world. Compared with foreign vessels, Chinese ones were bigger, stronger, faster and equipped with compasses. The many Arab and Persian merchants who entered China via Quanzhou and Guangzhou and stayed, preferred to sail on Chinese vessels, a fact that may have facilitated the invention of the dry-disc magnetic compass in Europe.

The primitive dry-disc magnetic compass was superior to the liquid magnetic compass in stability, though a fairly rough sea might disturb the needle, sometimes jamming it against the casing. The 16th century in Europe saw the mounting of the magnetic compass on gimbals — a simple combination of two copper rings and three pivot-axes maintaining an object in horizontal equilibrium and making it independent of motions of the ship and sea. This device had, however, already become quite popular in China when navigation between China and the West started early in the Han and Jin dynasties (206 B.C.-A.D. 420).

A 4th-century book entitled *Xi Jing Za Ji* (*Miscellaneous Records of the Western Capital*) tells of a tiny incense-burner being invented by a master handicraftsman named Ding Huan which did not spill the burning incense or the ashes no matter how the burner was turned. Used to perfume bedding, it was safe with the sleeper in bed as the inner container always kept its horizontal position. The principle of this incense-burner, produced in large quantities from the Han Dynasty, found better application in Europe in mounting the mariner's compass.

The Compass in Navigation

The magnetic compass played an important part in ancient and medieval China in military affairs, production, surveying, astronomy and people's daily life. Its most important use was in navigation, however.

Early in the Qin and Han dynasties (221 B.C.-A.D. 220) the Chinese travelled by sea to Korea and Japan. Navigation flourished in the Sui, Tang and Five Dynasties period (581-979) as trade between China and the Arab countries developed. In the Song Dynasty (960-1279), Chinese commercial fleets often sailed across the south Pacific and Indian Oceans. These developments in navigation were due partly to the use of the compass, its only alternative being the heavenly bodies. But rain or shine, day and night, the compass could be relied on to guide ships at sea.

The earliest Chinese report of the use of the compass at sea appears in *Ping Zhou Ke Tan* (*Pingzhou Table Talk*) written in 1119 by Zhu Yu of the Northern Song Dynasty. Zhu tells about Guangzhou's being a very busy seaport, and of the working conditions on board sea-going vessels. In his words:

> The sailors are sure of their bearings. At night they judge by the stars. In daytime they tell by the sun. When it is cloudy, they rely on the south-pointing needle.

The use of the compass only when heavenly bodies could not be seen indicates the sailors' long experience in navigation at that time, and that they were not yet quite accustomed to the instrument. The south-pointing needle was, however, more and more depended upon. Wu Zimu of the Southern Song Dynasty writes in *Meng Liang Lu* (*Records of a Dream of Grandeur*), 1275:

> When the weather is stormy or cloudy, only the compass can be relied upon. The navigator is put in charge of it so as not to make the least error. Everybody admits in awe that all lives on board the ship depend on the precise use of the instrument.

In the Yuan Dynasty (1271-1368), the mariner's compass was already regarded as necessary in all circumstances, as all important sea routes had been charted, with detailed marks on the compass to be taken at various places en route. These charts were called *luo pan zhen lu*, as mentioned in *Hai Dao Jing* (*Manual of Sailing Directions*) and *Da Yuan Hai Yun Ji* (*Records of Maritime Transportation in the Yuan Dynasty*) written in the 14th century. *Zhen La Feng Tu Ji* (*Description of Kampuchea*) written by Zhou Daguan in 1297 specifies the compass point to be taken by the ship upon leaving Wenzhou for Kampuchea aside from recording in detail things seen and heard abroad. "Steer the ship to sail along the *dingwei* needle direction," says Zhou. This is natural since Southeast Asia lies south by west of China, hence all ships leaving Wenzhou must sail southward and slightly west. The great Ming navigator Admiral Zheng He and his powerful fleet sailed to the western oceans seven times early in the 15th century, which greatly promoted China's trade and cultural relations with the Southeast Asian and East African countries, augmented international friendship and China's influence in the world. This would have been impossible without the compass. All the way from Liujiagang in Jiangsu Province to the north of Sumatra, Zheng's big ships were guided daily by the points of direction to be taken on the compass (i.e. *luo pan zhen lu*). From Sumatra westward, both *luo pan zhen lu* and the stars were used. The compass was important in charting a safe route for Zheng's fleet to sail to East Africa. Columbus' discovery of the Americas and Magellan's round-the-world voyages, both later than Zheng's adventures, would also have been inconceivable without the compass. These gallant voyages prompted the growth of a world economy, which was requisite for the development of capitalism in history.

Ancient China's Knowledge of Magnetism

Ancient China had ample knowledge of magnetism, judging from ancient classics. The Chinese first discovered magnetism in mining and refining iron ores, in which they encountered magnetite or magnetic iron ore (Fe_3O_4 or $Fe Fe_2O_4$). The earliest mention of magnetite is in *Guan Zi* (*The Book of Master Guan*) written by Guan Zhong and others in the 6th century B.C. or after. The essay says: "When there is magnetite above, 'copper-like gold' will be found below." The magnetic properties of this magnetite were gradually recognized in use.

Property of Attracting Iron

Magnetite (lodestone) has the property of attracting iron, and this easily distinguishes it from other ores. This property of magnetite was very early noticed in China and was often compared to the static electrical property shown in "the picking up of a mustard seed by a piece of rubbed amber". The ancient Chinese likened this to the affinity between mother and child. "The lodestone is the mother of iron. It attracts iron because of its motherly affinity for its child. A piece of stone without this affinity can never do the same," says Gao You in an annotation on the text of *Lü Shi Chun Qiu* (*Master Lü's Spring and Autumn Annals*) of 239 B.C. *Huai Nan Zi* (*The Book of the Prince of Huai Nan*) in 120 B.C. says: "The lodestone attracts iron, but not copper, to say nothing of tile or other earthenware." The attraction between lodestone and iron was further studied by Chen Xianwei and Yu Yan in the Song Dynasty (960-1279). They attributed it to an internal property of lodestone and iron, believing that a certain *qi* (physical force) in both brought the two together. "It is a kind of physical force that brings lodestone and iron together, as in the case of *yin* and *yang*, even when they are placed apart or there is something between them," said these two Song scholars. Liu Xianting (1648-95), a scholar of the Qing Dynasty, writes in *Guang Yang Za Ji* (*Miscellaneous Records of Guangyang*) that a unique property in the form of "an invisible force brings lodestone and iron together". Liu even mentions the phenomenon of magnetic screen in his book, saying: "When I was asked what could shield iron from being attracted by lodestone, my adopted son answered that only iron itself could do it." Liu adds that the reason for this can only be found in nature itself. That was the best Liu could do at a time when natural science was still in its infancy. But Liu's insistence on explaining natural phenomena by means of nature itself was materialistic.

In ancient and medieval China, the attraction of iron by lodestone was put to practical use. In a book *Tao Shuo (On Pottery)* Zhu Yan of the Qing Dynasty (1644-1911) says of a long-established practice in Chinese potteries and porcelain works: "When white porcelain is made, the glaze while in its liquid state has to be filtered through a layer of lodestone before application lest black blemishes form on the white surface." Traditional Chinese pharmacies have also from early times subjected every kind of powder that emerges from an iron roller or mortar to a treatment of lodestone to remove possible iron dust. Lodestone has also been long used in traditional Chinese surgery. In *Ben Cao Gang Mu (Compendium of Materia Medica)* written by Li Shizhen in 1596 appears an account of the application since the Song Dynasty (960-1279) of lodestone in such surgical procedures as extracting bits of iron from the eyes or throat. Today magneto-therapy has been developed into a modern branch of medical science effective for such diseases as arthritis.

Artificial Magnetization and Others

Two methods of artificial magnetization are mentioned in ancient Chinese classics: that mentioned by Shen Kuo of rubbing a steel needle on lodestone. In modern physics a ferromagnetic substance is regarded as an assemblage of small magnets called the domains. When the substance is unmagnetized the domains are arranged haphazardly. When it is magnetized, as after rubbing with lodestone, the domains inside the substance line up with their axes approximately parallel due to the magnetic field of the lodestone. Steel was used for the needle of a compass because steel was found to retain its magnetism and so could be made into a permanent magnet.

Another method was to magnetize an iron object in the magnetic field of the earth as described in *Collection of the Most Important Military Techniques*:

> In the fish method a thin leaf of iron is cut into the shape of a fish two *cun*[1] long and half a *cun* broad, having a pointed head and tail. The fish is heated in a charcoal fire. When red-hot, it is taken out by the head with iron tongs with the fish's tail due north. In this position it is partially quenched with water in a basin, submerging the fish's tail several tenths of a *cun*. The fish, so treated, is then kept in a tightly closed box.

Modern physics teaches that when iron is red-hot (i.e. above the Curie point of 600-700°C), the kinetic energy of its molecules is increased and the magnetic domains inside the iron are no longer fixedly oriented. Holding red-hot iron in the earth's magnetic field forces the iron domains to lie approximately parallel under the influence of the terrestrial magnetic field, and the rapid cooling in the quench fixes them that way, magnetizing the iron. Since China is situated in the earth's northern hemisphere, the terrestrial magnetic field dips at the north and so the iron fish was quenched with its tail dipped in water due north. In that position the iron was magnetized to the greatest extent by this method. The water was for rapid cooling as well as hardening the iron into steel. This second method of magnetization was entirely empirical, but it conformed with later findings in geomagnetism.

Magnetizing iron in the earth's magnetic field is also described in several books written in the Ming and Qing times. *Wu Li Xiao Shi* (*Small Encyclopaedia of the Principles of Things*) written by Fang Yizhi in 1664 quotes from a man named Teng Yi: "If an iron bar uniform in shape is horizontally suspended in air so that it pivots at its centre freely, it will stop in a south-north position." In 1570 Li Yuheng writes in *Qing Wu Xu Yan* (*Introduction to the Blue Raven Manual*):

---

[1] 1 (11th-century) *cun* = 3.19 centimetres.

Recently I met a geomancer named Wang Nongwan and learned from him that when any iron bar, thin or thick, is balanced horizontally in air by suspending it at the middle on a string, and made to rotate that way, it will always come to rest in a south-north direction like the needle of a compass. I have tried it several times and I am convinced.

The earth's magnetic field comprises magnetic deviation, magnetic inclination, and horizontal intensity. Magnetic deviation was discovered by Europeans in 1492 during Christopher Columbus' explorations at sea. Magnetic inclination was discovered in Europe even later, while in China both were noticed in the early years of the Northern Song Dynasty (i.e. not much later than A.D. 1000).

The making of the south-pointing fish as instructed in *Collection of the Most Important Military Techniques* (1040) showed some knowledge of geomagnetism and magnetic inclination.

When Shen Kuo said in 1086 that a needle could be made south-pointing by rubbing it with a lodestone, he added that "the needle often points slightly east of south and not strictly southward", creating the oldest record of magnetic deviation. Shen did not say that this was always the case, but stressed the word "often". Again when he told about the south-pointing needle being suspended horizontally in air, he said: "It *often* points to the south." Not much later Kou Zongshi, who compiled *Ben Cao Yan Yi* (*Dilations upon Materia Medica*) in 1116, cited Shen but with "slightly" deliberately deleted. The quotation became "The needle *often* points to the south by east" and "It *often* points to the south by the *bing* direction." Both Shen and Kou were careful in wording, especially with the word "*often*". Now it has been found that magnetic deviation varies with place and also with time, as the magnetic pole of the earth itself varies. Shen Kuo was probably writing from his personal observations at various places and over a long period of time. As stated, Columbus discovered magnetic deviation on his voyage from Spain to America in 1492. In 1634 other Europeans noticed that the deviation as recorded by Columbus at the same place was different. These discoveries were observed and recorded in detail, while Shen's observations were vague. However, the continuous variation of magnetic deviation was first recorded in *Dream Stream Essays* in 1086.

In the Southern Song Dynasty (1127-1279) the fact that magnetic deviation varies was more clearly recorded and was taken into consideration when geomancers used their compasses. Zeng Sanyi writes in *Discourse on the Cause of Things*: "Since heaven and earth are correctly placed on the *ziwu* meridian, one ought to use the *ziwu* line, but since there is what is now called the sloping off of the land south of the Yangzi River, it is difficult to use the *ziwu* system there, and *bingren* axis is preferred." This amounts to saying: When and where the difference between the magnetic north-south and the geographical north-south is not great, the *ziwu* line may be used. Since in the maritime provinces of southeast China there is a deviation between magnetic north-south and true north-south, *bingren* axis checks up better.

Magnetic deviation is shown on all geomancers' compasses made in the Yuan, Ming and Qing dynasties (A.D. 1271-1911), but is differently oriented. These can be regarded as original records of the variation of magnetic deviation at different places and time.

Ancient and medieval China had ample knowledge of magnetism and geomagnetism. The invention of the compass and its application in navigation indeed ushered in an epoch of magnetic navigation which was highly significant in world history.

Optics

Jin Qiupeng

Optics, an important branch of modern science, has proved its usefulness and pervaded all aspects of social life. Ancient China made important contributions to this branch of science in its long history.

Linear Propagation of Light

The sun provides mankind with light and heat, and is indeed man's source of light. When night falls, the earth is dark. Our ancestors in prehistoric times were helpless against the dark night. We do not know how many centuries elapsed before man discovered that fire could benefit him with light and warmth. Peking Man, who lived 500,000 years ago, already used natural fire, and a few tens of thousands of years ago he learned to make fire by drilling wood. For a long time fire was man's only artificial light source. Later, he invented oil lamps and wax candles, still using fire, and these were displaced only by the invention of modern light sources.

Over a long period of observing light phenomena man discovered that light rays filtering through tree leaves in a forest formed light beams, as is true for sunlight entering a room through a small window. Continual observations taught man that light travels in a straight line. To demonstrate this the outstanding Chinese scientist Mo Zi or his students performed the world's first experiment on pinhole inverse image formation nearly 2,500 years ago. Though they talked about "shadow" and not "image", the underlying principle is the same.

The experiment was to perforate the wall of a small dark room facing the sun. An inverse shadow was formed on the opposite wall of a person standing outside facing the hole. The Mohists explained this strange phenomena as light darting through the pinhole like an arrow taking a straight linear course. The head of the person blocks the light from above so that the shadow is formed below; his foot blocks the light from below so that this shadow is formed above, producing an inverse shadow. This was the first scientific explanation of the linear projection of light.

The Mohists further applied this property of light to explain the relationship between a body and its shadow. The shadow of a flying bird seems also to be flying. The Mohists discovered that the shadow, which remains fixed at any given instant, is formed because the light falling on the bird is blocked by it, since light travels in straight lines. When a bird is flying, the space darkened momentarily when the light is blocked will be illumined the next moment. That shadow will vanish and a new shadow is formed by the light that was blocked at the last moment. The Mohists concluded that "a shadow does not move". What they were saying is that a shadow is motionless at any particular instant. A shadow's apparent motion is due to the shadow of an object in motion constantly reproducing and changing place. The shadow of the bird therefore seemed to be following it in flight. It is really quite extraordinary that 2,400 years ago man made such a detailed study of the properties of light and explained the relationship between mobility and immobility of shadow. The Mohists, moreover, explained the phenomena of perspective and half-shadow by the principle of linear projection of light rays.

In the middle of the 14th century Zhao Youqin further investigated the relationship between pinhole and image formed by sunlight shining through a small hole in the wall. Zhao discovered that when the pinhole is tiny, the image is circular though the pinhole is not. During a solar eclipse, a notch also appears on the image corresponding to the blotting out of the sun. And although the size of the pinhole may vary, only the brightness varies, and not the size of the image. If one moves the screen nearer to the pinhole, the image will become smaller and the brightness will increase. After minute observation and speculation, Zhao Youqin arrived at the law of image formation through a pinhole. He maintained that when the hole is small enough the image formed is the inverse of the light source irrespective of the shape of the hole; in this case, the size of the hole is only related to the brightness of the image and unrelated to its shape. When the hole becomes bigger, the image will be the erect image of the hole, that is, not inverse.

To prove this conclusion, Zhao Youqin devised a further experiment. Two round wells about four feet in diameter were dug in the earthern floor of each of two downstairs rooms. The well on the right side was four *chi*[1] deep, that on the left eight *chi* deep. A table four *chi* high was placed in the well on the left so that the effective depth of the two wells was the same. He then made two round boards four *chi* in diameter and placed 1,000 candles on each board. After lighting all the candles, one board was placed in the bottom of the well on the right, the other board being placed on the table in the well on the left. He then covered the two well mouths with two round boards five *chi* in diameter with a square hole at the centre of each. The hole in the left board was about one *cun*[2] square, however, while the hole in the right board was about half a *cun* square. The images appearing on the ceiling were both round, only the one of the larger hole was brighter than the other. Zhao Youqin used the principle of linear projection of light to explain how candles on the east will form images on the west and vice versa, the same applying to north and south, every candle having a corresponding image. Because the 1,000 candles were massed into a round shape, the image formed by the interspersed separate images was also round. This illustrates that when the light source, hole and screen distances remain unchanged, the shape of the image also remains unchanged, only the illumination differs — the larger hole "will accommodate more light" and is therefore brighter, while the smaller hole "will accommodate less light" and is therefore dim-

---

1 1 (14th-century) *chi* =0.321 metre.
2 1 *cun* = $^1/_{10}$ *chi*.

mer.   If 500 candles on the east in the well on the right were extinguished, then the image formed on the ceiling of the right side room
would have a blank on the west.   This corresponds to the shadows
appearing during solar and lunar eclipses.   If, in the well on the left,
only 20 or 30 candles were lighted but spaced evenly, there would
be some unconnected dark squares though the images would be distributed in a round shape; if only one candle was lighted, the square
hole would not be small compared to the light source and so the
image formed would be square.   When all the candles were relighted,
the image on the left was round again, and when two large boards
acting as image screens were suspended on the ceiling parallel to the
floor, the distance observed between the screen and hole decreased,
and the image became smaller and brighter.   Then the two suspended boards were removed, the ceiling acting as the screen, the table
in the well in the left side was removed and the candles placed [in the
bottom of the well.   Now the light source of the left well was farther from the square hole and the image on the ceiling became smaller; it also became dimmer, because the candle-light was weak and
increasing the distance decreased the brightness.   These experiments
led Zhao Youqin to infer the law of pinhole image formation and
formulate the relationship between the distance and brightness of
the light source versus the distance of pinhole and image screen.
He pointed out that when the image screen is nearer to the pinhole
the image becomes smaller and vice versa; also that when the candle
is nearer to the pinhole the image becomes larger and vice versa,
and that the smaller the image the brighter it becomes, and vice versa.
He also pointed out that if the light is weak even when the candle
is near the hole, the image is dim; and that if the light is strong, the
image is brighter even when the candle is farther from the hole.
The last step of the experiment was to remove the two boards covering the wells and suspend from the ceiling above each well a round
board one *chi* in diameter, the board on the right having a square
hole four *cun* in width and the board on the left having an equilat-
eral triangular hole of five *cun*.   The lighting was then adjusted,
i.e. the distances between light source, hole and image screen were
changed.   The left image now formed on the ceiling was triangular

while the one on the right was square, demonstrating that when the hole is large the shape of the image formed is the same as the shape of the hole; that as the distance between hole and screen becomes smaller the image becomes smaller and brighter; and that as the distance between hole and screen increases, the image becomes larger and dimmer.

From the above experimental data, Zhao Youqin concluded definitely that the image of a small hole is the same as the shape of the light source and that the image of large hole is the same as the shape of hole. Zhao's work was at that time the sole proof of the linear projection of light and only demonstration of the principle of pinhole image formation by such sophisticated experiments.

China's shadow play was a very early application of this property of light. At the beginning of the Han Dynasty (206 B.C.-A.D. 220), Qi Shaoweng made paper into human figures and objects to manipulate in performances behind a stage curtain. Light shone on these figures cast its shadow on the curtain, and the audience witnessed the first shadow plays. This drama medium flourished during the Song Dynasty (960-1279) and was later taken to the West, where it created a sensation.

Principle of Image Formation of Mirror Surfaces

Light, however, demonstrates the phenomenon of reflection when its straight path is obstructed, and this too was known in China from ancient times.

Man had observed that at the beginning and end of the lunar month, the moon changes from the first quarter to the full moon and then to the last quarter and back again, changing cyclically. The ancient belief was that the moon itself emitted light. By about the 4th century B.C., the Chinese knew that this was not true, but that moonlight is light from the sun reflected from the surface of the moon. To demonstrate this, the Song scientist Shen Kuo conducted an experiment with an illuminated sphere. Then, using a ball to represent the moon, he sprayed white powder on half of the ball to represent that side of the moon which is lighted by the sun. Viewed from the side, "the part sprayed with white powder looks like a hook", said he, while viewed from the front "it looks circular". He thus explained how the full moon becomes crescent and then round again.

The technique of making mirrors was highly developed in ancient China. More than 3,000 years ago the Chinese were making and using bronze mirrors. They also investigated the principle of mirror image formation.

Man looked at himself in water before using a mirror. He knew that running water could not form images. As early as the 11th century B.C. bronze mirrors were used in China, and from the 3rd century their manufacture was improved so that they were exported as well as used locally. These mirrors remain to this day as valuable artifacts in world cultural history. The so-called "light-penetration mirrors" of more than 2,000 years ago attract great interest. These mirrors reflect on their polished surfaces designs which are executed in relief on their backs. The enigma of these mirrors occupied scholars in China and other countries for several hundred years, until only in modern times was the secret discovered. The reproduction of the design on the back of these mirrors on the polished surface was due to very slight inequalities of curvature, showing the highly sophisticated mirror making and remarkable understanding of light reflection in ancient China.

In the 2nd century B.C. China produced the world's first periscope by applying the principle of reflection of plane mirrors. In the book *Huai Nan Wan Bi Shu* (*The Ten Thousand Infallible Arts of the Prince of Huai Nan*) which appeared early in the Han Dynasty is this account: "Suspend a large mirror high up, put a basin of water underneath it and you can see the people around you." This simple though crude device was highly significant as the modern periscope is made on the same principle.

While using the plane mirror, people discovered certain strange phenomena of spherical mirrors, which are either concave or convex. China had for a long time studied the focusing property of concave mirrors and application of concave mirrors to kindle fire from the sun. The ancient name of China's concave mirrors was *yangsui* (sun-burning mirror), which means an implement to obtain fire from the sun. This was the first application of solar energy. As early as the 5th century B.C. elaborate studies on concave mirrors were carried out by Mo Zi or his pupils, their findings being recorded in the famous ancient scientific work *Mo Jing* (*Mohist Canon*). They discovered that when an object is placed in the centre of a sphere, the image is upright, the nearer the object is to the centre the larger the image will be, and the farther the object is from the centre the smaller the image will be. They also discovered that when an object is placed off-centre of the sphere the image is inverted, the nearer the object is to the centre the larger the image will be, and the farther the object is from the centre the smaller the image will be. Furthermore, when the object is placed at the centre of the sphere, the object and the image will coincide. The Mohists had already clearly distinguished between the focus and the sphere centre. They called the focus *zhongsui* (central fire).

The Mohists also studied convex mirrors and knew that wherever the object is placed relative to the convex mirror there will only be an upright image formed on the other side of the mirror's surface. This is a virtual image and always smaller than the object; the nearer the object is placed to the centre the larger the image will be, and the farther the object is placed from the centre the smaller it will be.

Ancient China's mirror makers applied the image-formation features of convex mirrors most skilfully, moulding the surface of large mirrors into a flat plane and making small mirrors slightly convex so as to reflect a person's whole face. The reflecting convex mirrors on vehicles and the large convex mirrors at road turns apply the same principle.

In the 11th century Shen Kuo correctly expressed the principle of image formation of concave mirrors, his findings based on the research of his forerunners. When you place a finger in front of a concave mirror to form an image, the image will change as your finger moves farther away or nearer to the surface of the mirror. Shen Kuo used this phenomenon to explain the relationship between the image of the concave mirror and its focus. As the finger moves nearer to the mirror, the image formed is upright; as the finger moves farther from the mirror the image disappears, because no image will be formed when the finger is at the focus. As the finger moves beyond the focus the image is inverted. He pointed out that concave mirrors "concentrate the light into a point", which he called *ai*. This is the so-called focal point in modern optics. He also used the window gap (image formed by a pinhole), fulcrum of an oar, and the thinnest part of a waist-drum as lively analogies to explain the formation of inverse images by concave mirrors.

Now we know that parallel beams of light reflected from a concave mirror will be focused at its focal point. The images formed by a concave mirror are the following: 1. when the object is placed beyond the centre of the sphere, the image is real and inverted but smaller than the object; 2. when the object is placed at the centre of the sphere, the image and the object are in the same position and have the same size but face opposite directions; 3. when the object is placed between the centre of the sphere and the focal point, the image is real and inverted but larger than the object; 4. when the object is placed at the focal point, there will be no image because light beams coming from the focus will be parallel when reflected from the concave surface; 5. when the object is placed inside the focal point, the image is virtual, upright and larger than the object. Convex mirrors will not form real images wherever the object is placed, the images will be upright, virtual and smaller than the object.

Mo Jing uses the sphere centre to differentiate the relation of object and image, but it makes no mention of the image formed when the object is placed between the centre of the concave mirror and its focal point. This is a defect, but we must remember that all this happened nearly 2,500 years ago when optics was still in its infancy, and a description of such specificity at that time is indeed admirable.

Refraction and Dispersion of Light

When light penetrates air into a transparent medium, refraction will occur, a fact known by mankind in past ages. For example, when a wooden bar is stuck obliquely into water it appears broken due to the refraction of light. Before glass, the Chinese people's knowledge about lenses was limited, yet they learnt about focusing convex lenses by using what they did have — ice. More than 1,000 years ago in the Jin Dynasty, Zhang Hua wrote in his book *Bo Wu Zhi* (*Record of Investigation of Things*): "Cut a piece of ice into a sphere, lift it in the sun and let its shadow fall on a piece of moxa; the moxa will be set alight." A remarkable discovery. Ice will melt in heat, but people in ancient times fashioned it into convex lenses to make fire by focusing light. We can see that the principle of focusing convex lenses was sufficiently understood at that early time but that its practical application was made possible only by repeated experiments and great effort.

The ancient Chinese people also explored what caused rainbows. Early in the Tang Dynasty (618-907) Kong Yingda stated: "When the sun shines through thin clouds onto raindrops, a rainbow will appear." This account is far-reaching, explaining as it does the conditions for rainbow formation: thin clouds, sunshine and raindrops. It says that the rainbow is the natural phenomenon when sun shines on raindrops. In the middle of the 8th century Zhang Zhihe experimented with making artificial rainbows. With his back to the sun he sprayed small water droplets and saw phenomena like a rainbow, demonstrating that a rainbow is the result of sunlight shining through water drops. He pointed out further that to see a rainbow you must stand with your back to the sun, that if you face the sun you cannot see it. Shen Kuo on his way to the Qidan (Khitan) in north China reached the same conclusion on the basis of actual observation. Shen Kuo quoted Sun Yanxian as saying: "A rainbow is the reflection of the sun in the rain; when the sun shines on rain, a rainbow will appear." On the basis of his predecessors' experience Zhu Xi (1130-1200) of the Southern Song Dynasty stated that when the rain is over and the sky begins to clear, a rainbow will appear. "It is not the rainbow that causes the rain to stop, but when the rain is thinning it is rain dispersed by sunlight that causes the rainbow to appear." These explanations of Sun Yanxian, Shen Kuo and Zhu Xi as "the sun's reflection in rain" and "rain dispersed by sunlight" are not accurate and even fallacious. We now know that rainbows result from sunlight dispersed in the raindrops through refraction twice and total reflection once (or twice). Still, their comparatively enlightened explanation at that time is impressive.

In the Song Dynasty, prolonged observation also led Chinese scientists to the conclusion that the red glow appearing in the morning and evening is the result of slanted sun rays. A poet of that time wrote: "The slanting sunlight brightens the glow. The rain divided by the rainbow." A lucid as well as poetic description of the relevant meteorological condition.

While the Chinese were gaining understanding of multi-colour rainbow formation by raindrops, they also discovered, not later than the 10th century, that natural transparent crystals when illuminated by the sun will emit spectral light. They called these crystals "five-colour stone" or "iridescent stone". Later they learned that transparent crystals are hexagonal, that "when sun shines on one, the light breaks into five colours like the rainbow", so linking light dispersion by transparent crystals with rainbow formation. This link stands today.

These statements about dispersion caused by sunlight through water drops and crystals, though primitive and crude, have important significance in the history of knowledge about light. It shows that man began very early to probe the mystery of the dispersion of light, and recognize it as a natural phenomenon — a great step forward in man's knowledge of light.

The Invention and Development of Papermaking

Pan Jixing

. The making of paper has contributed materially to the advancement of world science and culture. Its invention, along with that of the compass, gunpowder and printing, is indicative of China's high levels of science and technology in ancient times.

Paper was a brand-new type of writing material. Those used before had been tortoise shell, bone, metals, stones, bamboo slips, wooden tablets and silk. Records inscribed on bones or tortoise shells and bronzes during the Shang Dynasty (c. 16th-11th century B.C.) have come to light continuously since the turn of this century, and in recent years bamboo slips, inscribed wooden tablets and books and paintings on silk of the Warring States Period to the Qin and Han dynasties (5th century B.C.-A.D. 2nd century) have also been found in large numbers. None of these materials, however, suited the need. Tortoise shell was scarce, metal and stone cumbersome, silk costly and bamboo slips and wooden tablets took up too much space. The growth of China's economy and culture made more pressing the demand for a cheap and easily obtainable writing material. Experimentation over the years resulted in a vegetable-fibre paper. The raw material used was hemp, as in rope ends, rags and wornout fishing nets.

The historian Fan Ye of the 6th century in his "Biography of Cai Lun" in *Hou Han Shu* (*History of the Later Han Dynasty*) credits Cai Lun (d. 121), who served as a eunuch at the imperial court of the Eastern Han Dynasty (25-220), with the invention of paper in the year 105. However, no clear indication that Cai Lun invented paper can be found in his biography in *Dong Guan Han Ji* (*Han History Compiled at Dongguan*), which was completed earlier than *Hou*

176

Han Shu. Dong Guan Han Ji was written by a group of historians including Cai Lun's contemporaries Liu Zhen and Yan Du. It is strange that they should fail to record such an event if Cai Lun had in fact invented paper. Until the Tang Dynasty (618-907), *Dong Guan Han Ji* had always been regarded as the official history of the Eastern Han Dynasty. It mentions Cai Lun only as an official (*shangfangling*) in charge of papermaking in the imperial workshop. Although *Dong Guan Han Ji* was lost after the Song Dynasty (960-1279), some passages have come down through quotations in works of Sui (581-618) and Tang (618-907) dynasty authors. Cai Lun can thus scarcely stand as the inventor of paper, nor do archaeological excavations of this century substantiate the claim.

Remnants of Western Han Dynasty hemp paper of 49 B.C. were discovered in 1957 in Baqiao in an eastern suburb of Xi'an in Shaanxi Province. Microscopic study of these specimens by the author indicates that a small amount of ramie was combined with the hemp. It is the oldest vegetable-fibre paper extant today. In 1973-74 specimens of hemp paper of Western Han Dynasty (206 B.C.-A.D. 24) were found on the old site of Jinguan of the Han Dynasty, and in 1978 by the Zhongyan Production Team of Taibai People's Commune in Fufeng County of Shaanxi Province. These show that paper was invented by the Chinese labouring people more than 2,000 years ago. The role of Cai Lun appears rather as an improver of the coarse hemp paper of the Western Han Dynasty. He presented to the emperor in 105 a fine-quality paper which he produced by sponsoring and organizing the abundant man-power and natural resources of the court, which he as *shangfangling* administered. The technique of papermaking spread to all parts of China from that time. Meanwhile, the bark of paper-mulberry (*Broussonetia papyrifera*) and some other trees was added to the supply of raw materials in the Eastern Han Dynasty. The role of Cai Lun should not therefore be underestimated.

Numerous specimens of fine paper with characters written on them dating from the 2nd century have been unearthed in Xinjiang, Inner Mongolia and Gansu Province since the turn of this century. Simulated experiments indicate the processes in the manufacture of hemp paper to be roughly as follows:

Raw materials such as rope ends and rags are soaked in water (retted) and after expansion cut to pieces with an axe and then cleansed in water. The pieces are then cooked in a grass-stalk ash solution, which may be cited as the earliest alkaline treatment by chemical method. When such impurities in the raw materials as lignin, fructose, colouring matter and fat have been further removed, the pieces are washed in clear water and pounded in a mortar. The fine fibres after pounding are mixed with water to form a liquid pulp suspension. Paper moulds are used to scoop up portions of the liquid substance, which become paper sheets when dewatered and dried. If the surface of the paper is rough and creased, it requires calendering before it is fit for writing on.

The success of the labouring people as early as the Han Dynasty in making paper of vegetable fibres from cast-off textiles by chemical and physical methods, their recycling of fibrous material with crude apparatus, is an achievement worthy of note in chemical and technological history. Two technological cruxes present themselves: The removal of the noncellulosic constituents in the fibrous materials and cutting and fibrillating the macromolecules of the pure fibres by powerful pounding; and the use of a flat screen over which the pulp will run slowly and drain off most of the water while retaining the fibres. These moist fibres when dried become paper of specified tensile strength. The flat screen is called a mould, forerunner of modern Fourdrinier and cylinder papermaking machines.

After the spread of the papermaking technique throughout China in the 2nd century A.D. paper became a powerful competitor with silk, bamboo and wooden tablets. By the 4th century it had basically replaced these bulky materials for writing, a change which accelerated the propagation and growth of China's science and culture. During Wei, Jin, and the Southern and Northern Dynasties (3rd-6th century) many technical innovations were made. Mulberry and rattan were used as raw materials in addition to hemp and papermulberry. As for equipment, the advent of the "laid" transfer mould with movable bamboo screen set on a frame made it possible to scoop up thousands of moist paper sheets in rapid succession and in this way greatly raised work efficiency. With cooking in stronger alkaline solution and more efficient pounding in mortar the quality of paper improved, and processed papers such as coloured paper, coated paper and loaded paper appeared.

Examination of specimens of ancient paper of this period found in the Dunhuang grottoes and in the deserts of Xinjiang reveals the fibres evenly distributed and interwoven, white and with smooth surface, giving a really "beautiful and lustrous appearance". Jia Sixie of the 6th century devoted two separate chapters of his *Qi Min Yao Shu* (*Important Arts for the People's Welfare*) to describing the treatment of paper-mulberry bark as raw material for paper and the technique of dyeing paper with *huangbo* (*Phellodendrum amurense*) or yellow bark. It was in this period that the technique of paper-making found its way to Korea and Vietnam, China's near neighbours, and a wider popularization of papermaking technique began.

During the Sui, Tang and Five Dynasties (6th-10th centuries) *tan* tree-bark, daphne-bark and rice- and wheat-stalk papers and a new type of bamboo paper appeared in China in addition to those made of hemp, paper-mulberry, mulberry bark and rattan. In south China where bamboo is extensively cultivated, the manufacture of bamboo paper developed rapidly.

This bamboo paper was thought by some to have originated in the Jin Dynasty (265-420), but neither the literature nor artifacts substantiate this. Viewed technically, the appearance of bamboo paper should have occurred after a certain degree of development in the technique of making bark paper, as the treatment of such tough fibres as bamboo would have been quite beyond the possibility of bamboo paper appearing in the Jin Dynasty. Bamboo paper more likely originated after the Tang Dynasty, developing rapidly between Tang and Song.

Paper was now manufactured in all parts of north and south China. Then came the invention of block printing, which gave rise to book production which in turn speeded the expansion of the paper industry. Increased output, improved quality and steadily falling cost brought a variety of paper products into wide use in people's dialy life. Among the rare papers of this period were the hard yellow paper and Chengxintang paper of the Tang Dynasty. There were also water-marked and various processed art papers. The number of Tang Dynasty paintings done on paper reflects improvement in papermaking technique during that period.

From the 10th to the 18th century, during the Song, Yuan, Ming and Qing dynasties, bark paper of paper-mulberry and mulberry and bamboo paper became very popular and were in great demand. The screen for the forming of paper was usually made of fine polished bamboo strips densely interwoven, while the pulp was well pounded so as to produce paper of uniform texture. Starch-paste

Cutting and retting bamboo in a pond—an illustration from *Tian Gong Kai Wu* (*Exploitation of the Works of Nature*) by Song Yingxing, 1637.

was generally used in the Tang Dynasty as sizing, which also served as filler and to keep the fibres in suspension in the vat. In the Song Dynasty, a sticky vegetable juice called *zhiyao* (paper drug) was

蕩料入簾

**Dipping the bamboo pulp to form a sheet of paper
—an illustration from _Tian Gong Kai Wu_.**

commonly added to acquire an even distribution of fibres in the pulp. The most frequently used agents were a maceration extract of _yangtaoteng_ (_Actinidia chinensis planch_) and _huangshukui_ (_Hibiscus abelmoschus_). The use of such "glues", which had begun in the Tang Dynasty, became so popular during Song that the starch paste method was discarded.

Apart from the extensive use of paper for painting, calligraphy, printing and daily life in China, it was also first used here for the printing of paper money. Called _jiaozi_ (exchange medium) in the Song Dynasty, its issue continued in Yuan and Ming. Other countries followed China in issuing paper money. The wallpaper, paper flowers and paper-cuts of the Ming and Qing dynasties are most attractive and found markets within the country and abroad. The various coloured paper, _lajian_ (waxed paper), _lengjin_ (cold gold), _nijin_ (sprinkled gold) *luowen* (woven paper), *ni jin yin jia hui* (painted and flecked with gold or silver) and *yahua* (calendered pattern) were virtually reserved for the feudal ruling classes, being quite beyond the reach and requirements of the general user.

Books on papermaking were also written during this period, among them *Zhi Pu* (*Manual on Paper*) by Su Yijian of the Song Dynasty in 986, *Shu Jian Pu* (*Manual on Sichuan Writing Paper*) by Fei Zhu of the Yuan Dynasty, and the especially noteworthy *Tian Gong Kai Wu* (*Exploitation of the Works of Nature*) written in 1637 by Song Yingxing of the Ming Dynasty. In this latter work Song Yingxing devotes the entire Chapter 13: "Sha Qing" (Killing the Green or Papermaking) to a technical description and illustration of making paper from bamboo and tree bark.

This excerpt from Chapter 13 outlines the steps in the manufacture of bamboo paper.

In the fifth month, the bamboos on the mountains are cut into pieces five to seven feet long. After soaking in a pond for more than 100 days, they are carefully pounded and washed to remove the coarse husk and greenery. The inner fibres are then mixed with a high-grade lime solution and boiled for 8 days and nights in a cauldron. After the fire has been allowed to go out for one day, the bamboo fibres are taken from the cauldron and thoroughly washed in clear water until they are clean. Next they are soaked in a solution of wood ash and returned to the cauldron for boiling. When the mass reaches boiling point, it is strained. This is repeated for 10 days, when the bamboo pulp naturally becomes odorous and decayed. It is taken out and pounded in a mortar until it resembles clay or dough and then poured into a vat for moulding.

Song Yingxing's description of the processes of making bamboo paper is similar to later accounts by other authors.

The technique of papermaking found its way through Korea to Japan in the 7th century and was introduced into Arabia through Central Asia in the middle of the 8th century. The first workshops at Baghdad in Iraq, Damascus in Syria and Samarkand in Central

Asia were established with the knowledge imparted by Chinese craftsmen. The earliest hemp paper made from rags in Arabia was manufactured using methods and equipment quite similar to those used in China. When Arabian paper was being mass produced, it was exported to European countries in large quantities, and through Arabia the art of papermaking came to be introduced into Europe.

The first European countries to set up paper mills were Spain and France in the 12th century. They were followed by Italy and Germany in the 13th century. By the 16th century, paper was extensively used in Europe and finally completely replaced the traditional parchment and Egyptian papyrus. From then on, paper gradually spread throughout the world.

Gunpowder and Firearms

Zhou Jiahua

Gunpowder, a mixture of saltpetre, sulphur and charcoal, was invented in China more than 1,000 years ago. Some prefer to call it black powder because of its colour and civilian use today. But it is also known as brown powder, since its colour fades when composition varies.

In Chinese the term *huoyao* literally means "fire drug", or a "medicine" which easily catches fire. But why is it a "drug"?

Saltpetre and sulphur, indispensable ingredients of gunpowder, are important drugs in *Shen Nong Ben Cao Jing* (*Shen Nong's Materia Medica*). Written in the Han Dynasty (206 B.C.-A.D. 220), this is the earliest known pharmacopoeia in China. Since its invention, gunpowder has also been regarded as a medical agent in traditional Chinese pharmacology. Li Shizhen of the Ming Dynasty writes in *Ben Cao Gang Mu* (*Compendium of Materia Medica*) that gunpowder cures skin diseases and is an insecticide, desiccant and disinfectant. Gunpowder was a product of alchemy practised over a long period, named according to the Chinese alchemists' habit of calling every product or reagent a "drug".

The invention of gunpowder underwent a lengthy process. Perhaps arrived at fortuitously at first, it was deliberately improved to keep pace with social progress and production.

While pursuing various productive activities the ancient Chinese gradually grasped the properties of charcoal, sulphur and saltpetre. In the dynasties of Shang and Zhou (c. 16th-3rd century B.C.) charcoal was used in smelting metallic ores having been found to be superior to firewood as fuel for this purpose. Sulphur, a natural

184

substance, long ago caught people's notice because of the pungent odor of its gaseous compounds in smelting certain metallic ores, and was also noticed near hotsprings. Its chemical and pharmaceutical properties were learned over a lengthy period by the labouring people of ancient times, who later exploited it from mines. Alchemists found it unique in its properties. *Shen Nong's Materia Medica* says that it can ". . . transform gold, silver, copper and iron and is therefore a miraculous substance". In modern chemical parlance this means that sulphur reacts with metals like copper and iron to form compounds. *Zhou Yi Can Tong Qi* (*Kinship of the Three and the Book of Changes*), the earliest alchemists' classic written in the Eastern Han Dynasty (25-220), tells of sulphur and mercury combining to form red cinnabar. These properties of sulphur deeply impressed the alchemists, who set great store by them. Sulphur was used in almost all recipes for "potable gold" and "cyclically transformed gold elixir". In numerous alchemists' experiments it was found to be poisonous and highly volatile when heated. The alchemists soon found a method to "subdue the toxicity and volatility of sulphur by fire". They mixed sulphur with other inflammable substances (usually saltpetre), very carefully ignited the mixture till further ignition brought forth no more flames, then considered the toxicity and volatility "subdued". The mixing in of saltpetre was a step towards the invention of gunpowder. Saltpetre is also chemically active, inflammable on red hot charcoal and reactive with many other substances. For these reasons it was also much used in alchemy. Since like other salts such as *puxiao* (Glauber's salt $Na_2 SO_4$) it is white, alchemists learned to differentiate it from other salts by its peculiar violet flame over fire. Tao Hongjing, pharmacist-alchemist in the period of the Southern and Northern Dynasties (420-589), writes in *Ben Cao Jing Ji Zhu* (*Commentaries on Materia Medica*): "If saltpetre is ignited, it gives out a bluish purple flame." Modern chemistry bears this out. The flame test paved the way for accurate and large-scale exploration, mining and use of saltpetre.

Before the end of the Tang Dynasty (618-907) the Chinese had found out in the much often repeated practice of "subduing sulphur or saltpetre by fire" that a mixture of saltpetre, sulphur and charcoal is instantaneously combustible.   In Vol. 5 of *Zhu Jia Shen Pin Dan Fa* (*Methods of Elixir Preparations*) there is a "process for the subduing of sulphur" described by Sun Simiao, a pharmacist-alchemist of the Tang Dynasty, in his book *Dan Jing* (*Elixir Manual*). A similar process is recorded in Vol. 2 of *Qian Gong Jia Geng Zhi Bao Ji Cheng* (*Compendium on the Perfected Treasure of Lead, Mercury, Wood and Metal*).   The discovery of the inflammable and explosive properties of a mixture of saltpetre, sulphur and charcoal was of great scientific and historical significance. Another alchemists' classic, *Zhen Yuan Miao Dao Yao Lue* (*Essentials of the Truly Original Methods*) cautions against instantaneous combustion of a mixture of sulphur, saltpetre, realgar (As_2S_3) and honey as it will certainly be a calamity to persons and buildings nearby.   From various painful experiences the ancient Chinese gradually learned to mitigate or eliminate the toxicity and volatility of saltpetre and sulphur for their safe use in pharmacy and alchemy, and to exploit the violent capabilities of gunpowder to the full.   But when gunpowder was first invented, its wider use and significance were not sensed by its inventors.

Before gunpowder came in, fire had already been employed as a deadly weapon against the enemy in war.   There were some kinds of incendiary devices such as the *huojian* (fire arrow), which was an ordinary arrow smeared with inflammable materials such as grease, tar, resin or sulphur and ignited before shooting.   Weaponeers later found that gunpowder could be used for much more powerful incendiary purposes.   Near the end of the Tang Dynasty or the beginning of Song they started to attach small packages of gunpowder to arrows.   This marked the first use of gunpowder in war.   Later bigger packages of gunpowder with ignited fuses were hurled by catapults or ballistas which had formerly been used for hurling stones. The great potential of gunpowder in weaponry led to the invention of various other devices.   Tang Fu in the year 1000 presented to the Song emperor models of *huojian*, *huoqiu* (fire ball) and *huojili* (barbed fire package) of his own design.   In 1002 Shi Pu designed a number of fire packages and arrows which he was summoned to the court to demonstrate.

火砲

A Song Dynasty catapult used to hurl gunpowder packages.

Gunpowder weapons soon prompted larger-scale production of the inflammable material. *Wu Jing Zong Yao* (*Collection of the Most Important Military Techniques*) compiled by Zeng Gongliang and others in 1044 during the Northern Song Dynasty describes many kinds of gunpowder weapons and gives three precise prescriptions for making gunpowder as follows: for *du yao yan qiu* (gunpowder package which emits toxic fumes) use 30 *liang*[1] of saltpetre, 15 *liang* of sulphur, 5 *liang* of charcoal, plus croton seeds, arsenic, roots of *langdu* (*Euphorbia fisheriana*), bamboo and hemp fibres, wood oil, tar and other inflammable materials. For *ji li huo qiu* (barbed fire package) use 40 *liang* of saltpetre, 20 *liang* of sulphur, 5 *liang* of charcoal, plus bamboo and hemp fibres and other inflammable materials. For the *huopao* (gunpowder package for incendiary as well as concussion purposes) use 40 *liang* of saltpetre, 14 *liang* of

---

[1] 1 *liang* = 50 grammes.

sulphur, 14 *liang* of charcoal plus bindings and other inflammable materials.

Saltpetre, sulphur and charcoal are the indispensable ingredients in all the formulae. The proportion of saltpetre is greater than that of sulphur and charcoal combined. The proportion of saltpetre in gunpowder is much greater than that in "subduing saltpetre or sulphur by fire".

The Song rulers were constantly busy waging war against external invaders and internal peasant insurgents, and this greatly expanded the production of gunpowder and fire weapons. A special government bureau was assigned to take charge of the munitions industry, under which there were 11 large workshops and more than 40,000 workers. Gunpowder production ranked first in importance. *Xin Si Qi Qi Lu* (*Tearful Records of the Battle of Qizhou*) written in 1221 by Zhao Yurong states: ". . . On the same day there were produced 7,000 gunpowder cross-bow arrows, 10,000 gunpowder ordinary arrows, 3,000 barbed gunpowder packages and 20,000 ordinary gunpowder packages," indicating the scale of firearms production in the Song Dynasty.

Firearms at this stage were still limited to incendiary weapons. Later the explosive property of gunpowder was exploited by altering the percentages of ingredients in the recipe to increase the speed of combustion. The barbed gunpowder packages already developed would certainly explode, though the explosion was moderate and served merely to scatter harnassing barbs on men and animals that trod on them. Towards the end of the Northern Song Dynasty new types of bursting fire weapons appeared and were called in awe *pilipao* (thundering gunpowder charge) and *zhentianlei* (heaven-shaking thunder). The former was a factor in raising the siege of Kaifeng in 1126, while the latter signified the development of iron shells and splinters. The iron shell as a container for gunpowder and the splinters marked the use of gunpowder chiefly as an explosive and the whole *zhentianlei* as a fragmentation bomb which took a very heavy toll of the enemy. *Kin Shi* (*History of the Kin Dynasty*) describes in vivid terms the deadly might of the bomb:

When it went off, it made a report like sky-rending thunder. An area more than half a *mu*[1] was scorched on which men, horses and leather armour were shattered. Even iron coats of mail were riddled.

Another important advance in fire weaponry was achieved in the Song Dynasty by peasant insurgents in armed struggles against feudal rulers. This was the barrel firearm called *huoqiang* (literally "fire lance") which first appeared in 1132. In 1259 there appeared *tuhuoqiang* (fire-spitting lance). *Huoqiang* was a long bamboo tube into which gunpowder was packed. When set off, flames were thrown from the tube as from the barrel of a modern flamethrower.

Tuhuoqiang was made of a thicker bamboo tube into which *zike* (bullets) were packed with gunpowder. When fired, flames came forth followed by bullets. This was a primitive musket or gun.

Greater shooting power demands greater pressure in the barrel, which rendered bamboo useless. Bronze and iron gun barrels appeared not later than the Yuan Dynasty (1271-1368). They were then called *huochong* (fire gun) in general. Because of the great awe these firearms inspired they were also called *tongjiangjun* (bronze generals). The oldest *tongjiangjun* now reposing in Beijing's Historical Museum was cast in 1332. It is the world's earliest surviving gun.

During the dynasties of Song and Yuan a primitive type of rocket weapon was developed — a gunpowder arrow propelled by a separate charge of gunpowder. Somehow this came into fairly

Illustration of a Song Dynasty *tuhuoqiang* (fire-spitting lance).

---

[1] 1 *mu* = 1/15 hectare or roughly 1/6 acre.

wide use only in the Ming Dynasty (1368-1644). Illustrations of this type of rocket appear in *Wu Bei Zhi* (*Treatise on Armament Technology*) written in 1628. They were called *feidaojian* (flying-dagger arrow), *feiqiangjian* (flying-spear arrow) and *yanweiqiang* (swallow-tailed arrow). The names indicate that all were propelled by gunpowder charges and that their arrowheads were unusual. Soon multiple or repeated rockets were developed, called *huo nu liu xing jian* (meteoric fire arrows) which fired 10 arrows, *yiwofeng* (hornet nest) which fired 32, *si shi jiu shi fei lian jian* (volley of forty-nine flying arrows) which fired 49, *bai shi hu jian* (hundred curving arrows) and *bai hu qi ben jian* (hundred tigers set free) which fired 100 arrows.

Treatise on Armament Technology also describes prototype rockets called *fei kong ji zei zhen tian lei pao* (sky-flying anti-personnel cannon) and *shen huo fei ya* (magic flying fire crow). The former

A Ming Dynasty illustration of *shen huo fei ya* (magic flying fire crow).

was mortal, the latter incendiary, both were set off over the heads of the enemy by time fuses connected to the propellent charges.

In the Ming Dynasty prototypes of automatic mines and other fairly advanced rockets were developed. Among them a two-stage rocket called *huo long chu shui* (fiery dragon emerging from water) merits special attention. A dragon-shaped main rocket propelled by four first-stage propellent rockets set off the main dragon-shaped second-stage rocket upon exhaustion of the four charges. The "fiery dragon" then soared faster and farther.

A Ming Dynasty illustration of *huo long chu shui* (fiery dragon emerging from water).

China had established sea trade with India, Persia and the Arab countries early in the Tang Dynasty. Saltpetre, the most important ingredient of gunpowder, was introduced to these countries through the transmission of alchemical and pharmaceutical techniques. Saltpetre was therefore called "China snow" in the Arab countries and "China salt" in Persia. The Arabs and the Persians knew only of its use in metallurgy, medicine and glass-making at first, and learned to use it in gunpowder only between 1225 and 1248, when gunpowder was introduced in the Arab countries by merchants via India. Of the Europeans the Spaniards were the first to learn about gunpowder through translating Arabic classics. Firearms, on the other hand, were introduced to the West in the course of the Mongolian expeditions. History tells us that *huojian*, *duhuoguan* (poison fire flask), *huopao* and *zhentianlei* were taken to the Middle East by the Mongolian expeditionary forces. The Arabs learned the use and making of gunpowder and firearms from the invaders and were soon producing their own. The Europeans learned the techniques later, in wars against the Arabs. The earliest records concerning gunpowder and firearms appeared in England and France only in the 14th century. But in Europe gunpowder and firearms played a remarkable role in the bourgeoisie's victory over the aristocrats.

Porcelain

Hong Guangzhu

Porcelain is another Chinese invention, developed on the basis of steady improvement in pottery-making technique. Its beginning can be traced to the Shang Dynasty (c. 16th-11th century B.C.).

As early as in the Neolithic period of primitive society over 6,000 years ago the ancestors of the Chinese people were making and using pottery. Kiln-fired at the relatively low temperatures of around 500—600°C after the clay was shaped by hand, the early pottery was rather coarse in texture. During the periods of the Yangshao Culture[1] and the Longshan Culture[2] the nature of pottery clay as a viscous and plastic substance was better understood. The firing was also better controlled and used. Artisans in pottery already knew about making the bodies of vessels with clay that was finely washed. Most of the bodies were still fashioned by hand or they were pressed into moulds, while some were thrown on the wheel. Although the surfaces of the objects were generally calendered, some were decorated with red or black patterns producing what archaeologists term "painted pottery". The type that attracts the greatest attention, however, is the so-called "egg-shell pottery" of that early period. Its pure black body of considerable hardness is so thin as to be somewhat translucent. Improved pottery-kiln structure is mainly responsible for its high consistency. Flue, channel, inside

---

[1] Yangshao Culture of the Neolithic period, relics of which were first unearthed in Yangshao Village, Mianchi County, Henan Province, in 1921.

[2] Longshan Culture of the Later Neolithic period, relics of which were first unearthed in Longshan, Shandong Province, in 1928 and are distributed in the middle and lower reaches of the Huanghe, or Yellow River.

cavity and fire chamber are all seen in a pottery kiln of the Longshan Culture period discovered at Miaodigou in Henan Province. Such kilns could be heated to fairly high temperatures because of efficient flues, and the firing could be easily controlled. Pottery of that period is not only consistent in texture, but items range from the ordinary red and grey ware to fine white and black products.

Though porcelain and earthenware are essentially different, the firing processes of the two are similar so that the technique of porcelain manufacture can be regarded as derived from that of pottery-making. Between the Later Neolithic period and the Shang Dynasty, incised white pottery and hard pottery with stamped geometric designs made their appearance. These were made of porcelain clay and fired at temperatures above 1,000°C. Out of these types emerged primitive porcelain.

Since 1953 China has discovered a great variety of glazed wares of the Shang and Zhou period (c. 11th century-221 B.C.) at Erligang in Zhengzhou, Henan Province, Tunxi in Anhui Province, Dantu in Jiangsu Province and Xi'an and Fufeng in Shaanxi Province. They include *zun* (wine vessels), bowls, vases, jars and stamped cups. Fashioned of kaolinic earth, all are of compact texture and lustrous appearance. Having characteristics of early porcelain in appearance or ingredients, these glazed utensils are sometimes called "primitive celadons" or "primitive porcelain".

By "porcelain" is meant ware the raw materials of whose moulded or shaped unfired body are a combination of kaolin (also called porcelain clay), orthoclase and quartz; the surface of a porcelain base is coated with vitreous glaze and the object baked at a temperature of about 1,200°C; the finished product has a very low rate of water absorption and becomes hard and solid after agglomeration. Chemical analysis of the above-mentioned glazed ware specimens unearthed at the Yin ruins of the 14th-11th century B.C. near Anyang in Henan Province revealed the main raw material used for the bodies of primitive porcelain to be kaolinic clays. Primitive porcelain was also shown to be quite different chemically from pottery, the former containing relatively more acid oxide such as silicon dioxide and less alkaline oxide such as calcium, amgnesium and sodium oxides. By increasing the acid and decreasing the alkali content of the body the firing temperature of this primitive porcelain could be raised to about 1,000°C without the body melting. Water absorption of primitive porcelain agglomerated at this high temperature is slowed by a thin surface coating of green vitreous glaze. Study of the primitive porcelain specimens unearthed at Xiaotun Village in Anyang shows their average rate of water absorption to be only about 0.4 per cent. The development of primitive porcelain of the Shang and Zhou period can therefore be correctly regarded as marking the beginning of a new era in China's ceramics manufacture. The use of kaolin, invention and improvement of glazing and raising of firing temperature all prepared the way for the advent of modern porcelain.

In 1924, specimens of early celadons of A.D. 99 were discovered at Leigutai in Xinyang, Henan Province. Since the founding of the new China in 1949 many important discoveries of early celadons have been made, among them celadon cups and ink-stones of the year 241 found at Shimenkan in Nanjing, Yue kiln celadon *shuizhu* or water pot made by Yuan Yi of Shangyu in 251 found in a tomb at Zhaoshigang outside Guanghua Gate in Nanjing, and *aiqing* or mugwort green (*Artemisia argyi*) wares of 297 used as burial objects found in the tomb of Zhou Chu in Yixing, Jiangsu Province. Besides their fine and consistent bodies, these celadons have coatings of thick dark-green glaze, different from the thin pale-green glaze of the primitve celadons of earlier periods. The definite dates of these funerary objects sets the time of the Chinese labouring people's invention and perfection of porcelain as the Eastern Han period (25-220) at the latest, while signs of further development in the preparation of glaze had become unprecedentedly marked.

The beauty of porcelain is due largely to its body being coated with glaze of one or more colours. The so-called *piaoci* misty or pale-green porcelain of the Jin Dynasty (265-420), *qian feng cui se* or the green of a thousand peaks of the Tang Dynasty (618-907), *yu guo tian qing* or porcelain the colour of the sky after rain of the

Chai Zhou period (954-959), the *mise*[1] of the Wu Yue period, the powder and jade greens, black gold, turtle and motley of the Song Dynasty (960-1279), *qing hua you li hong* or blue-and-white in red glaze of the Yuan Dynasty (1271-1368) are all enchanting names given to porcelain types displaying distinctive styles in the prepartaion of glaze.

The first of these glazes invented early in the Shang and Zhou period was the green glaze. Like the body, glazes are made from mineral substances, their chief components being silicate, calcium oxide, borate or phosphate. The colouring agents of glaze in ancient times included iron, copper, cobalt, manganese, gold, antimony and other mineral elements. The so-called multi-coloured glaze of the Han Dynasty (206 B.C.-A.D. 220) was produced by using molysite or copper salt in lead glaze or lead oxide. Iron as a colouring agent exists as two iron oxides: ferrous oxide, which gives a green colour, and ferric oxide, which produces dark brown or terra-cotta. The iron in the glaze becomes ferrous oxide when heated in a reducing flame, and ferric oxide when heated in an oxidizing flame. Analysis shows that if the ferrous content of the porcelain glaze reaches 0.8 per cent, the fired porcelain will develop a pale-green colour, and if it increases over 0.8 per cent, the green colour will turn dark. An iron content exceeding 5 per cent makes reduction difficult, and the colour will also turn dark brown or nearly black. Rapid progress in porcelain-making technique enabled Tang Dynasty artisans of the Yue kilns (near today's Yuyao in Shaoxing, Zhejiang Province) to maintain an appropriate ferrous-oxide content (1-3 per cent) in the glaze and produce the famous *qian feng cui se*. It was no small accomplishment, as it meant not only that the iron content was accurate and the substance appropriate. The temperature and ventilation in the kiln also had to be strictly controlled so that the porcelain was baked in a reducing fire. Continuous practice based on experience through the ages further developed the technique of celadon manufacture, resulting in truly ex-

---

[1] *Mise*, the famous "secret" or forbidden colour, said to have been called this because it was reserved for imperial use.

quisite products.

China's white-glaze porcelain which originated in the Southern and Northern Dynasties (420-589) was perfected in the Sui Dynasty (581-618). By the Tang Dynasty, the white porcelain of the Xing kilns (in today's Neiqiu, Hebei Province) had become one of the two main items of blue wares and white wares. During the Tang Dynasty the white porcelain kilns of Jingdezhen in Jiangxi Province and of Dayi in Sichuan Province ranked along with the famous Xing kilns. The body of a Tang Dynasty white porcelain bowl unearthed at Shengmeiting in Jingdezhen in 1958 was found to have a fairly high calcium oxide content, its firing temperature had reached 1,200°C and its whiteness was assessed at 70 per cent, specifications close to the standard for modern high-grade fine porcelain. This level of achievement provided the foundation for developing the blue decorated porcelain of later periods.

The Song Dynasty (960-1279) was another classical age in the

Illustrations from *Tian Gong Kai Wu* (*Exploitation of the Works of Nature*) by **Song Yingxing, 1637.**
Left: Making pots. *Right:* Dipping in the glaze.

development of porcelain, and it has always been described as a period of maturing in the technique of porcelain manufacture. The wares of the Song Dynasty show new improvements in the substance of the body, glazing and manufacturing technique. A division of labour was instituted, with the various technological processes clearly defined. There was maintenance of firing temperature, preparation of ingredients, body making, and glazing. This too marked a step forward in porcelain making. The wares of the Ding, Ru, Guan, Ge and Jun kilns, perhaps the five best known, and other kilns of the Song Dynasty, each have their distinguishing features in the colour of glaze and decorative patterns. There were the *baijisui*, or hundred-fold crackled ware of the Ge kiln at Longquan in Zhejiang Province glazed with materials of different heat expansion coefficients, the *fenqing* or grey-green of the Di kiln, the *yingbai* or jade-white, *tianbai* or sweet white, and "embroidered", carved and pressed wares of the Ding kiln, "the brown moth and iron-coloured foot" of the Guan

Illustrations from *Tian Gong Kai Wu* (*Exploitation of the Works of Nature*) by Song Yingxing, 1637. *Left:* Porcelain kiln. *Rignt:* Decorating wares.

kiln, the moon-white or *yingqing* of Jingdezhen, "the black hare's fur" and "partridge spots" of the Jian kiln, the black glaze with carved designs and the motley of Cizhou kiln — all well-known and highly valued throughout the world.

Of the many celebrated kilns of the Song Dynasty, the Jun kiln in today's Yuxian County in Henan Province is noteworthy for its *yaobian* or variations which originally resulted from accidents in firing and which caused the glaze to come out mottled, streaked or splashed. Multi-coloured and entirely different from the monochrome porcelain of blue or white of the past, the wares of the Jun kiln were innovative for their strongly heated red-and-blue glaze and the derivative purple articles. Reduced copper is shown by analysis to produce the red tinge of the Jun kiln glaze, the colouring effect of copper being quite similar to that of iron. The copper content of the Jun kiln red glaze is about 0.33 per cent. Although small quantities of other mineral elements also produce colouring effects, it was remarkable that the labouring people of the Song Dynasty made use of copper salts and controlled firing to obtain several glaze colours.

Structural innovations in Song Dynasty kilns are important. The Longquan kiln, for instance, is dragon-shaped and built along the curve of a hill. Its enormous cavity held more than 170 rows of articles, each row with a capacity of up to 1,300 pieces, which meant a total of 20,000-25,000 pieces at one firing. Full utilization of kiln heat is achieved by the curve in the middle which serves as a damper, reducing the speed of the flames through the kiln. The glaze of wares which are evenly fired in such kilns is of uniform colour. At this time also the quality of the wares of the north was much improved when direct-fire kilns using wood as fuel were converted into down-draft kilns burning charcoal.

During the Yuan Dynasty (1271-1368), the red derived from copper as used to produce the unique red-tinged underglaze porcelain of the north was improved by the southern artisans of Jingdezhen, who developed blue-and-white porcelain with cobalt compounds for underglaze colours.

Further improved firing during the Ming Dynasty (1368-1644) is evident first of all in the successful baking of fine white glaze. Of high aluminium oxide and silicon dioxide content and small proportion of solvent, this white glaze has a jade-like smoothness and translucency and cream-like lustre and colour. Improving the quality of white glaze paved the way for the development of single glaze and painted porcelain.

Painted porcelain is commonly considered as either underglaze or overglaze. A ware is classified as underglaze when the decoration is done on the body before the glaze is fired; it is overglaze when the decoration is done on the fired glaze and the piece is re-fired. The celebrated Chinese blue-and-white porcelain is underglaze, a ware with blue decoration on white ground which was one of the main porcelain products during the Ming Dynasty. Cobalt oxide is found in the blue glaze substance, the colour varying greatly with variations in temperature and the type and conditions of firing. A reducing fire is required for the cobalt glaze to reveal its beautiful blue colour. The blue decoration will lose lustre if the temperature is not right. Complete control must therefore be maintained over the firing and preparation of the glaze ingredients. Porcelain workers of the Ming Dynasty achieved wonders in this craft, turning out exquisite blue decorated porcelain for both the home and foreign markets.

The Ming Dynasty produced a variety of lovely porcelains, among them single-glaze using the red derived from copper. There is the fresh red and emerald green of the Yongle period (1403-24), the delicate yellow of the Hongzhi period (1488-1505), the peacock hue and Mohammedan blue of the Zhengde period (1505-21) and the peacock blue of the Jiajing period (1522-66). Though copper-red glaze began with the *yaobian* of the Jun kiln during the Song Dynasty, it continued to develop in the Yuan, and by the Ming Dynasty it had been fired into underglaze red of such unique shades as fresh red and ruby red. This was by no means accidental, but was the result of mastering the technique of the reducing fire and of the chemical process of converting the oxide of copper into the metal's free state during the firing so as to distribute it evenly in a colloidal state in the glaze material.

The variety of painting methods used in decorating Ming

wares is noteworthy in China's porcelain manufacture technique. The clashing colours of the Chenghua period (1465-87) and the "five colours" of the Jiajing and Wanli (1573-1619) periods are world-famous types. The clashing-colour type is fired with red, yellow, green or purple painted on the already fired blue-and-white ware, while the "five-colour" does not necessarily include that number of colours. It is polychrome including red.

Qing Dynasty (1644-1911) porcelain developed on the foundation of the remarkable achievements of Ming, and attained a very high level indeed.

The best monochrome glazes are the sky blue, emerald green, blue-green, apple green, delicate yellow, and the red, purple and green soufflés fired during the Kangxi period (1662-1723). There are also the various Song-type and five-colour glazes produced in the Qianlong period (1736-96). The rouge water, oil green, sky blue and nine imitations of ancient products including Ru, Guan, Jun and Longquan wares duplicate the originals exactly because of accurately following the recipes and perfect control of firing temperatures.

The fresh red and Lang kiln red of the Kangxi period of Qing imitated the cloudless red and red underglaze of the Xuande period (1426-36) of Ming. All these red glazes resulted from adopting and improving Ming Dynasty porcelain-making technique.

In coloured porcelain, the plain three-colour and five-colour of the Kangxi period and the powdered and enamel colours of the Yongzheng (1723-36) and Qianlong periods are well-known in China and abroad. Both powdered and enamel colours are overglaze colours. Powdered colour is achieved by adding powdered lead to the colouring material, or applying it on the surface to produce different shades of colour on the glaze through temperature control. Powdered colour, pleasing in tone and of subdued lustre, is universally appreciated for its three-dimensional, contrasting contour appearance. The technique of enamel-colour ware is the same as for powdered colour. Like powdered-colour wares, enamel decorated bodies are exquisite in substance, pattern and style.

A feature of the blue decorated glaze of the Qing Dynasty is white often suffused with the blue due to a fairly high calcium oxide content.   Ferrous oxide in the body and glaze is responsible for the blue colour in the glaze.   In fact, the amount of ferrous oxide in the glaze reaches over 90 per cent of the total iron content.

In order to minimize deformation of the body, the method of using a large proportion of kaolin earth in the formula was adopted in the Qing Dynasty.   With the raw materials very carefully washed and processed, the quartz particles are finer than in past ages and are evenly distributed.   This allows good development of the silicate hydrate of aluminium crystals in the body, given an appropriate and stable temperature and optimal control of the firing throughout. The high-quality porcelain made in this way is translucent and the whiteness of the painted plates of the Yongzheng period exceeds 75 per cent with a firing temperature as high as 1,310°C.   This high temperature no doubt plays its part in the hardness and beauty of the body and glaze.   Microscopic structural study reveals the quality of this Qing Dynasty porcelain to have reached various standards for modern hard wares.

As long ago as in the Tang Dynasty, Chinese porcelain was transported by sea and land via the Silk Road and sold abroad together with tea and silk.   It continued to be exported without interruption in the following dynasties.

In the 11th century, China's porcelain-making technique found its way to Persia and afterwards to Arabia, Turkey and Egypt.   It was not until 1470, however, when it was introduced into Venice that Europe began to make porcelain.

Lacquer and Lacquer Technique

Pan Jixing

Like porcelain, lacquer is an important invention in chemical technology and industrial arts and crafts of ancient China's labouring people. Anti-corrosive, resistant to acid and alkali, it is a durable material besides being light in weight and attractive in appearance. Lacquer is widely used in daily life and also in various branches of industry. The varieties of lacquer which China exports have a distinctive style that makes them popular throughout the world to this day. The lacquer produced in China is also called *daqi* (major lacquer). This is a physiological secretion of the indigenous lacquer tree, *Rhus vernicifera*, its chief ingredient being urushiol. The sap taken from the lacquer tree, called raw lacquer, is watery. Stirring in the sunshine to evaporate excess moisture reduces this raw lacquer into a dark viscous liquid called mellow lacquer.

Vegetable oils such as tung oil were mixed into lacquer as a drying agent when lacquerware was made in ancient China. Colour paints composed of tung or other drying oils and various colours and dyes were also used for painting decorative patterns, establishing a lacquer technique with distinct national style. Tung oil, a product of China as its Chinese name indicates, is extracted from the seeds of the tung-oil tree, its chief ingredient being elaeostearic acid ($C_{17} H_{29}$. COOH). The Chinese knew its properties already in ancient times and used it in combination with paint, a remarkable and singular undertaking in the history of chemical technology.

Although the production of lacquerware has been outlined in chemical terms only quite recently, the Chinese labouring people long ago empirically applied the chemical laws involved in inventing lacquerware and understood paint-coating and coat-forming conditions.   Records document Chinese lacquerware as originating over 4,000 years ago.   According to the books *Han Feizi* (*The Book of Master Han Fei*) and *Yu Gong* (*Tribute of Yu*) completed in the Warring States Period (475-221 B.C.), China had already begun to use lacquer in the manufacture of dining and sacrificial utensils and in preparing colour (vermilion) paint in the Later Neolithic Age when the clan commune disintegrated and gave place to slave society.

The lacquer tree is easily identified by the shiny black paint coat that forms when its sap is exposed to the sun.   The Chinese labouring people in ancient times ingeniously used this natural phenomenon, increased the amount of sap flowing from the tree and applied it after sunning to utensils, producing primitive lacquerware.   When red colour was mixed into the sap, the result was colour paint. *Han Feizi*'s record appears to be a factual account handed down through the ages, as such records have been confirmed by archaeological excavations in recent years.   It is interesting that the black pottery pieces with lacquer decorations unearthed in the remains of the Late Neolithic Age in Wujiang County, Jiangsu Province, in the fifties correspond nicely in time with the period recorded in *Han Fei zi*.   The laterite-coloured carved furniture stamps found among the Yin Dynasty (14th-12th century B.C.) ruins are the earliest extant lacquerware decorations.

Great importance was attached to the cultivation of lacquer trees during the Spring and Autumn Period (770-476 B.C.).   "Since there is lacquer in the mountains, chestnuts on the moist lowlands and you have wine and food, why not play the *se*[1]?"   appears in the Chapter "Guo Feng" in *Shi Jing* (*Book of Odes*).   "Small lacquered tables are also used" is seen in "Gu Ming" Chapter in *Shu Jing* (*Book of History*).   Exquisite colour-decorated lacquer objects such as tea tables, desks, *zu* (ancient sacrificial utensils), drums and *se*, dagger-axe handles and tomb-guarding animals of the end of the Spring and Autumn Period have been unearthed, providing material data to complement ancient records.

Lacquer-tree plantations run by special officials already existed in

---

[1]   *Se*, a 25-stringed plucked instrument, somewhat resembling the zither.

the Warring States Period. The ancient philosopher Zhuang Zhou (c. 369-286 B.C.) is described in the "Biographies of Lao Zi and Zhuang Zhou" in *Shi Ji* (*Records of the Historian*) as having been a lacquer-tree plantation official. Furthermore, artisans of the Warring States Period had a rudimentary knowledge of the anti-corrosion protective property of paint coat on utensils. *Kao Gong Ji* (*Artificers' Record*) states that lacquer is a substance that can withstand frost and dew. Decorated lacquer vehicles, weapon handles, tea tables and desks for daily use, trays, *lian* (vanity case) as well as musical instruments of the period from Western Zhou (c. 11th century-770 B.C.) to the Warring States have been excavated in large quantities. Examination reveals their base to be most often wood, leather or inserted hempen cloth. For anti-corrosion, some of the wooden buildings and metal objects in later periods were also given coats of

Lacquer *lian* **(vanity case) of the Warring States Period (475—221 B.C.).**

lacquer, while many lacquer articles were adorned with coloured patterns. As the lacquer artisan Huang Cheng of the Ming Dynasty (1368-1644) put it, "It is preferred because of its durability and rich colour."

It has been ascertained that the decorative patterns on some of the lacquerware of the Warring States Period were drawn with oil paint made up of tung oil or other drying oils variously coloured. Though oil paint is brighter, it is not as preservative as lacquer. But the output of lacquer could not compare with that of tung oil. It was also more costly to produce lacquer. The addition therefore of drying oil into lacquer as a thinner made up for the deficiencies of each. The combined use of superior properties continues today. The ancient Chinese also knew about applying albumen and litharge (PbO) or *tuzi* (containing MnO_2) as a drying agent for a high polymer membrane of "major lacquer" and tung oil respectively.

The five colours (red, yellow, blue, white and black) and their combinations used for coloured lacquer decorations during the Warring States Period are probably mineral pigments such as cinnabar, stone yellow, realgar, orpiment, red or white clay and vegetable dyes such as indigo.

By the Qin and Han dynasties (221 B.C.-A.D. 220), the technique of lacquer entered a new phase of development and spread to all parts of China. "Biographies of Comic Characters" in *Records of the Historian* gives an account of a "damp chamber". This was probably for the production of lacquer, as urushiol is easily polymerized into membrane in the presence of moisture. Drying in a moist atmosphere further prevents cracking. Among the unearthed lacquerware of the Han Dynasty (206 B.C.-A.D. 220) are ladles, trays, tables, *lian*, boxes, eared cups, pillows and inner and outer coffins with wood and hempen cloth bases. Lacquerware with inlay of gold, silver or copper is called *kouqi*, and was costly. Writing about the different kinds of lacquerware in his *Yan Tie Lun* (*On Salt and Iron*), Huan Kuan of the Western Han Dynasty (206 B.C.-A.D. 24) pointed out:

Rich people use such utensils as lacquerware with a silver mouth and gold ears, gold and jade pots, while the middle class use hempen lacquer decorated with jade and Shu cups decorated with gold.

He also said, "One decorated lacquer cup can be exchanged for ten copper ones." Lacquerware with inlay of gold or silver was certainly much more valuable than the decorated type.

Industrial officials were appointed in the Han Dynasty (206 B.C.-A.D. 220) to supervise the lacquer industry in the prefectures of Shujun (now Chengdu) and Guanghan, the chief lacquer centres in Sichuan Province. "Biography of Gong Yu" in *Han Shu* (*History of the Han Dynasty*) states that as many as three industrial offi-

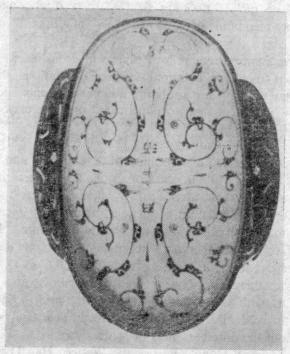

Lacquer eared cup of the Western Han Dynasty (206 B.C.- A.D. 24) found at Mawangdui near Changsha.

cials were appointed in a single prefecture of Guanghan, spending 50 million *qian* (copper coins) annually and consuming a huge amount of manpower and natural resources. Rich families contended with each other in using lacquerware. Lacquer objects with gold, or silver inlay have been found in excavations, and words like *huangtugong* (gilders) which are inscribed on them may have something to do with their production.

An excavated lacquer object of the Han Dynasty gives the date and place of production and the name of the artisan. Also inscribed on it is a chronological record of events. The division of labour in a lacquer workshop is obviously rather complicated. Excavated objects also furnish valuable data in understanding lacquer technique of that period. A government-owned lacquer workshop of the Han Dynasty employed the following type of workers: elementary (who made the basis), brushers and appliers (who put on the lacquer), gilders (who did the gilding), painters (who painted the oil colour decorations), inscribers (who carved the inscription), cleaners (who did the finishing touches), and the constructor. Among the officials there was also division of responsibility. Apart from being a government-owned business, the lacquer industry was a flourishing enterprise among the people. "History of Agriculture and Commerce" in *Records of the Historian* says:

> In the states of Chen and Xia the natives who own 1,000 *mu*[1] of lacquer trees, in Qi and Lu states ones who own 1,000 *mu* of mulberry trees and hemp, and those of Weichuan who own 1,000 *mu* of bamboo are equal in wealth to a marquis supported by 1,000 families.

Along the same line is the folk saying: "He who owns 1,000 tung-oil trees will never suffer from want as long as he lives." The lacquerware unearthed recently at Mawangdui near Changsha in Hunan Province typify the excellent work done in lacquer art in the early Han Dynasty.

---

1 *Mu* — A Chinese measure of land which now equals $1/15$ hectare or roughly $1/6$ acre.

Hempen cloth statue-making was invented in the Jin and the Southern and Northern Dynasties (3rd-6th century). The base of such a statue is a clay mould on which is applied a coat of hempen cloth fixed on with paste and lacquer. The statue is then painted in colour and allowed to dry, after which the clay mould is removed leaving the completed statue hollow inside. Artists of those early times were already making lacquer statues six metres high, their art taking an important place in ancient lacquer technique. Jia Sixie, a scientist during the Northern Dynasties (386-581), devoted an entire chapter of his *Qi Min Yao Shu* (*Important Arts for the People's Welfare*) to lacquer. Discussing ways to increase the durability of paint coat and its care, he pointed out, "A lacquer object will get wrinkled and thus damaged if allowed to come in contact with salt and vinegar for any length of time." Also that it would go mouldy in the presence of moisture and so "should be exposed to sunshine" during the rainy summer. This was using the ultra-violet rays of the sun as a bacteriocidal agent. This early scientist noted that "cinnabar mixes well with oil so that it can endure sunshine". This knowledge could only have been the result of long practice.

Developing on from lacquer with gold and silver inlay is gold and silver *pingtuo*, done by pasting patterns of sheet gold or silver on the lacquer object after which it is given another coat of lacquer. It is then rubbed smooth and polished until the bright gold or silver patterns emerge. It was in the Tang Dynasty (618-907) that *tihong* (carved red lacquer) was invented. The wood or metal basis of such objects is coated many times with vermilion paint and the carving is done after each coat. The pattern is three-dimensional. *Luodian* (mother-of-pearl inlay) with shell or jade encrusted on the surface was also quite common during that period. Zhu Zundu of the Five Dynasties (907-960) wrote *Qi Jing* (*Book of Lacquer*) summing up lacquer technique experience of the preceding dynasties. This important work, the earliest monograph on lacquer technique, has unfortunately been lost.

The Tang Dynasty *tihong* continued in vogue during the Song and Yuan dynasties (960-1368). Precious metals were used as basis and it was called *diaohong*. Zhang Yingwen of the Ming

Dynasty (1368-1644) comments in his *Qing Mi Cang* (*Secret Collections*):

> The majority of carved red lacquer made by the lacquer artists of the Song Dynasty for the court chose gold or silver for their bases. These were given a number of very bright vermilion coats. The cutting was done with such skill after each coating that the scenes of hills, rivers, towers, pavilions, people, birds and animals on them look extremely lifelike.

Some of these articles have come down to us today and they are truly exquisite.

"Rhinoceros skin" lacquer, of which the best examples were produced in the Song Dynasty, was painted vermilion, black and yellow, the resulting colour like that of rhinoceros skin. In the Song Dynasty also, Li Jie gave a far more detailed account of the boiling and purification of tung oil than his predecessors when he discussed architectural paints in his *Ying Zao Fa Shi* (*Architectural Methods*).

Of the carved lacquer of the Yuan Dynasty (1271-1368), the products made by Zhang Cheng and Yang Mao in Jiaxing of Zhejiang Province were best known for their superb carving. Peng Junbao, also from Jiaxing, was noted for his *chuangjin*, a type of *tianqi* in which powdered gold is filled into the carved pattern on a *chuangjin* object. After polishing it has a distinctive style comparable to gold and silver *pingtuo*. Mother-of-pearl inlay is a high-quality lacquer that was produced for the rich in the Yuan Dynasty. Apart from shell, it is also ornamented with pearls, jewels and various colours of jade.

Lacquer technique continued to develop in the Ming (1368-1644) and Qing (1644-1911) dynasties. Lacquer-tree and tung-oil tree plantations were established in Nanjing in great numbers during the reign of Hongwu (1368-98) of the early Ming Dynasty as an example to follow. In the reign of Yongle (1403-24), a governmental bureau was established at Guoyuanchang in Beijing for the manufacture of carved red lacquer for imperial use. Famous artisans including Zhang Degang, son of the celebrated lacquerer Zhang

Cheng of the Yuan Dynasty, and others worked at this factory. They used yellow copper, wood and tin for bases. The reign of Xuande (1426-35) was especially noted for its *tihong* and *tianqi*. During the reign of Longqing (1567-72), the wares of Huang Cheng, a folk *tihong* artist in Xinan, can be compared with the products made at Guoyuanchang belonging to the governmental bureau.

Huang Cheng wrote *Xiu Shi Lu* (*Records on Lacquer*), which was later annotated by the lacquer worker Yang Ming of Jiaxing in 1625. The book consists of two parts, Part 1 dealing with the

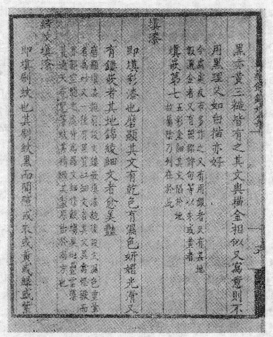

Page from the book *Xiu Shi Lu* (*Records of Lacquer*)
by Huang Cheng of the Ming Dynasty (1368-1644).

materials, tools and methods of lacquer production. It also lists possible defects in various kinds of lacquerwork and their causes. Part 2 deals with the taxonomy of lacquer and the several dozens of decorative techniques used in various kinds of lacquerwork. *Records on Lacquer* is the only extant comprehensive monograph on lacquer technique.

Since the Qing Dynasty, the technical tradition of the past has been followed. In the reign of Jiaqing and Daoguang (first half of the 19th century), the lacquer artistry of Lu Kuisheng and his works were typical of the time. Some of his works including inlays, carvings and statues have been handed down to the present day. Since that period no progress in lacquer technique has been made, while some of the methods have been lost.

China's lacquer and lacquer technique spread abroad early in the Han, Tang and Song dynasties, being passed on to such East Asian countries as Korea, Mongolia and Japan, to the Southeast Asian countries of Burma, India, Bangladesh, Kampuchea and Thailand as well as various countries in Central and West Asia. Lacquer soon became a special Asian handicraft industry. Many pieces of the *jinianming* lacquerware (bearing chronological inscriptions) made by the government-owned lacquer works in Guanghan Prefecture of Sichuan Province have been found in north Korea, while quite a number with copper inlays have been discovered in the ancient tombs of Noyin Ula in Mongolia. The Imperial Treasury (Shoso-in) at Nara in Japan still preserves a collection of gilded lacquer and gold and silver *pingtuo* of the Tang Dynasty.

It was the Persians, Arabians and later the Central Asians who carried Chinese lacquer westward to the Europeans. With direct contact between China and Europe following the discovery of a sea route, Portuguese and Dutch traders one after another carried Chinese lacquer to Europe where it was highly valued. Chinese lacquer was successfully imitated in various parts of Europe in the 18th century, the lacquerwork of Robert Martin of France being well known on the European continent at that time. Germany and Italy were next in starting lacquer industries. The style of the early products, which the Europeans called Rococo, is actually a blending of Chinese and European styles. As with porcelain, the lacquerwork in these countries was benefited by the Chinese invention.

From the 16th century, Chinese tung oil was transported to Europe by the Portuguese together with porcelain. In fact the Europeans had already heard about tung oil from the accounts of the Venetian Marco Polo who travelled in China in the 13th century. Because tung oil is faster drying than linseed oil, it replaced linseed oil in making paint in the United States from the second half of the 19th century when it was imported from China. The United States began to cultivate tung-oil trees in 1902.

Alchemy in Ancient China

Wang Kuike

Chemistry today, highly developed, has not only unravelled the mystery of transformation of matter, but has also produced a legion of substances never brought forth by nature. Our ancestors' dream of "wresting from the Creator his power of creation", so to speak, has come true.

But chemistry, like other sciences, has had an embryonic past —alchemy, which in ancient China was a thaumaturgy for preparing elixirs of life. The origin of Chinese alchemy can be traced back to time immemorial. As recorded in *Zhan Guo Ce* (*Records of the Warring States*), early in the Warring States Period (475-221 B.C.) the king of the state of Chu was presented with an "elixir of deathlessness" by thaumaturgical technicians. Qin Shi Huang, the First Emperor of the Qin Dynasty (221-207 B.C.), having unified the country and subdued all enemies mortal, recruited a number of thaumaturgists in making a "miraculous drug ensuring immortality", and dispatched thousands of virgin boys and maidens headed by an adept Xu Fu over the seas in quest of it. For a life-giving recipe Liu Che, the Emperor Wu Di (reigned 140-87 B.C.) of the Han Dynasty (206 B.C.-A.D. 220), took part in the experiments himself and sought for it from among the people. Sponsored by emperors and kings, lords and gentry, alchemy began to thrive with the pyrogenation of cinnabar and other minerals.

The practice of alchemy in ancient China was quite extensive but did not include *neidan* "internal" or physiological alchemy[1],

---

[1] Or macrobiotics "inside" the human body, in contrast to *waidan* (laboratory alchemy or pharmaceutical macrobiotics), which was conducted "outside" the human body.

which was limited to respiratory, dietetic, gymnastic and other body exercises. Ancient Chinese alchemy was composed of three distinct fields. First, researches in subjecting metals and other minerals to proto-chemical processes with the aim of discovering an elixir of life. Second, researches in metallurgical production of artificial gold or silver as "therapeutic" metals. Third, pharmaceutical-botanical researches for macrobiotic plants.

Ancient alchemists never succeeded in producing any elixirs of life or genuine gold or silver by transmutation. Nevertheless, they did a large amount of creative work in their fields, drawing on the experiences of the labouring people in metallurgy, pharmacy and other handicraft production, and made achievements important to the later development of science and technology, especially to the birth of a true science — chemistry.

Pyrogenic Methods

Ancient Chinese alchemical methods fall into two categories: pyrogenic and aqueous. Pyrogenic methods are methods of direct heating (without water as the medium) of the object as generally employed in metallurgy. Wei Boyang, an alchemist of the Eastern Han Dynasty (25-220), writes in *Zhou Yi Can Tong Qi* (*The Kinship of the Three and the Book of Changes*) that in his days 600 essays had been written on the subject of pyrogenation. But this collection of essays has long been lost and their contents must remain unknown. According to the "Chapter of Basic Principles" of *Bao Pu Zi* (*The Book of Master Baopu*) written by Ge Hong (284-364) of the Jin Dynasty (265-420), and some later works on alchemy, pyrogenic methods generally consisted of *duan*, or prolonged calcination; *lian*, or transformation of a dry substance by roasting; *zhi*, or baking; *rong*, or fusing; *chou*, or distillation; *fei*, or sublimation; *fu*, or subduing the toxicity or volatility of a substance by heating, etc.

The first thing studied by Chinese alchemists was *dansha*, or cinnabar, a bright red mercuric sulphide. Alchemists tried the pyrogenic methods. They found that when heated, cinnabar decomposes into sulphur dioxide and mercury. The latter combines directly with sulphur to form mercuric sulphides again, usually the black metacinnabar, which can be sublimed into its original state, the bright red cinnabar, when once more heated. Mercury is the only metal in liquid state at ordinary temperatures. Silver-white, heavy, fluid and volatile, it aroused the common people's interest as a miracle in itself and caught the alchemists' attention as it could be pyrogenated over and again into *huandan*, a cyclically transformed regenerative elixir, which was also called *shendan* (the "miraculous elixir"). "The miraculous elixir not only ensures longevity, but is also capable of turning other substances into gold," says the "Chapter on the Metallous Enchymoma" of *Bao Pu Zi*. That amounts to saying that the "miraculous elixir" was a panacea which would bring long life to those who took it and could "turn iron into gold by projection" (by adding a grain of the "miraculous elixir" to a large crucible of molten iron). For obtaining a "ninefold cyclically transformed elixir", the alchemists repeated the pyrogenation of cinnabar over and again till they became quite familiar with the process. Liu An (179-122 B.C.) of the Western Han Dynasty (206 B.C.-A.D. 24) writes in *Huai Nan Wan Bi Shu* (*The Ten Thousand Infallible Arts of the Prince of Huai Nan*): "Cinnabar is actually mercury." *Zhou Yi Can Tong Qi* gives a vivid description of the properties of mercury as volatile and easily reactive with sulphur. It also describes the process of the "obviously turning into vermilion" of metacinnabar when sublimed in the alchemists' reaction-vessel. *Bao Pu Zi* sums up the long experienced process in a single sentence: "Cinnabar when heated yields mercury, which after many transformations turns into cinnabar again." In his *Jiu Huan Jin Dan Miao Jue* (*Wonderful Instructions on the Ninefold Cyclically Transformed Gold Elixir*) Chen Shaowei of the Tang Dynasty (618-907) describes in detail the process of the "metamorphosis of mercury" (the combination of mercury and sulphur to form cinnabar). There is an exact proportion between the mercury and sulphur used, a clear description of the duration and temperature of heating, and an operation sequence right up till the two ingredients "turn into a purple pigment without the slightest loss of the total weight".

The process described here is not much different from that in a chemistry laboratory today. There are two kinds of red mercuric sulphide, natural and synthetic. The former is called *dansha*, or cinnabar, as stated; that produced in Hunan Province is generally considered to be of the best quality and is called *chensha* after the obsolete geographical place name Chenzhou. The latter is called *yinzhu* or vermilion and is probably one of the earliest products of chemical combination conducted by man in a laboratory. It reminds us of what should have been considered a remarkable achievement in proto-chemistry.

Ancient Chinese alchemists were also quite familiar with other mercuric or mercurous compounds. *Tai Qing Shi Bi Qi* (*Records in the Rock Chamber*) written in the Tang Dynasty describes the method of making "frost of mercury" (*shenggong* or corrosive sublimate): Heat mercury and tin together till an amalgam is formed; then pulverize it and mix it with table salt (sodium chloride), *tai yin xuan jing* (selenite), *dun huang fan shi* (ordinary gypsum) or *jiangfan* (gypsum with ferrous sulphate). Cover the mixture with powder of *puxiao* (sodium sulphate) and heat the lot for seven days and nights running. Chemically, mercury and sodium chloride heated together for a long time could possibly form mercuric chloride, which if made to react further with the superfluous mercury might form mercurous chloride. But this process would have been very slow and complicated. It has since been greatly simplified.

The dissolving of other metals in mercury to form amalgams was made use of in the handicraft of *liujin*, or amalgamation gilding of metal objects, early in the Warring States Period. There is no mention in ancient classics whether the invention of *liujin* had anything to do with alchemy. But the property of mercury as solvent of metals obviously arrested the attention of Chinese alchemists who studied it assiduously as shown in their various writings since the Eastern Han Dynasty (25-220). "Finally 'gold flower' was obtained," says Wei Boyang. "Gold and mercury merged completely into a homogeneous silver-white pasty substance which, if allowed to harden, became most rigid." Alchemists succeeded in making extremely fine dust of gold, silver and other metals through amalgamation.   In *Zhu Jia Shen Pin Dan Fa* (*Methods of the Various Schools for Magical Elixir Preparations*) written in the Song Dynasty (960-1279) a method is given of making gold dust through amalgamation: First dissolve gold into mercury to form an amalgam.   Add table salt to it.   Then the mercury is volatilized and driven out by heat.   Finally wash the salt away with water.   Gold dust is thus obtained.

Exploiting other properties of mercury, alchemists made a series of further interesting experiments.   In *Gan Qi Shi Liu Zhuan Jin Dan* (*The "Responding to the Qi Method" in Preparing the Sixteen-Fold Cyclically Transformed Gold Elixir*), also written in the Song Dynasty, we find a method of obtaining a "14-fold transformed *ziheche* (a reddish-brown solid solution, not the placenta mentioned in the pharmaceutical treatises)".   The method is:

> Pound into fine powder in a mortar four *liang*[1] of cinnabar, four *liang* of realgar and two *liang* of mercury.   Put the mixture in a reaction-vessel and make the vessel air-tight.   Heat the vessel for 60 days running and *ziheche* is obtained.   A few grains of this *ziheche* is capable of "drying up one *liang* of mercury, tincturing it yellow".

This process is now construed to be a possible formation of a reddish-brown solid substance which, when added to mercury, could possibly turn the latter into another solid solution of a yellow-ish colour.   Clearly the alchemists were motivated by an illusion about discovering a "miraculous elixir" a small amount of which added to mercury would transform it into "gold".   Although their illusion had not the slightest chance of being realized, their futile attempts certainly broadened the horizons in the quest of knowing more about nature.

"Gold will never be corroded, so it is the most precious of all things," says Wei Boyang.   Chinese alchemists had the hope that the wonderful essence of durability of gold and silver could be communicated to man so that he could acquire a similar "immuta-

---

[1]   1 *liang* = 50 grammes.

bility". They tried to find a way of administering these metallic elements and to refine "pharmaceutical gold and silver" through alchemical processes, thinking that the miraculous man-made products were more effective than natural ones. Consequently, aurifaction and argentifaction became an important component part of alchemy. Beginning with the Emperor Wu Di of the Han Dynasty, rulers of various dynasties recruited alchemists in an attempt to produce artificial gold. The products obtained were of course never genuine. Nevertheless the alchemists succeeded in producing alloys similar to gold and silver in appearance. The "pharmaceutical gold or silver" produced were actually yellow or white alloys of base metals and chemicals. According to *Bao Zang Chang Wei Lun* (*Discourse on the Contents of the Precious Treasury of the Earth*) written in the period of the Five Dynasties (907-960), there were 15 kinds of "pharmaceutical golds" at that time; 11 were "prepared from chemicals" (*shuiyinjin* or mercury "gold", *danshajin* or cinnabar "gold", *liuhuangjin* or sulphur "gold", *heiqianjin* or black lead "gold", etc.); four were "prepared from chemicals by projection" (*tongjin* or copper "gold", *toushijin* or brass "gold", etc.). There were thirteen kinds of "pharmaceutical silvers"; nine were "prepared from chemicals" (*shuiyinyin* or mercury "silver", *caoshayin* or "silver" from under-the-plants sand, *zengqingyin* or chalcanthite "silver", *liuhuangyin* or sulphur "silver", etc.); four were "prepared from chemicals by projection" (*tongyin* or copper "silver", *baixiyin* or white tin "silver", etc.). Some of the products could hardly be told from gold or silver and had been generally regarded as a marketable currency. Aurifaction and argentifaction thrived in the dynasties of Tang and Song, lingering even into the Ming Dynasty (1368-1644). That was why in his *Ben Cao Gang Mu* (*Compendium of Materia Medica*) Li Shizhen (1518-1593) warns against artificial or counterfeit "gold" and "silver" prepared from base metals and chemicals. The distinctive advances made in ancient China in the preparation of yellow and white alloys were obviously related to the alchemists' activities.

Lead and its compounds were early in common use in China. *Hufen*, ceruse or basic lead carbonate, a cosmetic, was produced before the Han Dynasty. Wei Boyang noticed that when thrown into fire, *hufen* "loses its white colour and becomes lead". This caught the attention of the alchemists who made it one of the important phenomena to be watched and studied. Besides dissolving lead into mercury to form an amalgam, they used lead in preparing *qiandan*, minium or lead tetroxide. As stated in the "Chapter of Yellow and White" of *Bao Pu Zi*, "Ceruse is white; it becomes red when it turns into minium. Minium is red; it becomes white when it turns into ceruse." That is to say, lead can be used not only in making *hufen* which is white, but also in making *huangdan* which is red. If minium is thrown into fire, it also "loses its colour and becomes lead". Alchemists of later generations learned ways of making other compounds of lead. Qing Xu Zi, Master of Simplicity and Easiness, of the Tang Dynasty gives an account of the "method of making elixir" in *Qian Gong Jia Geng Zhi Bao Ji Cheng* (*Compendium on the Perfect Treasure of Lead, Mercury, Wood and Metal*). Lead, sulphur and saltpetre fused together and subjected to a vinegar treatment yield a powder called *huang dan hu fen*, probably impure lead acetate.

Some metallic and non-metallic minerals are highly toxic, such as sulphur and arsenic. Before use, they were often subjected to a fire treatment called *fuhuo*, or "subduing toxicity by fire". Sun Simiao (581-682) of the early Tang Dynasty told of a method of "subduing the toxicity of sulphur by fire": Grind to powder in a mortar, two *liang* each of sulphur and saltpetre. Put the powder mixture in a cauldron to be fired directly with three dried pods of soap-bean tree (which when burned provide carbon). When the flames subside, pour in three *jin* of charcoal powder of various extent of charring and stir till the charcoal shrinks to two-thirds of its original volume. During the reign of the Emperor Xian Zong (806-820) in the Tang Dynasty, Qing Xu Zi invented "the method of subduing the toxicity of alum [*sic*] by fire": Two *liang* of saltpetre and two *liang* of sulphur were ignited with three and a half *qian*[1] of dried birthwort. The prescription is almost the same as that of Sun Simiao in subduing the toxicity of sulphur. In both prescriptions a burned plant is a source of carbon. Sulphur and saltpetre, both highly

---

[1] 1 *qian* is equivalent to 1/10 *liang* or 1/100 *jin*.

combustible, were believed by alchemists to have the power of subduing the toxicity of each other in fire. However, it is not easy to justify such treatment from the viewpoint of chemistry. The fire disaster imminent in such crude and rash practices, often leading to the burning down or blowing up of the alchemists' laboratories, inspired the alchemists in the Tang Dynasty as an important experience: that sulphur, saltpetre and charcoal powders mixed together formed *huoyao* (literally "fire drug", or gunpowder) which, instantly combustible, could be used as an incendiary or explosive. During the later years of the Tang Dynasty this experience, together with the prescription, were passed on from the alchemists to military experts. Gunpowder was one of the four most important inventions in ancient China.

Aqueous Methods

In alchemical practices, metallic and non-metallic minerals were not only pyrogenated into solid elixirs, they were also made to dissolve in liquid solvents. A considerable amount of empirical knowledge concerning complicated reactions of minerals in water solution had been accumulated. In "The Chapter on Grotto Gods" of *Dao Zang* (*Taoist Patrology*) the "Thirty-Six Methods for Bringing Solids into Aqueous Solution" is included, which could have been written before the Jin Dynasty. In this work also the ancient alchemists' 54 recipes for bringing 34 minerals and two non-mineral substances into aqueous solution are well preserved in a clear text. Similar recipes are given in the "Chapter on the Metallous Enchymoma" of *Bao Pu Zi*. These, together with similar records written in the Tang and Song dynasties, sum up in sketchy form the aqueous methods used in Chinese alchemy.

The aqueous methods employed in alchemy consist of: *hua* or dissolving (including fusing), *lin* or rinsing (to obtain the solution of a part of a solid substance), *feng* or buried underground, *zhu* or boiling (in a large amount of water), *ao* or decocting (long-time, high-temperature boiling), *yang* or simmering, *niang* or fermentation (long-time exposure in humid conditions or in air containing carbon dioxide), *dian* or projection (transformation of a large quantity of substance by adding a small amount of an elixir), *jiao* (pouring out from a container or repeatedly ladling in the container to cool a hot liquid), *ji* or dipping (to dip a container into cold water to cool the substance inside it), filtration, recrystallization, etc.

A solvent bath, called *huachi* in alchemy, is generally needed in aqueous processes. It is a ceramic trough or jar in which the chief solvent is concentrated vinegar with saltpetre and other chemicals dissolved in it. Saltpetre is called *xiaoshi* (stone dissolvent) in Chinese alchemy classics, since it was believed to be able to "dissolve 72 kinds of stones". It therefore plays an important role in alchemical aqueous methods. In an acid solution it provides nitrate ions like dilute nitric acid, capable of oxidizing many metallic and non-metallic minerals. The purposeful adding of saltpetre to vinegar in the view of chemistry is to form a solvent in which many substances dissolve because of oxidation-reduction. This is still a useful method today and may be regarded as a practical invention in the history of chemistry.

Some of the reactions involved in dissolving metals and non-metallic minerals into the *huachi* bath are very complicated. Their applications prior to the emergence of chemistry were remarkable, as shown in the following two examples:

Dissolving gold. There is a recipe in the "Chapter on the Metallous Enchymoma" of *Bao Pu Zi* for preparing "potable gold". One of the substances used is *xuan ming long gao* (mysterious bright dragon's fat), which according to *Shi Yao Er Ya* (*Synonymic Dictionary of Mineral Drugs*) written by Mei Biao of the Tang Dynasty can mean mercury or vinegar and raw wild raspberry. According to the "Chapter on the Metallous Enchymoma", gold is soluble if sealed for 100 days in *huachi* in the presence of the substances prescribed in the aforesaid recipe. Chemistry tells us that gold is chemically inactive and insoluble in ordinary solvents with the exception of the following:

1. Aqua regia, anhydrous selenic acid or other strong acid liquids in which chlorine, bromine, or iodine is produced;

2. Chlorine water (to form auric chloride, $AuCl_3$);

3. Mercury (to form a liquid amalgam when gold is less than 15%);

4. Dilute aqueous solution of cyanides of alkali metals in the presence of air (to form aurocyanid ions $[Au(CN)_2{}^-]$).

Neither a mixture of strong acids like aqua regia nor chlorine water is formed in the recipe for "potable gold". But if we take *xuan ming long gao* to mean mercury, gold will be soluble in it. Suppose we take the term to mean raw wild raspberry. In its juice will be CN^- ions which in acetic acid will form hydrocyanic acid HCN. The saltpetre and other substances will provide alkali metallic ions such as K^+ and Na^+. In a solvent like that gold will dissolve, though the process will be slow. To piece out the recipe and master the skills the alchemists must have gone through lengthy throes of trial and error.

Dissolving cinnabar. "Cinnabar solution" is mentioned in *Dao Zang*, and in the recipes provided in later alchemists' writings. *Shidan* (copper sulphate) is generally used besides vinegar and saltpetre. "*Shidan* is necessary in dissolving cinnabar", remarks *Huang Di Jiu Ding Shen Dan Jing Jue* (*The Yellow Emperor's Canon of the Nine-Vessel Spiritual Elixir*) written in the Tang Dynasty. This comment merits attention. Mercuric sulphide is hardly soluble in a mixture of vinegar and saltpetre but soluble in the presence of copper sulphate, which could have played the role of a catalyzer.

The aqueous methods in alchemy are by no means limited to dissolving metals or other minerals in vinegar and saltpetre. They have many other applications.

The displacement of metallic ions is an important discovery in aqueous methods. Alchemists had long worked out a theory about "transmutation of one metal into another". They were under the illusion that the displacement of metallic ions in water solution was the transmutation they had long dreamed of. *Huai Nan Wan Bi Shu* says: "*Zengqing* (chalcanthite) turns into copper when it meets iron." Ge Hong noticed: "When iron is smeared with *zengqing*, it assumes a copper-like reddish lustre. . . . Although the outside is copper-coloured, the inner substance remains to be iron." Tao Hongjing (456-536) of the Southern and Northern Dynasties (420-589) discovered that *jishifan* (chicken dropping alum, basic copper carbonate or copper sulphate) could be used instead of *zengqing* in the process in which iron "assumes the lustre of processed copper". Although alchemists made various experiments in the displacement of metals and wrote down detailed records, they failed to give a satisfactory explanation. Later their experiments became the precursor of *shui fa ye jin dan tong fa* (aqueous metallurgical method of obtaining copper from *danfan* [copper sulphate]) which was initiated during the Song Dynasty.

Huang Di Jiu Ding Shen Dan Jing Jue gives a recipe for preparing what chemistry calls potassium sulphate. The recipe says: Dissolve a mixture of *puxiao* (sodium sulphate) and *xiaoshi* (saltpetre) in hot water. Then boil the solution till it becomes quite concentrated. Place the container in cold water and let it remain overnight. Potassium sulphate was thus allowed to crystallize and separate out. It is remarkable that the principles of double decomposition and different solubilities of matters should be so early (though empirically) exploited in the preparation of what we now call a chemical compound.

In alchemy, aqueous methods constitute an extensive field of study. Even organic acids, plant alkaloids and animal hormones are explored. One is amazed at the wide range of subjects studied and the complicated processes involved.

Raw Materials and Reagents in Alchemy

What chemical products or reagents has chemistry inherited from alchemy? Some explanation is needed before giving an answer to this question. Alchemy and traditional medicine and pharmacology in China were once interwoven. Alchemists were often practitioners of traditional medicine and pharmacy, and vice versa. No clear demarcation line can be drawn between the written works in these two categories. *Shen Nong Ben Cao Jing* (*Shen Nong's*

Materia Medica), completed in the Eastern Han Dynasty, is the oldest Chinese pharmacopoeia in existence. Instead of being exclusively a medical book, it rates cinnabar as the topmost precious drug. Altogether 40 alchemical metallic and non-metallic mineral drugs are rated into three grades of importance in the book. "The grading," says the book, "warrants the upper-grade drugs as capable of making man live in health, live long, even ascending to heaven and becoming immortal. . . ." These statements smack strongly of alchemy. Medical works written by alchemists such as Ge Hong's *Zhou Hou Bei Ji Fang* (*Handbook of Medicines for Emergencies*), Tao Hongjing's *Ming Yi Bie Lu* (*Informal Records of Famous Physicians on Materia Medica*) and Sun Simiao's *Qian Jin Yi Fang* (*Supplement to the Thousand Golden Formulae*) are even more conspicuous in this trend. In such circumstances, it is very hard to tell which drugs are primarily therapeutic and which are exclusively alchemical reagents. The following list is based on incomplete statistics made by Prof. Yuan Hanqing. This is a conspectus of more than 60 organic and inorganic substances mentioned by Chinese alchemists in their classical works without specifying individual sources, since many of them were not first discovered or produced by alchemists.

Elements: mercury, sulphur, carbon, tin, lead, copper, gold, silver, etc.

Oxides: *sanxiandan* (mercuric oxide, HgO), *huangdan* (litharge, PbO), *qiandan* (minium, Pb_3O_4), *pishuang* (arsenolite, As_4O_6), *shiying* (quartz, SiO_2), *zhishiying* (amethyst, quartz containing manganese compounds), *wumingyi* (pyrolusite, MnO_2), *chishizhi* (haematite, Fe_2O_3), *cishi* (magnetite, Fe_3O_4), *shihui* (quicklime, CaO), etc.

Sulphides: *dansha* (cinnabar, HgS), *xionghuang* (realgar, As_2S_2), *yushi* (mispikel, $FeAsS$), etc.

Chlorides: *yan* (table salt, including *rongyan*, *bingshi*, $NaCl$), *naosha* (sal ammoniac, NH_4Cl), *qingfen* (calomel, Hg_2Cl_2), *shuiyinshuang* (corrosive sublimate, $HgCl_2$), *luxian* (table salt containing $MgCl_2$), etc.

Nitrates: *xiaoshi* (saltpetre or Chile saltpetre, KNO_3 or $NaNO_3$).

Sulphates: *danfan* (blue vitriol, $CuSO_4 \cdot 5H_2O$), *lüfan* (green vitriol, $FeSO_4 \cdot 7H_2O$), *hanshuishi* (gypsum, $CaSO_4 \cdot 2H_2O$), *puxiao* (Glauber's salt, $Na_2SO_4 \cdot 10H_2O$), *mingfanshi* (alum, $K_2SO_4 \cdot Al_2 (SO_4)_3 \cdot 2Al_2O_3 \cdot 6H_2O$), etc.

Carbonates: *shijian* (soda, Na_2CO_3), *huishuang* (potash, K_2CO_3), *bai'e* (chalk including *shizhongru* $CaCO_3$), *shizeng* (azurite, $Cu(OH)_2 2CuCO_3$), *kongqing* (malachite, $Cu(OH)_2 \cdot CuCO_3$), *qianbai* (white lead, $Pb(OH)_2 \cdot 2PbCO_3$), *luganshi* (calamine, $ZnCO_3$), etc.

Borates: *pengsha* (borax, $Na_2B_4O_7$), etc.

Silicates: *yunmu* (mica, $H_2KAl_3 (SiO_4)_3$), *huashi* (talc, $H_2Mg_3 (SiO_3)_4$), *yangqishi* (actinolite, $Ca(Mg, Fe)_3(SiO_3)_4$), *changshi* (felspar, $K_2O \cdot Al_2O_3 \cdot 6SiO_2$), *buhuimu* (*shimian*, asbestos, $H_4Mg_3Si_2O_7$), *baiyü* (white jade, $Na_2O \cdot Al_2O_3 \cdot 4SiO_2$), etc.

Alloys: *toushi* (copper and zinc), *baijin*, *baitong* (copper and nickel), *baila* (lead and zinc), and the various amalgams.

Sand and soil: *gaolingtu* (kaolin, SiO_2, Al_2O_3, etc.), *yuyuliang* (haematite, Fe_2O_3 with sand and soil), *shizhonghuangzi* (flint), etc.

Organic solvents: *cu* (vinegar, CH_3COOH), *jiu* (alcohol, $CH_3 CH_2OH$), etc.

The statistics are incomplete not only because of the total absence of plant and animal drugs but also because the number of minerals employed in alchemical processes would have been much greater. However, the list may create a sketchy impression of the scale on which alchemical experiments were conducted.

Alchemical Apparatus and Equipment

More than 10 kinds of apparatus and equipment are mentioned in alchemical classics. They include furnaces and vessels specially designed for alchemical purposes, water basins, distilleries for the extraction of mercury, *huachi* baths, mortars or grinders, sieves, etc.

Alchemists' furnaces are called *danlu* or *danzao*. Described in *Dan Fang Xu Zhi* (*Indispensable Knowledge for the Alchemical Laboratory*) written by Wu Wu in 1163 are two kinds of alchemists' furnaces: the *jijilu* and the *weijilu*, or furnaces with or without a water bath attached. The reaction chamber built in the furnaces is called *danding*, which has a number of other names such as *shenshi* (magical chamber) *gui* (cabinet), and *danhe* (elixir box). Some of the vessels are in the shape of a bottle gourd, some in the shape of a crucible. Some are made of metal (gold, silver or copper), some of porcelain. *Xuantaiding* (suspended inner vessel) is introduced in *Jin Dan Da Yao* (*Essentials of the Metallous Enchymoma*) of the Yuan Dynasty (1271-1368). *Jin Hua Chong Bi Dan Jing Yao Zhi* (*Confidential Instructions on the Manual of the Heaven-Piercing Golden Flower Elixir*) gives an account of a silver water basin on top of the *shenshi* for the purpose of cooling. In *Xiu Lian Da Dan Yao Zhi* (*Essential Instructions for the Preparation of the Great Elixir*) is mentioned *shuihuoding*, probably a double-layer vessel with water in between. The reaction-vessels are the most important apparatuses of the alchemists. They can be heated on furnaces with drugs inside to bring about the expected reactions or sublimation.

Besides the *danding* mentioned above, alchemists are provided with special apparatus for the extraction of mercury from cinnabar. That recorded in *Jin Hua Chong Bi Dan Jing Yao Zhi* is a rather simple device consisting of two parts. The upper part is in the shape of a round-bottomed flask and hence is called the "pomegranate" pot; the lower part is a tub-shaped crucible called *ganguozi*. When in use, the *ganguozi* is buried in earth with cold water inside it. Cinnabar and charcoal are loaded into the pot before its mouth is covered with a small tile which is secured with iron wire. It is then placed on the *ganguozi* and heated. The vaporized mercury thus produced goes into the pot and is condensed to liquid mercury by the cooling effect of water. In 1970 four silver pomegranate pots were discovered among the mass of pharmaceutical and alchemical cultural relics of the Tang Dynasty excavated at Hejia Village in the southern suburbs of Xi'an. This shows that alchemists were already using such simple retorts during that period. The distillery in *Dan Fang Xu Zhi* is of a more complicated type. Although no specifications are given, an illustration shows that it is an air-tight vessel with cinnabar or other drugs inside heated in a furnace. A side tube of the vessel leads the vapour into a condenser alongside the vessel and the furnace. It is of a rather advanced design even by modern standards. Its appearance could be earlier than the date of writing *Dan Fang Xu Zhi* (1163). Western writers on the history of science always attributed the invention of the distillery to the Arabs. Actually the Chinese have a very long tradition in producing such apparatus.

Relations Between China and the West in Alchemy

It is universally acknowledged that chemistry was developed on the basis of medieval European alchemy, which had its sources in Arab alchemy. But it was little known three or four decades ago that Arab alchemy, which first appeared in the eighth century, was probably related to Chinese alchemy.

There were many similarities between Chinese and Arab alchemy: Chinese alchemy was motivated by the quest for *shendan* (magical elixir) which was believed to be capable of giving man long life and of turning iron into gold by projection. Arab alchemy was the same, in which the magical elixir was called *aliksīr* or philosopher's stone. Aurifaction constituted an important part of both Chinese and Arab alchemical attempts. In *Huai Nan Zi* (*The Book of the Prince of Huai Nan*) written in the 2nd century B.C. and in other alchemists' classics are to be found theories about the formations and transformations of metals underground. Jābir ibn Hayyan, the Arab alchemist, said something about similar formations and transformations. Since Arab alchemy appeared much later than its Chinese counterpart, and since there had long been cultural contacts between China and the Arab countries, it seems reasonable to claim that the above similarities were consequences of the transmission of Chinese alchemy to the West.

What is further noticeable is that Chinese alchemists had long been making use of saltpetre and *naosha* (ammonium chloride) when these were still unknown in ancient Greece and Egypt but were used in contemporary Arab countries and Persia. Saltpetre was called "China snow" in Egypt and other Arab countries, and "China salt" in Persia. Among the seven leading metals listed in the Arab and Persian alchemists' classics appears "China metal" or "China copper". These can be regarded as evidences of the transmission of Chinese alchemy to Central Asia and Egypt. The Arabic for alchemy is *al-kīmivā*. It has been noted that *kīmivā* was how the two characters *jinye* (potable gold) were pronounced in ancient China. This seems a relevent testimony since in the Tang and Song dynasties China had already established contacts with Central Asia. This was especially true during the Song Dynasty, in which trade with the Arab countries by sea had become quite flourishing. Quanzhou of Fujian Province was then an important trade port. In modern Quanzhou dialect, *jinye* is still pronounced "kim-ya".

Since the beginning of the 20th century the above historical facts have been quite ignored by many Western scholars, who have tried to trace the sources of European alchemy to ancient Greece and Egypt, while some have even alleged that Chinese alchemy was an import from Greece via the ancient Arab countries. The distorted historical facts have been partly righted since the 1930s due to the findings of some Western historians of science who have made profound studies. Dr. Joseph Needham puts it, ". . . One of the most important roots of all chemistry, if not indeed the most important single one, is typically Chinese."[1] The Chinese nation has certainly contributed its share in blazing the trail for chemistry, as for other sciences.

---

[1] *Science and Civilisation in China*, Cambridge University Press, London, 1954, Vol. 1, Preface, p. 9.

Phenological Calendars and Knowledge of Phenology

Cao Wanru

With Chinese agriculture beginning to develop as early as 6,000-7,000 years ago, and with the ancient Chinese eager to do farm work in the right season, their knowledge of phenology in production long ago led to the appearance of the phenological calendar in China.

The earliest extant work recording phenological phenomena in China is *Xia Xiao Zheng* (*Lesser Annuary of the Xia Dynasty*[1]). Though there are different views as to the date of its writing, the language and contents lead us to believe that it dates from China's slave society.

Of only about 400 words and written in archaic and simple style, the document carries a great deal of information. Phenological, meteorological and celestial phenomena and also such major administrative matters as those related to agriculture (for example, sericulture and horse-raising) are outlined in the order of the 12 lunar months respectively. Most remarkable is the observation on phenological phenomena, which indicates that China was founded in antiquity on the basis of agriculture. Phenological data were collected and recorded month by month for arranging agricultural production at the optimal season. Being mainly a monthly record of phenology and farm activities, the document is the oldest phenological calendar — a valuable Chinese cultural heritage.

---

[1] The Xia Dynasty existed approximately between the 21st and the 16th century B.C.

First Month and Ninth Month in *Xia Xiao Zheng* deal with phenology:

First Month (corresponding to February in the Gregorian calendar, beginning of spring):

Phenological phenomena: Hibernating insects wake up; wild geese fly north; pheasants drum and call; fish rise beneath the ice; leeks appear in the garden; field rodents emerge; otters seek for fish as food; hawks become turtledoves;[1] catkins burst forth on the willow; the plum, apricot and mountain peach blossom; the cyperus flowers; hens begin to lay.

Meteorological phenomena: In this season sweep gentle breezes; though chilly still, they melt the frozen soil.

Celestial phenomena: *Ju* star (i.e. *Xu*, 11th *xiu*) now is seen; at dusk *Shen* (Orion, 21st *xiu*) culminates; *Doubing* (the "handle" of the seven stars of the Great Bear) hangs downward.

Farm tasks: Farmers repair and maintain their *leisi* (a farm implement used for tilling); the farming inspector proceeds to mark off the fields; yellow rape is picked as sacrificial flowers for the ancestors.

Ninth Month (corresponding to October in the Gregorian calendar, when the weather becomes cold in China):

Phenological phenomena: Wild geese fly south; swallows rise to hibernate; bears, brown bears, panthers, racoon dogs, *si* (a rodent) and weasels enter their caves; chrysanthemums, queen of flowers, are in their glory; sparrows dip the sea and become clams.[2]

Meteorological phenomena: (No record for Ninth Month.)

Celestial phenomena: The sun is close to the star *Huo* (Antares, central of *Xin*); *Chen* (another name for Antares) is joined with the sun.

---

[1] Turtledoves which came in spring were mistaken for hawks.

[2] A mistaken notion by which sparrows were thought to have become clams, for the streaks on the shells of the latter resembled the feathers of the former.

Farm tasks: Wheat is sown; fur gowns are provided for the king.

Although the above-quoted passages form only a part of *Xia Xiao Zheng*, they do suffice to show the scope of phenological observations made by the Chinese about three millennia ago. In botanical observation, they mention herbaceous and woody plants; in the animal world they note the behaviour of wild birds, beasts, poultry and fish. In short, the linking of phenological phenomena with farming indicates that from the very beginning phenological knowledge developed in tandem with the demands of agriculture.

It must be pointed out that *Xia Xiao Zheng* not only records plum, apricot and mountain peach blossoming in the first month, but also mentions the Huaihe River,[1] the sea and Chinese alligators[2]. Obviously, phenological data in the document mainly concern the coastal areas up the Huaihe and Changjiang rivers.

Shi Jing (*Book of Odes*) is a collection of verses from the era of slave society in China which contains quite a number of records of phenological phenomena. Outstanding are those described in "Qi Yue" ("Seventh Month") in the chapter "Bin Feng" ("Folksongs from the Bin State"). We find such vivid descriptions as, "In the fourth month the milkwort is in spike", . . . "in the fifth month cicadas sing", "in the sixth month grasshoppers shake their wings", . . . "in the 10th month crickets go under my bed", etc. "Qi Yue" appears to be the earliest poem concerning phenology.

Observation and recording of phenological phenomena at each *jieqi* (fortnightly solar term) is also of long standing in China. As early as in the Spring and Autumn Period (770-476 B.C.) there was the tradition of recording phenological phenomena and weather conditions at Spring Equinox, Autumn Equinox, Summer Solstice and Winter Solstice.

---

[1] Tenth Month: "Black pheasants enter into the Huaihe River and become bigger clams."

[2] Second Month: "Skinning alligators" referring to Chinese alligators as found in the middle and lower reaches of the Changjiang (Yangzi) River.

The chapter "Twelve Ji" of *Lü Shi Chun Qiu* (*Master Lü's Spring and Autumn Annals*) of the Warring States Period (475-221 B.C.) follows on the whole a pattern similar to *Xia Xiao Zheng* in recording celestial and phenological phenomena and administrative affairs in the order of the early, middle and late phases of the four seasons, i.e., following the order of the 12 months of the year. A difference between the two books is that records given in the first work are numerous and jumbled, many having nothing at all to do with phenology, and parts that do have are mainly taken from the latter. New materials are added, however, especially in the field of meteorology.

"Yue Ling" ("Monthly Ordinances") of *Li Ji* (*Book of Rites*) and "Shi Ze Xun" ("On Times and Seasons") of *Huai Nan Zi* (*The Book of the Prince of Huai Nan*) and some other books appearing in the Han Dynasty (206 B.C.-A.D. 220) have been renowned for their phenological records, but the contents are almost the same as those in *Lü Shi Chun Qiu*.

It is noteworthy that *Yi Zhou Shu* (*Lost Books of the Zhou Dynasty*[1]) written in the Han Dynasty has two chapters on phenology: "Yue Ling Jie" ("Notes on Monthly Ordinances") with almost the same contents as "Yue Ling" of *Li Ji*, and "Shi Xun Jie" ("Interpretations of Times and Seasons") in which phenological phenomena are recorded according to the order of the 24 fortnightly periods and the 72 *hou* (1 *hou* = 5 days), marking a big advance in China's phenological calendar. In the Northern Wei Dynasty (386-534) the calendar with 72 *hou* was adopted as the national calendar, to be followed in succeeding dynasties with phenological contents approximating those in "Shi Xun Jie" of *Yi Zhou Shu*. In the 19th century, the Taiping Heavenly Kingdom (1851-1864) issued a "Heavenly Calendar" (solar calendar) to correct the defect of former calendars which neglected the difference of phenological areas. The Heavenly Calendar is attached with a section called "Meng Ya Yue Ling" ("Preliminary Monthly Ordinances") compiled from phenological observations made in Nanjing, capital of the Taiping Heavenly

---

[1] The Zhou Dynasty existed approximately between the 11th and the 3rd century B.C.

Kingdom. It also includes monthly observations of the previous year as a reference for farmers.

Progress made in the astronomical almanac plays an important role in agricultural production, and phenological knowledge had always figured in agricultural advance in ancient times when frequent mention is made of suitable ploughing and sowing times by referring to phenological phenomena. For instance, in *Fan Sheng Zhi Shu* (*The Book of Fan Shengzhi*), a famous treatise on agriculture of the Western Han Dynasty (206 B.C-A.D. 24), there appears this passage about agricultural production in Guanzhong District, Shaanxi: "Light soils are to be ploughed when apricot trees begin to blossom. Plough again when the blossom fades." And for sowing beans: "In the third month, when elm trees are fruiting, sow soya beans in the upland fields whenever it rains. When mulberries turn dark, sow subsidiary beans after rain." Now let us turn our attention to *Qi Min Yao Shu* (*Important Arts for the People's Welfare*) written by the famous agriculturist Jia Sixie of the Northern Wei Dynasty. Jia said that in growing cereals people should do what was appropriate to the season and consider well the nature and conditions of the soil, and that then and only then would the least labour bring best results. Relying on one's own ideas, not on the orders of Nature, he cautioned, would make every effort futile. He divided the sowing time into three periods — optimal, medium and the latest possible. Since the region Jia referred to was the North China Plain, the best time for sowing spiked millet was the first 10 days of the second lunar month, till the male hemp plants puff and poplar leaf-buds swell; the second best time was the first 10 days of the third lunar month, till about Qingming (Pure Brightness Day, in early April) when the peach blooms; and the latest time was the first 10 days of the fourth lunar month, till jujube leaf-buds swell and mulberry blossoms fall. Records related to phenology appear also in other ancient books on agriculture such as *Si Min Yue Ling* (*Monthly Ordinances for the Four Classes of People*) by Cui Shi (-c. 170) of the Eastern Han Dynasty (25-220), *Nong Sang Ji Yao* (*Fundamentals of Agriculture and Sericulture*) and *Wang Zhen Nong Shu* (*Agricultural Treatise of Wang Zhen*) of the Yuan Dynasty (1271-1368) and *Nong Zheng Quan Shu* (*Complete Treatise on Agricultural Administration*) by Xu Guangqi (1562-1633) of the Ming Dynasty (1368-1644).

In the Northern (960-1127) and Southern (1127-1279) Song dynasties and the latter Ming and early Qing (1644-1911) dynasties, progress continued being made in phenological observation and study.

Very few first-hand observations on phenology have been handed down from ancient times. The earliest extant record worthy of mention is *Geng Zi Xin Chou Ri Ji* (*Diary Written in the Years 1180-1181*) by Lü Zuqian, a celebrated man of letters from Jinhua, Zhejiang Province. For two years before his death while ill at home he kept a diary of phenological phenomena as he carefully observed them every day. The diary begins on the first day of the first month of the year of 1180 and ends with the 28th day of the seventh month the following year. Recorded are the blossoming times of more than 20 plants including the wintersweet, cherry, apricot, peach, Chinese redbud (*Cercis chinensis*), plum, Chinese flowering crabapple, pear, rose, hollyhock (*Althaea rosea*), tawny daylily, lotus, cottonrose hibiscus and chrysanthemum. He also noted the time of the first cry of spring birds and autumn insects. As the records are based on the lunar calendar, this must be taken into account when using them. They are treasured as the earliest extant records of actual observations in the world.

Shen Kuo (1031-1095), a scientist of the Northern Song Dynasty, also studied various aspects of phenology, such as the effect of altitude and latitude on plant cultivation. In his *Meng Xi Bi Tan* (*Dream Stream Essays*) he observed that owing to the difference in altitude, plants which blossomed in the third month on the plain would blossom in the fourth month in mountain areas. He appreciated very much the famous lines of Bai Juyi:

Most blossoms have faded in the fourth month on the plain

While peach trees in the mountain temple are still in full bloom.

He considered that these implied "a universal truth" of phenology. He also remarked that the grass along the Nanling Mountains "remains green in winter," but that trees in the Fenhe River valley of Shanxi "wither and fall with the coming of autumn". All this is due to "differences in *diqi*" (soil temperature). He analysed the causes for the different ripening times of rice strains — "some ripen in the 7th month, some in the 8th or 9th while others in the 10th"— as due to their "different characteristics". More important is his explanation of the human factor, i.e., cultivation, as playing an important role aside from phenology, which is restricted to natural factors. He said that since "human efforts are not the same," "how can we confine ourselves to a definite month?" This was a truly brilliant and progressive view indicating that cultivation can make crops ripen earlier and yield more.

Ben Cao Gang Mu (*Compendium of Materia Medica*) by Li Shizhen of the Ming Dynasty records nearly 2,000 medicines and many phenological phenomena. Xu Xiake (1586-1641), a geographer and a contemporary of Li Shizhen also refers in his *Xu Xia Ke You Ji* (*Travels of Xu Xiake*) to the differences in phenological phenomena between the south and the north as due to "the further northward, the colder it is", which is a more scientific explanation of Shen Kuo's theory of the *diqi*.

The geographer Liu Xianting (1658-1695) of the early Qing Dynasty also carefully observed the phenological phenomena of different places. He criticized the contemporary calendar as indiscriminately copying the 72 *hou* in the ancient *Yi Zhou Shu*. He remarked that "different" places have a "different" 72 *hou*. "For example, the plum south of the Five Ranges[1] blooms in the 10th month whereas the peach and the plum in Hunan blossom in the 12th. . . ." He suggested that detailed phenological phenomena should be recorded both in the south and in the north and "handed down to later generations" so that the laws of Nature could be discovered.

From *Xia Xiao Zheng* to Liu Xianting, phenological knowledge over more than 2,000 years in China is one of the country's splendid achievements in science.

---

[1] Roughly Guangdong and Guangxi.

Water Conservancy Projects and Knowledge of Hydrology

Song Zhenghai

Irrigation is known as the lifeblood of agriculture. For several thousand years the industrious, brave and ingenious Chinese people have struggled dauntlessly with rivers, lakes and seas, constructing countless water conservancy projects, effectively promoting agriculture and achieving a corresponding development in the knowledge of hydrology.

Several Major Water Conservancy Projects

Not a few water conservancy projects constructed in ancient China have been world-famous. Aside from their colossal size, these projects were expertly designed even by modern standards.

Dujiangyan. Situated on the Minjiang River near Guanxian County on the Chengdu Plain, Sichuan Province, the project was built under the direction of Li Bing, who assumed office as Governor of Sichuan in 250 B.C. Dujiangyan consists of three component units: Fish Snout (the primary division-head), Flying Sands Spillway and Cornucopia Channel. The river-diverting Fish Snout is a midstream weir dividing the Minjiang River into Inner River on the east and Outer River, its mainstream, on the west. Cornucopia Channel is the upper end of a canal built by splitting Jade Rampart Mountain. Flying Sands Spillway regulates the volume of flow into the canal.

Below Cornucopia Channel, Inner River feeds into irrigation channels that crisscross the Chengdu Plain. Because of Dujiang-

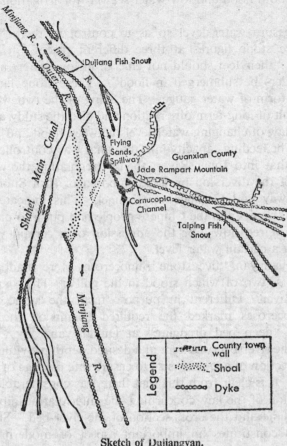

Sketch of Dujiangyan.

yan, the Chengdu Plain "had the benefit of irrigation against drought, and of prevention of flood in the rainy season", according to classical records. The area became a rich granary and was called "Land of Abundance".

The planning, designing and building of Dujiangyan were remarkably scientific, original and nearly perfect. Fish Snout, Flying Sands Spillway and Cornucopia Channel co-ordinate to handle the flow in both flood and low-water seasons for irrigation and flood prevention.

To measure water level so as to control it, there "were stationed three stone figures at three different places" at the canal intake, and "their feet should not emerge in low water, nor should their shoulders be submerged in flood". These stone figures were a primitive form of water gauge. The fixing of the two water levels was the result of long-term observation and also the study and cognition of the law of changing water levels between flood and low-water seasons. The feeding capacity into the canal was controlled by manipulating that into Inner River, then by regulating the water level by means of the diversion project composed of Fish Snout, Flying Sands Spillway and Cornucopia Channel. This indicates that as early as 2,300 years ago the Chinese people had put into practice the "theory of weired flow", or the admission of a certain volume of discharge at a certain water level.

At Dujiangyan "five stone rhinoceroses were made upon Li Bing's orders, two of which stood in the gulf". The "gulf" refers to Inner River. Different in purpose from the stone men, the stone rhinoceroses marked the required height of the riverbed for regular "deep bed dredging" at Dujiangyan. By "deep bed dredging" the river-bed was kept at a desired depth providing a riverflow cross-section of adequate size to ensure safe passage of relatively big floods. It is clear from this that there was considerable mastery and utilization of the law concerning the interrelationship between flowing volume and the cross-section of the river-bed. Such interrelationship constitutes an important aspect of modern formulae about flowing capacity.

Zheng Guo Canal. Built in the first year of the reign of Qin Shi Huang, the First Emperor of the Qin Dynasty, (246 B.C.), this waterway was designed by hydraulic engineer by the name of Zheng Guo, who was also in charge of construction. Starting from the present-day Jingyang County, Shaanxi Province, the 150-kilometre canal directs the water of the Jingshui River eastward into the Luoshui River, providing irrigation for the Guanzhong Plain. The construction of the canal made possible what was known as "fertilizing irrigation", that is, silt precipitation from the Jingshui River water which carried with it fertile silt. As a result, two million mu[1] of alkaline soil in central Shaanxi became fertile. From then on, "central Shaanxi became rich land, free from drought". Following is a song then popular in central Shaanxi:

> Zheng Guo sprang forth,
> Bai Qu following,
> Each measure of water
> They brought from the Jingshui
> Yielded measures of rich soil
> To reward our toil.
> Moist and fertile, our land
> Produces lush crops.
> Abundantly ever
> It feeds and clothes the millions.

The "rich soil" is of course the silt deposited by the two canals, which contributed to the developed agriculture and prospering economy of the central Shaanxi Plain.

The building of Zheng Guo Canal, with its head at the bank of the Jingshui in the mid-lower reaches of the river, is scientific indeed and shows a knowledge of hydrology. Where the river makes a turn, there is a cross current to the mainstream. The water in the upper stratum flows from the convex to the concave one, the rate of flow reaching its maximum in the mid-lower reaches close to the lower bank and directly opposite the entrance to the canal. This ensures abundant feeding into the canal of water carrying with it large quantities of fine silt which is deposited by irrigation. The cross current at the lower stratum, contrary to that above it, flows from the lower bank to the higher, in which direction it flushes the coarser sand moving near the river-bed due to its greater weight, preventing the coarse material from getting into the canal and choking it.

---

1 mu = 1/15 hectare or roughly 1/6 acre.

Longshou (Dragon Head) Canal. A major waterway was dug by 10,000 soldiers by order of the Emperor Wu Di of Han, for the irrigation of Zhongquan, an area lying 20 kilometres southeast of present-day Pucheng County, Shanxi Province. The builders discarded the idea of an open canal dug round the foot of the mountain, which would have been liable to destruction by the collapse of the eroded loess. In place of this the working people developed the "pit-canal", which "conducts water by connecting pits". This innovation enabled Dragon Head Canal to be led underground through the three and a half kilometres wide Shangyan Mountain (Tielian Mountain today). The idea of the pit-canal was adopted in Xinjiang where it was developed into the technique of "pits-on-slopes".[1]

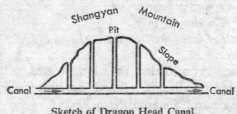

Sketch of Dragon Head Canal.

Shaopi (Anfeng Pond). From 613 to 591 B.C. the people of the state of Chu built Shaopi in the southern part of present-day Shouxian County, Anhui Province. It was built on the basis of a natural pond, with dykes erected in low places. Round the pond are arranged 36 sluice-gates and 72 culverts. Receiving the floods from the Lu'an mountainous region it becomes a reservoir over 60 kilometres in circumference, providing irrigation for more than 100,000

---

[1] The "pits-on-slopes", an irrigation system in which the underground water is made to flow through an underground canal, is widely employed in the dry districts of Turpan and Hami in Xinjiang. An underground canal is dug, starting from the spring-head in a mountainous district. Such canals may be 20-30 kilometres long. Above the canal is a row of vertical pits 20-30 metres apart and extending to the ground. The depth of the pits depends on their positions on the slope. The water feeds into open channels for the irrigation of farmlands.

mu of cropland. Now it has become a component part of the Pi-Shi-Hang Water Conservancy Network.

The Great Yellow River Dykes. Flowing through the loess plateau, the Yellow River (Huanghe) carries with it enormous quantities of mud and silt. Downstream from Mengjin, Henan Province, where the Yellow River enters the plain, its sloping bed suddenly levels up, resulting in rapid precipitation of the mud and silt. This became the cause of frequent floods, dyke-breaching and shifting of river course in flood seasons. The Yellow River has left a record of extremely serious natural calamities.

For generations the Chinese people won illustrious successes in their fight against Yellow River floods, developing a high level of technique in harnessing it. The present 500 kilometres of dykes along its banks are the result of their efforts. They were built and improved through the ages. In the Spring and Autumn Period (770-476 B.C.) there were already small dykes along the lower reaches of the Yellow River. By the time of the Warring States (475-221 B.C.) these had increased. On some sections of the river there appeared what was known as the Great Revetment and Thousand-Zhang[1] Dyke. Qin Shi Huang, the First Emperor of Qin, having brought China under his unified rule, gave the order to "tear down all barricades and eliminate all obstacles", starting construction of the first great dyke uninterrupted by state boundaries. The dykes of the Yellow River were improved in the ages that followed. Before the Ming (1368-1644) and Qing (1644-1911) dynasties, when dykes began to be built with the idea of "narrowing the stream to attack the sand", dyke construction had been undertaken with the sole aim of flood-prevention. Now it began to have the more positive aim of preventing silt deposition. This was a great step forward and led to co-ordinated function of various types of dykes along the Yellow River. There appeared also many projects of a supplementary nature, such as draining dams, dams parallel to the stream, dams with sluice-gates and dams for retardation. In critical sections the dykes were strengthened by various kinds of revetments.

---

[1] *Zhang* — A Chinese measure of length which in the Warring States Period equalled 2.29 metres.

The great dykes on the Yellow River should have prevented flood disaster, protected the people's lives and property and promoted agriculture on the North China Plain. Actually, the reactionary ruling classes turned a blind eye to the people's plight and lined their own pockets with the funds for river control. The dykes fell into disrepair and failed in their purpose, writing into history records of incessant havoc wrought by "China's sorrow", the Yellow River.

The idea of "narrowing the stream to attack the sand", remarkable in classical Chinese hydrology, is still a theoretical tenet in controlling and improving the construction of the Yellow River dykes. Pan Jixun of the Ming Dynasty got the idea and formulated it on the basis of earlier illustrations of the law governing the movement of mud and sand in rivers, plus his own experiences in river-control. He wrote:

> When the stream is wide, its flow is slow, the sand stays, and the river-bed rises. . . . When the stream is narrow, the flow is rapid, the sand is washed away and the river deepens. . . . Build dykes to narrow the river, for the water to attack the sand. The water, prevented from running sideways, cannot but rush ahead along the river-bed. This is a matter of course, and it is the reason why narrowing is preferable to widening.

These quotations are from *He Yi Bian Huo* (*On Rivers: Some Erroneous Representations*). The theory of "narrowing the stream to attack the sand" has had its place in world hydrological history.

Chen Huang of the Qing Dynasty added to this theory the suggestion that timely outlets of water should be made at certain sections, depending on the width of the river-bed, in order to maintain a steady water level and flow speed. "It must be seen," he wrote, "that the volume discharged equals the volume increased." Chen Huang devised a means to measure the volume of water, drawing on the method then popular for measuring excavated earth. Comparing the flow of a river to a pedestrian walking, he said that a fast-flowing stream, like a "fast walker", covers 100 kilometres in a day, while a slow-flowing river, like a "slow walker", covers 35-40 kilometres. Taking the cube of 10 as a volume-unit, he calculated the river's flowing capacity from the number of such volume-units of water per day — the earliest known method for river-discharge calculation.

Sea-Walls. The earliest record of sea-walls dates from the Western Han Dynasty (206-B.C.-A.D. 24), about 2,000 years ago. They are mentioned in the chapter on the Jianjiang River in *Shui Jing Zhu* (*Commentary on the "Waterways Classic"*) quoting from *Qian Tang Ji* (*Notes on the Qiantang*). This "protective wall against the sea" was one of the largest such embankments of the time. Earth in time gave way to stone for sea-wall construction, and there was continual increase in size.

The provinces of Jiangsu and Zhejiang were then sources of tribute grain transported by water to the capital. Bordering the East Sea, they were vulnerable to typhoons and disastrous tidal waves. Worst affected was the area around Hangzhou and Jiaxing. Sea-wall construction was therefore an important part of water conservancy in Zhejiang, and much effort went into sea-wall building in the Tang (618-907), Song (960-1279) and Yuan (1271-1368) dynasties. Later, during the reigns of the emperors Kang Xi, Yong Zheng and Qian Long of the Qing Dynasty, many more sea-wall construction projects were launched.

The 150-kilometre-long sea-wall between Jinshanwei in Jiangsu on the north and Hangzhou in Zhejiang to the south is world-famous. Like a Great Wall along the coast, it wards off the tidal waves, protecting the vast expanse of the fertile delta and coastal plain.

Two Ancient Navigation Canals

Besides the above water conservancy projects directly benefiting agriculture, the ancient Chinese dug canals to link up rivers for navigation.

Lingqu Canal. Situated in Xingan County, Guangxi Province, its construction was directed by Shi Lu on the decree of the First Emperor of Qin who, having conquered all the other six states of China, unified the whole country and needed the canal for supplying his armies across the natural barrier of the Wuling Mountains. About 15 kilometres long and five metres wide, the canal connects the Xiangjiang River, a tributary of the Changjiang (Yangzi), and the Lijiang River, a tributary of the Zhujiang (Pearl River). The construction of Lingqu Canal started with the building of stone dams known as Plough-Share and spillways called Datianping and Xiaotianping mid-stream of the Xiangjiang, halting its flow. Southern and northern canals were dug below Plough-Share. The northern canal linked the Xiangjiang River, while the southern canal, the Ling-

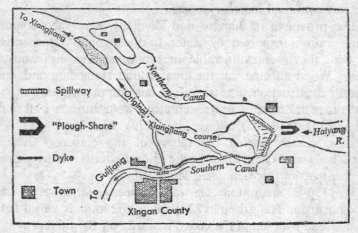

Sketch showing Lingqu Canal.

qu, let to the Lijiang River, which is the upper reaches of the Guijiang.

The flow of the Haiyang River, of the upper reaches of the Xiangjiang, is divided by Plough-Share and proceeds in the two canals. Thus are connected the Xiangjiang and the Lijiang, two rivers of the two major waterway systems feeding respectively into the Xiangjiang and the Zhujiang. When the Haiyang River is in spate, Lingqu Canal discharges the flood water over the spillways into its original Xiangjiang course, preventing flood.

The Lingqu Canal site was at a place where the Xiangjiang and Lijiang rivers are nearest to each other, and where the difference in water level is not great.   Its course moreover is winding in order to reduce the slope of the riverbed, steady the flow and so facilitate navigation.

Scientifically conceived and designed, Lingqu Canal is prominent in world inland navigation history.   Having undergone large-scale reconstruction, the canal today affords irrigation while continuing to serve as a transport waterway.

The Grand Canal.   This is the canal earliest in the date of construction, largest in scale and longest in mileage of all navigation canals in the world.   Starting from Beijing in the north and reaching Hangzhou in the south, it has a total length of 1,794 kilometres.   It channels through the five great water systems of the Haihe, the Yellow, the Huaihe, the Changjiang and the Qiantang rivers.   The advent of the Grand Canal concluded the absence in China of a major north-south waterway.   Its benefit was immense, serving as a trunk

The Grand Canal.

line of transportation between northern and southern China until the completion of the Beijing-Guangzhou Railway.

A canal dug as far back as 2,400 years ago, known as the Hangou, is a component part of the Grand Canal.

From then on it was continuously developed and enlarged, till the Emperor Yang Di of the Sui Dynasty (581-618) who, desirous of speeding up the shipment of tribute grain to the capital, launched into the construction of the canal, completing its total length of 2,400 kilometres in six years. The section of the Grand Canal between the Huaihe and the Haihe was different in the Sui Dynasty from what it is now. It went southeastward and northeastward from Luoyang, then the Eastern Capital of the country. The Yuan Dynasty made Beijing its capital, and it was to Beijing that the shipment of tribute grain was directed from Jiangsu and Zhejiang. To avoid detouring via Luoyang, the canal was straightened, resulting in the present-day Beijing-Hangzhou Grand Canal.

Traversing a vast area with a great variety of geographical conditions, the canal involved very complicated efforts in building, the crucial problem being that of procuring a water source and maintaining the water level. The canal builders, surmounting tremendous technical difficulties and resolving many a complicated problem, finally succeeded in making the canal navigable.

The section of the canal through the hilly regions of Shandong, which was dug in the Yuan Dynasty, was most difficult. It was a key point in the through traffic along the canal, yet sailing through this section was precarious due to the turbulence caused by the drop in water level as the canal crossed the Yellow River. This was resolved during the reign of Yongle in the Ming Dynasty thanks to a rational plan suggested by a civilian by the name of Bai Ying. This was the choosing of a place at which the bed of the canal was highest. At this point was concentrated the entire discharge of the Wenhe River, where the water was brought to flow separately southward and northward. Along the canal were constructed a number of reservoirs, as the topography permitted, to assure water supply. The flow of the water was retarded to a degree of navigability by over 30 sluice-gates all along the canal.

Considering that the canal crossed the Yellow River at this point, the builders made the canal feed into the Yellow River instead of the reverse, eliminating the danger of the canal being choked up with silt carried by the Yellow River.

The Grand Canal that resulted from the two major construction schemes undertaken in the Sui and Yuan dynasties served as the artery of north-south traffic in China. In the Tang Dynasty over two million tons of grain was shipped north yearly via the Grand Canal, a tonnage that increased to seven million by the Song Dynasty. The Grand Canal proved an asset in developing China's economy.

For long periods certain sections of the Grand Canal were not navigable due to silt deposited by Yellow River floods. However, dependent as they were for their subsistence on rice from the south, the rulers in the north never failed to order timely dredging. During the reign of the Emperor Qian Long (1736-1795) of the Qing Dynasty, when the decline of China's feudal system accelerated, the dykes and sluice-gates along the canal fell into disrepair and the channel was choked with silt, seriously impeding through traffic. Beginning from the 20th century (especially after the commissioning of south-north railway trunk-lines), the silting up worsened and there was no longer uninterrupted passage, though various sections were navigable.

In the new China, sponsored personally by Premier Zhou Enlai, the giant programme of "diverting water from the Changjiang to northern China" was mapped out. This envisaged the abundant water of the Changjiang Valley being channelled to the thirsty north China, an unprecedentedly great undertaking for the reformation of nature. The part of the project now underway is its easternmost section, namely, drawing the Changjiang water at Jiangdu, near Yangzhou, to feed the lower reaches of the Yellow and Haihe rivers, linking the four great water systems of the Changjiang, Huaihe, Yellow and Haihe rivers. This waterway is routed along the Grand Canal, which is being dredged and widened. When it is completed, the canal will not only supply large quantities of water from the Changjiang, but will provide round-the-year navigation for 1,000-2,000-ton vessels. The ancient Grand Canal will be restored to fresh life.

Early Hydrological Observations

Advancing along with the development of water conservancy, hydrology was to water conservancy as theory to practise or science to production. Aside from the water conservancy projects described above, the achievements of ancient hydrology worthy of special mention include the Peiling Stone Fish and the Wujiang River Water-Mark Tablets.

The Peiling Stone Fish. Since low water had always been the deterrent to navigation, irrigation and water supply, great attention was paid to low river level from ancient times. To understand the laws governing the recurrence of low water in the Changjiang, an investigation was conducted in recent years into historical records of low water. Markers recording low-water levels were found in 11 places along the section from Yichang in Hubei to Chongqing in Sichuan. The most important of these are Lotus

Stone markers recording high and low water levels along the Chong-qing-Yichang section of the Changjiang and its tributaries.

Stone at Jiangjin, Bumper Harvest Tablet at Chongqing, White Crane Ridge in Peiling, Dragon-Back Stone at Yunyang and Water Recording Tablet at Fengjie. The best preserved and most valuable of these inscriptions are those carved on a stone fish at White Crane Ridge in Peiling. It bears 163 items recording intermittent low-water levels over a period of 72 years. Thus we know that as long as 1,200 years ago, the Chinese working people created the country's earliest "water-level observatories" in our own national style. Archaeologists have traced inscriptions of low-water records to the reign of the Emperor Guang Wu of the Han Dynasty, two millennia ago.

The Wujiang River Water-Mark Tablets. In the Song Dynasty two water-level tablets were erected on the Wujiang River, one for recording changes in water level within the year (from month to month and from season to season), the other to record changes from year to year. On one of the tablets is inscribed:

> Seven horizontal notches represent the water level at different times. They serve as a water gauge. When the water remains below the first notch, fields both high and low are safe. When it rises above the second notch, the low fields will be inundated. . . . When the water rises above the seventh mark, the highest land will be submerged. (*Wu Jiang Kao* [*Notes on the Wujiang River*] by Shen Qi, Vol. 2.)

On the other tablet was inscribed: "Water reached here in the fifth year of the reign of Shaoxi of the Great Song Dynasty" (A.D. 1194). Also, "Water reached here in the 23rd year of the age of Zhiyuan of the Great Yuan Dynasty" (A.D. 1287). From this it is evident that as long ago as in the Song Dynasty a system of observation of water levels was established for the statistical recording of the inundation of farmlands during flood seasons. This amounted to the prototype of the water-level observatory in the direct service of agricultural production as recorded in Chinese history.

Maps 2,000 Years Ago and Ancient Cartographical Rules

Cao Wanru

China's legends say that in the Xia Dynasty (c. 21st-16th century B.C.) nine bronze cauldrons were cast with pictures of maps showing the various regions of the country with their mountains and rivers, plants and animals, as a guide for those who intended to travel to distant places.   We have as yet found no historical evidence of the existence of the cauldrons, though the drawing of maps on utensils 4,000 or more years ago is quite possible.   Few such maps have survived the long years of vicissitudes.

Maps are mentioned in a number of ancient Chinese classics written before the Qin Dynasty (221-207 B.C.), such as *Shang Shu* (*Book of History*), *Shi Jing* (*Book of Odes*), the *Zhou Li* (*Rites of the Zhou Dynasty*), *Guan Zi* (*The Book of Master Guan*), and *Zhan Guo Ce* (*Records of the Warring States*).

In the chapter concerning maps in *Guan Zi* it is stated:

Before a military campaign the commander must read the maps most carefully for terrains so steep or waterlogged as to hamper or damage carts, for possible valleys or passes through impassable mountain ridges, and for dense forests of bushes where enemy ambuscades threaten.   He must learn all the necessary details such as the length of routes, sizes and possible strongholds of cities, and even the causes of prosperity or decline of towns.   He must memorize all these geographical features and set great store by them.   Only thus can he succeed in manoeuvres and raids, in unfolding his every strategic or tactical step logically, and in exploiting the topographical advantages to the full. Maps are used oftenest for military purposes.

Guan Zi, of about the Warring States Period, stresses the study of maps as important to a military strategist or tactician. Maps were in fact attached to all important ancient Chinese military classics such as *Sun Zi Bing Fa* (*Master Sun's Art of War*) written in about the 5th century B.C., and *Sun Bin Bing Fa* (*Sun Bin's Art of War*) written in about the 4th century B.C. The military use of maps demanded on the part of ancient map-makers fairly clear ideas of orientation, distances and scales to locate mountains and rivers, and indicate lengths of routes and sizes of cities and towns.

By good fortune, three marvels of ancient cartography were unearthed in 1973 in the No. 3 Han tomb at Mawangdui near Changsha. All three are painted on silk fabric. The first is a topographical map of contemporary Changsha Guo, an early Han Dynasty (206 B.C-A.D. 220) marquisate. The second is a military map showing how armed forces were stationed throughout the marquisate. The third points out all cities and towns of the marquisate. However, no titles, scale specifications or dates of drawing are given on any of the maps. Judging from the archaic names of places mentioned and from the time when the tomb was built (168 B.C.), we are sure the maps were made in the early years of Western Han (206 B.C.-A.D. 24), over 2,100 years ago. They are the oldest maps based on actual surveying ever found in China.

Contrary to modern practice, the three maps put the south at the top of the sheet. The focal part of the topographical map was drawn on a scale approximately 1:180,000. The military map uses a scale approximately between 1:80,000 and 1:100,000.

The topographical map, about 96 centimetres square, depicts southern Changsha Guo (Tai Marquisate), which is today the Xiaoshui River basin in the upper reaches of the Xiangjiang River. All mountains, rivers, population centres and routes at that time are clearly located; even contour lines of slopes and contrasting heights of mountain peaks are partially marked. However, the upper part of the map fades out on an abruptly diminishing scale to the

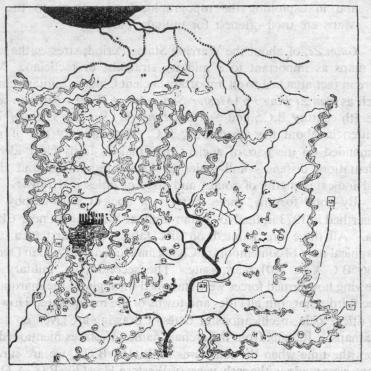

Reconstruction of the topographical map unearthed in the No. 3 Han tomb at Mawangdui near Changsha.

extent that the South Sea coastlines are grotesquely represented. This is because the upper part of the map extends beyond the border of the Tai Marquisate and so enters the domain of another prince. The surveying was obviously limited to within the boundaries of the marquisate.

The military map, which is 98 centimetres long and 76 centimetres wide, shows only part of the area represented by the topographical map. It marks off the garrison districts with castles and strongholds where troops were stationed at that time. The map is very useful in studies of Han Dynasty local military practice and thought.

Comparing the topographical map with a modern map, we

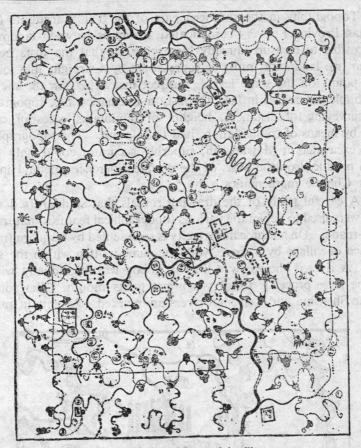

Reconstruction of the Mawangdui military map.

find that the courses of rivers, for example the Shenshui (now Xiao-shui) and its tributaries, largely coincide.   Fairly accurate also are the locations of such county seats as Yingpu (now Daoxian), Nan-ping (now Lanshan), Chongling (now Xintian) and Lengdao (now Ningyuan). The towering Jiuyishan Mountain range and the Dupang-ling Ridge, which lies north-southwards, are exquisitely drawn. All these would have been impossible without a fairly scientific and accurate method of surveying 2,100 years ago.

Topography in the southern part of the Tai Marquisate (now Hunan Province) is very complicated, distances between landmarks impossible to measure directly by pacing, so that the distance between two landmarks must be represented by a straight line joining two points on the same level. Indirect measuring, or the method of double differences in Han Dynasty mathematics, i.e., topographical surveying (of distances, heights of mountains, sizes of cities, etc.) by means of triangulation, had to help out. The topographical map excavated at Mawangdui, superbly accurate, could only have been produced on the basis of the double differences method in ancient Chinese mathematics.

A careful study of the drawing reveals that standardized symbols were used to depict the various topographical features, natural or man-made. On all three maps a city is represented by a square frame, a town or village by a circle, with the name written within the symbol. A river is shown by a line or ribbon which gradually widens as the river proceeds from its sources to its mouth. The sources of the Shenshui and Lengshui rivers are marked. Mountain ranges

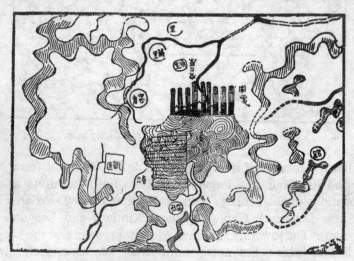

Details of the Jiuyishan mountain region on the reconstructed topographical map of the Tai Marquisate unearthed at Mawangdui.

are represented by closed lines following their courses. The Jiuyi-shan mountain region is wonderfully depicted by hachuring, a system of thickening the lines as the slope sharpens, very similar to the contour lines used today. Contrasting heights of peaks are shown by imagined pillars of proportional heights on top. Roads are indicated by thin lines without other specifications. All sketchy lines were drawn with a deft hand. Compared with the topographical map, the military map is less exquisite in depicting the mountains, but its waterways and important military places and fortifications are in bright colours, a feature which does not appear on the topographical map.

The excavation of the 2,100-year-old maps at Mawangdui is a great contribution to the study of the history of cartography. The magnificent maps add new lustre to ancient cartography of China and the world.

Of course, very few maps were produced from actual surveying (such as those for military use) in the Han Dynasty. Most demarcated administrative districts and, not soundly based on surveying, were of inferior quality. Some were later examined and criticized by Pei Xiu (223-271) of the Western Jin Dynasty. "None ... employs a graduated scale and none ... is arranged for orientation .... One cannot rely on them."

In 267 Pei Xiu was appointed minister of works by the Jin emperor and was soon charged with the compilation of *Yu Gong Di Yu Tu* (*Yu Gong Regional Maps*). In the preface to this work Pei summed up the experiences of earlier cartographers and formulated six rules in map-making: First, there was to be a graduated scale to which the map was to be drawn. Second, it had to show the correct directional relations between the various parts of the map. Third, the distance between two landmarks was to be measured by man's pacing. Fourth, fifth, and sixth, in case two landmarks were not on the same level, they had to be first reduced to the same level. Curves had to be reduced to straight lines. Pei Xiu's six rules, written in the 3rd century, were the earliest guidelines in map-making.

These rules were highly valued and followed by Jia Dan (730-

805) of the Tang Dynasty, who compiled *Long You Shan Nan Tu* (*Map of Longyou and Shannan Prefectures*) and *Hai Nei Hua Yi Tu* (*Map of Chinese and Barbarian Peoples Within the Four Seas*). The latter was a map about 10 metres long and 11 metres high; on a grid scale of 1 *cun*[1] to 100 *li*[2]. This huge map, long lost, was well-known for two features: First, it depicted an area of 30,000 *li* from east to west and 33,000 *li* from north to south, including China and many "barbarian" states.   Jia collected material about the "barbarian" states from their envoys to China.   Second, historical changes in the names of states, provinces, prefectures, and counties, and changes in the different areas that had been under their jurisdiction or administration were marked down in the map.   Jia used black for names which had been abolished, for current names he used vermilion — a practice followed by later cartographers.

Shen Kuo (1031-1095) of the Northern Song Dynasty writes in his *Meng Xi Bi Tan* (*Dream Stream Essays*), about "the map of the flying bird" and the method of finding out "the distance as covered by a flying bird", which is "as the crow flies" or unmistakably straight.   Shen applied the method to the production of *Shou Ling Tu* (*Map of the Counties and Prefectures*).   His "flying bird" measuring of a distance was the same as described by Pei Xiu in the last three of his six rules.

Before the excavation of the Mawangdui maps, the greatest marvels of ancient Chinese cartography in existence were *Hua Yi Tu* (*Map of China and the Barbarian Countries*), *Yu Ji Tu* (*Map of the Tracks of Yu the Great*), and *Di Li Tu* (*General Geographical Map*). Carved in stone in the Southern Song Dynasty (1127-1279), all are well preserved.   The maps cover a vast area including the valleys of the Changjiang (Yangzi), Huanghe (Yellow) and Zhujiang (Pearl) rivers, and regions north of the Great Wall.   All river courses and coastlines are highly accurate.   *Hua Yi Tu* and *Yu Ji Tu* were carved in the same year (1136) and are on opposite sides of one stone stele. However, there are some differences in the two maps due to the dif-

---

[1]   1 (Tang) *cun* = 2.94 centimetres.
[2]   1 (Tang) *li* = 18,000 *cun.*

ferent data collected in surveying. *Yu Ji Tu*, in which the grid scale is 100 *li* to each side of a small square, is obviously more accurate than *Hua Yi Tu*.

Zhu Siben (1273-1333) of the Ming Dynasty was an avid traveller. He made use of his personal findings to check the maps produced by earlier cartographers and found many mistakes in them. It took him 10 years to prepare his own *Yu Di Tu* (*Terrestrial Map*), with of course the grid scale. Zhu's successor two centuries later, Luo Hongxian (1504-1564), thought Zhu's single map too unwieldy and went on to revise and enlarge it into a book of many sheets. Zhu's grid scale was naturally of great help to Luo in dissecting the original map into handy sheets. The enlarged parts are mainly of the border regions and river valleys. Luo's map book is called *Guang Yu Tu* (*Enlarged Terrestrial Atlas*). There is definitely a network of squares on the sheets, and on nearly all maps drawn in the Ming and Qing dynasties the grid scale is to be found. Pei Xiu's rules of indicating distances by a network of tiny squares have become features on traditional maps of China.

During the reign of the Emperor Kang Xi (1662-1722) in the Qing Dynasty, a complete geographical survey of the empire was carried out in preparing *Huang Yu Quan Tu* (*Complete Atlas of the Imperial Domain*). The European method of regarding the earth's surface as a sphere was adopted. Latitude and longitude surveys were done and the projection method was used in drawing the map. The traditional Chinese rules of map-making — accurate proportion, correct orientation and precise distances — were closely adhered to by cartographers of those days.

Rocks, Mineralogy and Mining

Yang Wenheng

Rocks and Mineralogy

Minerals and rocks play a vital part in people's life. Over half a million years of obscure prehistory the ancestors of the Chinese gradually learned the use of 21 kinds of minerals and more than 30 kinds of rocks, according to archaeological findings.

The Chinese were mining copper ores and refining them 4,000 years ago. Towards the end of the Yin time (14th-11th century B.C.) the Si Mu Wu Ding, a bronze quadripod weighing 875 kilogrammes, was cast at the order of a certain Si in memory of his mother Wu. Tin and lead, components of bronze, had long been mined. Metereoric iron was found and put to use by the Chinese more than 3,000 years ago. This prompted the exploitation of iron ores. During the Spring and Autumn and Warring States periods (770-221 B.C.) the Chinese were already masters at refining iron ores. Carburizing steel appeared. Iron tools came into common use in agriculture and the people's daily life. The development from using stone to using bronze and iron tools epitomized the entire history of prospecting, mining and refining a number of ores and non-metallic minerals. "Treatise on Mountains" in *Shan Hai Jing* (*Classic of Mountains and Rivers*) and *Yu Gong* (*Tribute of Yu*) are the earliest summing-up of ancient Chinese knowledge and uses of rocks and minerals from the Stone Age to the Iron Age. "Treatise on Mountains", written in about the 5th century B.C., tells of 89 ores, non-metallic minerals, special rocks or clays, and 309 places where they were produced. These are fairly detailed statements, specifying topographical surroundings of mines such as the top or foot, whether on the sunny or shady side of a mountain; the outward appearance and properties of a rock such as its hardness, colour, lustre, type of aggregation (clay lump, nugget, oval-shaped agglomeration, grains, massive rock), etc. It is also told whether the minerals can be smelted. A nomenclature based on the above findings lists such names of different rocks as magnetite, multi-coloured rock, soft rock, red grain, purple rock, white jade, black rock, etc. "Treatise on Mountains" expounds also on the paragenesis and association of minerals such as that of copper and "grindstone", iron and garnet, silver and "grindstone"; iron, jade and "black chalk"; gold and silver, white gold and iron; gold, silver and iron; "golden jade" and ochre. *On Stones*, written between 371-286 B.C. by the Greek scholar Theophrastus and classifying 16 minerals as metals, stones or clays, is believed by many to be the earliest lapidary in the world. Actually, "Treatise on Mountains", much richer in content, had been in existence for more than 200 years.

Mineralogy and pharmacology were closely related in ancient China. In all the more than 30 pharmaceutical classics since time immemorial, from *Shen Nong Ben Cao Jing* (*Shen Nong's Materia Medica*) down to *Ben Cao Kang Mu* (*Compendium of Materia Medica*), a large number of mineral drugs are listed and described as to their pharmaceutical, chemical and physical properties, their distinguishing features, production sites and mining methods. As a whole their statements on mineral drugs form a fairly rich compendium of mineralogy in themselves. *Ben Cao Yan Yi* (*Dilations upon Materia Medica*) written in 1116 by Kou Zongshi of the Song Dynasty, qualitatively analyses minerals according to their chemical changes, forms of crystals, cleavages and colours. *Ben Cao Kang Mu* (1596) by Li Shizhen of the Ming Dynasty sums up all essential information of previous pharmacists plus his own erudite knowledge from meticulous field investigations and is also a concise but comprehensive work on minerals. Of the 266 kinds of minerals listed in the book, more than 160 are metals or rocks. Li Shizhen classifies them into four groups: metals, stones, jades and gemstones, and salts. His descriptions of their source, appearance, properties and tests are fairly detailed and are verified by his personal textual and field findings.

Chinese monographical classics on rocks and stones include *Yun Lin Shi Pu* (*Yun Lin's Manual on Stones*) (1133) by Du Wan of the Song Dynasty, and some of *Yan Pu* (*Manual on Inkstones*). Classics partly devoted to lapidary including *Zhui Geng Lu* (*Talks in the Intervals of Ploughing*) (1366) by Tao Zongyi and *Tian Gong Kai Wu* (*Exploitation of the Works of Nature*) (1637) by Song Yingxing were both written in the Ming Dynasty.

Yun Lin Shi Pu focuses on rocks for decorative purposes in rockeries and stones for ink-slabs and artistic carvings. Each kind of rock or stone is clearly described as to shape, colour, sound on striking, hardness, veins, lustre, crystal form, magnetic property, translucency, moisture absorption, weathering and erosive process. Rocks and stones are classified into the following nine groups: 1. Limestone of some 20 kinds eroded by water into fantastic shapes and used in rockeries; 2. stalactites and stalagmites; 3. limestone or sandstone that contains feldspar; 4. limestone or sandstone that contains manganese or iron ores; 5. fairly pure quartzites, sandstones and agates; 6. pyrophyllites, micas and talcs; 7. shales and inkstones; 8. ores and jades; 9. fossils. *Yun Lin Shi Pu*, with emphasis on metamorphic rocks, is a lapidary true to its name.

Tao Zongyi in his book is mainly interested in gemstones.

Tian Gong Kai Wu on the contrary dwells on metals, salts, clays, limestones, coal, alums and other ores or minerals which were important in the munitions industry and people's lives. It gives minor consideration to gemstones, reflecting the author's attitude of thinking highly of useful materials while belittling decoration. The distinguishing feature of *Tian Gong Kai Wu* is that it brushes aside the outward appearance and classification of rocks and minerals while dealing at length with their prospecting, mining and refining techniques. *Tian Gong Kai Wu* is a scientific work in encyclopaedia form.

Minerals are also mentioned in numerous other records such as local historical and natural science records and some personal notes.

China is among the countries that exploited and used petroleum the earliest. "Chapter on Geography" in *Han Shu* (*History of the Han Dynasty*) written in the 1st century tells of definite places where petroleum was discovered and came into use. In the Qin (221-207 B.C.) and Han (206 B.C.-A.D. 24) dynasties a liquid that burned easily was found on top of the flowing water of the Weishui River, a tributary of the Yanhe. The local inhabitants of Gaonu County (now Yanan) ". . . scooped it up and used it (as a fuel)". During the Eastern Han Dynasty (25-220), *shiyou* (literally "rock oil") was found in a spring flowing out of a mountain south of Yanshou County of Jiuquan. This "rock oil" changed from a brownish yellow to black when collected in a pot and so was also called *shiqi* (rock lacquer). When lit, it was found to burn with an extremely bright flame. It was also an excellent lubricant for cart-axles according to "Chapter on States and Prefectures" of *Hou Han Shu* (*History of the Later Han Dynasty*). In the Northern Wei (368-581) and Tang (618-907) dynasties, petroleum was also discovered in Kuqa County of Xinjiang and in Yumen County of Gansu Province, and knowledge spread as to its exploitation and various uses. In the Northern Song Dynasty (960-1127), work shops processing petroleum were formed. Petroleum was an important material used in making munitions, though the term *shiyou* was coined by Shen Kuo (1031-1095) of the Northern Song Dynasty, who collected carbon black from petroleum flames and made ink-sticks of it. During the Southern Song Dynasty (1127-1279), *shizhu* (rock candle) appeared. This was the solid paraffin extracted from crude petroleum and used for lighting, as indicated by its name. Pharmaceutical uses of petroleum are mentioned in Li Shizhen's *Ben Cao Gang Mu*. In the Ming Dynasty (1368-1644) its indigenous primitive refining began in Sichuan and Shaanxi provinces as told in local historical, scientific and other records, which are of great interest to us.

Prospecting Theories

Theories on mineral prospecting were formed very early in China. A well-known paragraph on the subject is included in

"Essay on Numerical Data Concerning the Earth" in *Guan Zi* (*The Book of Master Guan*). In a summary of the law of distribution of ores in ore-bearing formations, the essay points out that ores are indicated by the outcrops of their lodes, parageneses and associations. It says:

> When there is haematite above, iron will be found below. When there is lead above, silver will be found below. Where there is cinnabar above, "yellow gold" will be found below. Where there is magnite above, "copper-like gold" will be found below. Mineral deposits in a mountain are revealed by their respective signs on the surface.

The upper layer of iron ores are often composed largely of ferric oxide (haemite) of brown colour, while lead and silver often exist in paragenesis, as taught in modern theory of ore-bearing formation. The essay accords in general with modern theory and practice of mineralogy, except that the text confuses copper and sulfides of iron with "yellow gold" and "copper-like gold". It should be noted that the Chinese character *jin* may mean both "gold" and "metal". The text should be interpreted in the latter sense of the word.

Prospecting by association or paragenesis of one ore with another formed an empirical creed in the mind of traditional Chinese miners. Examples can be further found in *Tian Gong Kai Wu*.

New prospecting theory emerged in about the first half of the 6th century during the period of the Southern and Northern Dynasties. The theory ushered in a new epoch of geobotanical and biogeochemical prospecting, and is embodied in the well-known *Di Jing Tu* (*Illustrated Mirror of the Earth*), which unfortunately survives only in fragments. In the theory, connections between plants and ores are established — an important original scientific idea which had been in formation for a long time. The "Essay on Encouraging Studies" in *Xun Zi* (*The Book of Master Xun*) of about 240 B.C. says that "jade can be found in a mountain on which trees and grasses grow luxuriantly". *Bo Wu Zhi* (*Notes on the Investigation of Things*) of 270-290 by Zhang Hua of the Jin Dynasty (265-420) says essentially the same. However, *Di Jing Tu* gives much fuller and more substantial statements such as:

> During the second month, where newly grown plants are drooping, there is precious jade underground. During the fifth month where leaves of plants are green and thick but juiceless, and where branches of trees are drooping, there must be precious jade underground too. It is also said that where plants have early withered after the eighth month there is also precious jade underground. Where in the mountains there is green onion, there is silver below, shining with a white lustre. Where stalks of grasses are red and elegant, there is lead below; where they are yellow and elegant, copper is below.

Duan Chengshi of the Tang Dynasty in Vol. XVI of *You Yang Za Zu* (*Miscellany of the Youyang Mountains*) written in 863 sums up:

> When in the mountains there is green onion, there is silver below. When in the mountains there is shallot, there is gold below. When in the mountains there is ginger, there is copper and tin below. If the mountain has precious jade, the branches of the trees on it are drooping.

These statements are not necessarily all true but the idea of finding minerals with the help of plants nearby has been proved correct.

Dr. Joseph Needham, outstanding English historian of science, comments on the ancient Chinese theory and practice of prospecting minerals with the plants nearby as indicators: "The medieval Chinese observations were the forerunners of a vast and rapidly growing body of modern scientific theory and practice."[1]

Mining Technology

Growing demands for minerals in ancient China spurred the development of mining and mining techniques.

---

[1] J. Needham, *Science and Civilisation in China*, Cambridge University Press, 1959, Vol. III, p. 680.

The Mining of Copper Ores

Mining of copper ores was already done on a fairly large scale before the 11th century B.C. as borne out by excavations of large numbers of bronzes on which the dates of casting are often visible. In 1974 an ancient copper mine of the Spring and Autumn Period (770-476 B.C.), the first of its kind, was discovered at Tonglü Mountain in Daye of Hubei Province. With shafts and galleries largely intact, the ancient mine sheds light on the whole process of mining 2,500-2,700 years ago, and merits extensive and intensive researches. The deposits are rich and mainly of malachite, copper brilla and cuprite. Bright green, orange and red, these are easily recognized and mined. The formation is of a contact metamorphic iron and copper ore-bearing type. The ancient mine was discovered during topsoil stripping of a modern open pit. Two groups of vertical shafts, called *laolong* (old wells) in the local dialect, are 300 metres apart along a narrow south-north region on Tonglü Mountain. The 12th row of these shafts was discovered more than 40 metres underground. Within an area of 50 square metres eight vertical shafts and an inclined shaft have been found, each about 80 centimetres in diameter. The timber props are 5-10 centimetres in diameter. The 24th row was discovered more than 50 metres underground. Within an area of about 120 square metres five vertical shafts, an inclined shaft and ten horizontal galleries have been found. The shafts are 110-130 centimetres in diameter. Known as "raises" in mining, the vertical shafts served the purposes of communication between the working levels, water drainage and ore extraction. Each 50 metres deep, the raises are intersected by horizontal galleries which radiate from them. Timber props in the galleries are 20 centimetres in diameter. At each level where the galleries meet the raises are found signs of a windlass installed for ore and buckets of underground water to be hauled in alternation to the surface. The galleries are of various sizes. The largest is the No. 2 gallery, which is 1.6 metres high and 1.95 metres wide. The other galleries range from 1.3 to 1.5 metres in both dimensions. The inclined shaft was perhaps meant for exploring ores. Ores were extracted from the lowest level up, roughly selected where they were dug with the help of a boat-like wooden tray (by gravitation). Poor ores and waste materials were used to fill up the void left by earlier excavations. This method resulted in the ores that were hauled up being of higher quality, thus greatly reducing haulage. In short, the ancient mine is found to be fairly complete with shafts and galleries, timber props and cross-members, as well as primitive mechanical devices for water-draining and ore hauling. The ventilation is interesting, as the different atmospheric pressures at different levels and different points were certainly made use of to facilitate the natural flow of air. Abandoned galleries are found closed so that the air flow was always in the direction of the digging-in and would reach the depth desired. Recent exploration also reveals this ancient mine to be advantageously located at a point where the deposits were most likely rich between faulted rock layers. It is indeed an ideal museum of China's mining techniques two and a half millennia ago. In fact, few written records exist on ancient Chinese copper mining, though *Tan Yuan* (*Garden Corner of Talks*) by Kong Pingzhong of the Song Dynasty says:

> Censhuichang of Shaozhou [now Shaoguan] was formerly rich in copper. Ores could be excavated 20 *zhang*[1] underground. Now much less copper is produced in this area, as ores are found only at depths of 70 or 80 *zhang*. Miners say that it is risky working at such great depths, where lurks "cold smoky gas" fatal to man. A long bamboo pole with a lamp suspended at the end is first thrust to the bottom of the pit. If the flame turns blue, it shows that toxic gas is present. Those who flee it are saved.

This "cold smoky gas" is certainly carbon monoxide. Kong, however, says nothing about the construction of the mine. *Tian Gong Kai Wu* states: "In Wuchang of Hubei and Guangxin of Jiangxi, copper mines are many.... Ores mixed with sand and

---

[1] 1 (Song) *zhang* = 3.19 metres.

stones are excavated from a depth of several *zhang* in the mountains."
Dian Nan Xin Yu (*New Talks About South Yunnan*) by Zhang Hong
of the Qing Dynasty (1644-1911) gives an account of the preparatory
workings of a copper mine at Xiangyang Mountain of Yunnan
Province.   It tells about vertical shafts, horizontal galleries and timber props.   "Every two and more *chi*[1] there are four props to shore up
the galleries, constituting a compartment.   The length of a gallery is
counted by the number of compartments."   But this description is far
less informative than the actual mine found at Tonglü Mountain.

Salt, Natural Gas and Petroleum Mining

The sinking of brine wells in Sichuan Province is said to have
been initiated by Li Bing in the period of the Warring States (475-
221 B.C.).   Li succeeded in organizing the local people to dig a number of brine wells and brine ponds at Guangdu.   These were probably all primitive, shallow wide wells.   The dynasty of Eastern Han
(25-220) was a period of shallow and wide brine wells.   Brine was
brought out not only from underground where its indications were
clear on the gound surface; it was also brought out where there
were no surface indications of any kind.   After the Eastern Han
period the depth of wells gradually increased.   In the first year of the
reign of Yongkang (A.D. 300) in the Jin Dynasty, the depth of wells
reached 100 metres.   In the Tang Dynasty 250-metre wells were dug.
Labour at brine wells 2,000 years ago can be seen depicted on a
number of bricks made in the Han Dynasty.

During the Han and Jin dynasties fire wells (natural gas wells)
were sunk alongside the brine wells.   The natural gas wells went
down to 200 metres, and the gas was used to evaporate the brine.
Natural gas wells appeared in China more than 1,300 years earlier
than those in England, where the first well was sunk in 1668.

The small but deep well was named "bamboo pipe well",
and was first drilled between 1041 and 1053 during the reign of the

---

[1]   1 (Qing) *chi* = 0.315 metre.

Emperor Ren Zong in the Song Dynasty. The drill and bits then used were made of an iron shaft and round iron blades, which could reach hundreds of metres below while the well was but 20 centimetres in diameter. Thick bamboo was used as outer piping to stop fresh water seeping in. Smaller bamboo sections were used as containers to haul up the brine. The hauling was done with the aid of an animal-drawn wheel. The iron shaft and round blades, indispensable to deep-well drilling, were the forerunners of modern drills. The drilling of brine wells is dealt with in detail in *Tian Gong Kai Wu*.

The increasing depth of brine wells eventually reached the oil-bearing formation in the Sichuan basin. In the 16th year of the reign of Zhengde (1512) of the Ming Dynasty, the first petroleum well was completed in Jiazhou (now Leshan) at the foot of Mount Emei. The well, several hundred metres deep, ushered in a new epoch in oil-well drilling in China 300 or so years earlier than any of its kind in North America and Europe. During the reign of Daoguang (1821-50) of the Qing Dynasty, drillers of Ziliujing in Sichuan, still using the simple drilling rig of bamboo and wood and the

Sinking a brine well in pre-Ming Dynasty (1368-1644) times.

indigenous round-blade drill bits, penetrated the main natural gas formation in the basin and completed the first 1,000-metre gas well. The exploitation of natural gas reached a new level.

Coal Mining

Coal was used in refining iron ores early in the Western Han Dynasty (206 B.C.-A.D. 24) where coal mines were in operation, as in Henan Province. In the Song Dynasty (960-1279), coal-mining techniques were fairly well developed and systematized. From the relics of an ancient coal mine at Hebi in Henan the original preparatory workings and the active phase of mining can be traced. A main vertical shaft was first dug to a depth of 46 metres. Seam-traversing galleries were then dug at various levels to reach the angled seam. The galleries are found to be more than a metre high with a cross-section in the shape of a trapezium one metre wide at the roof and 1.4 metres wide at the floor. The coal deposit was then divided into checkered sections. Excavation was carried out in two stages. In the first stage coal was excavated from the remotest sections towards the main shaft, and from one of every two neighbouring sections crisscross, leaving the intact sections as natural props for the whole roof. In the second stage the remaining sections were excavated with great care, again from the remotest towards the main shaft. Underground water was drained by a two-pronged channel. Windlasses in the vertical shaft were used to haul it up to the ground surface, while lower voids left by earlier excavations accommodated part of the water. Coal-mining techniques are also described in *Tian Gong Kai Wu* which says:

> Experienced coal miners can tell from the colour of sur-
> face earth whether there is coal below. But as a rule no coal
> can be found till a well has been sunk to a certain depth, say
> five *zhang* underground. One must guard against poisonous
> gas that escapes with the uncovering of the first batch of coal.
> A long bamboo pipe, hollowed full length and sharpened at the

thinner end, is thrust into the coal to exhaust the poisonous gas. Only after this is done can the coal round the pipe be excavated with a pick. The number of galleries round a main vertical shaft varies according to where and how a seam can be reached. The galleries usually radiate from the shaft to shorten hauling distances. Timber props and cross-members are used to shore up the galleries against possible cave-ins.

Coal mining in south China as illustrated in *Tian Gong Kai Wu* (*Exploitation of the Works of Nature*).

The extraction of carbon monoxide, as described also in *Tan Yuan* written in the Song Dynasty, was an indigenous but ingenious method of ancient times.

Earthquake Forecasting, Precautions Against Earthquakes and Anti-Seismic Measures

Tang Xiren

China has experienced many earthquakes. For thousands of years the Chinese people have fought perseveringly and bravely against earthquake disasters, and have left earthquake records covering long historical periods. And China is the land of the world's first seismograph, her people observing and recording many foreboding earthquake phenomena accumulated no little experience in precaution against earthquakes and anti-seismic measures.

A Jin Dynasty excavation revealed records in *Zhu Shu Ji Nian* (*The Bamboo Annals*) that in the reign of King Shun (2179-2140 B.C.) "the earthquake cleft the earth so deep that water wells were exposed", and that in the reign of King Jie (1763-1711 B.C.) of the Xia Dynasty "the earthquake broke down the temples". These are probably the earliest records of earthquakes. The book *Lü Shi Chun Qiu* (*Master Lü's Spring and Autumn Annals*), written in the 3rd century B.C., says:

> The sixth year of the reign of King Wen (1177 B.C.) of the Zhou Dynasty, in the sixth month the monarch lay in bed ill for five days. The earth was moving west, east, north and south, within the city and suburban area.

This states explicitly the time and scope of the quake, and is the earliest specific and verifiable earthquake record in China. Such classics written before the Qin Dynasty (221-207 B.C.) as *Shi Jing* (*Book of Odes*), *Chun Qiu* (*Spring and Autumn Annals*),

Guo Yu (*Discourse on the States*) and *Zuo Zhuan* (*Zuoqiu Ming's Chronicles*), recorded and preserved many descriptions of earthquakes in ancient times. Beginning from the Han Dynasty earthquakes were recorded in *Wu Xing Zhi* (*Records of the Five Elements*) and in all subsequent dynastic chronicles as disasters. Earthquake data greatly increased after the Song Dynasty, when regional chronicles began to flourish, and in these, earthquakes were also recorded as disasters. Apart from these official and regional chronicles there were notes, commentaries, novels and poems written by individuals which also mentioned earthquakes and often described them in vivid terms. Series of books of different dynasties, such as *Tai Ping Yu Lan* (*Taiping Imperial Encyclopaedia*) compiled during the Song Dynasty (960-1279), *Gu Jin Tu Shu Ji Cheng* (*Collection of Books, Ancient and Modern*) compiled in the Qing Dynasty (1644-1911), also collected and classified many earthquake data. Even inscriptions on tombstones recorded earthquakes in different ages.

These earthquake records accumulated throughout China's history are a valuable legacy indeed. After the establishment of the People's Republic of China, Chinese historians and seismologists began jointly to collect and arrange these data. They examined more than 8,000 manuscripts and books from which they extracted more than 15,000 items relating to earthquakes. After sifting these materials, they chose 8,100 earthquake records from 1177 B.C. to A.D. 1955. The number of these records and the millennia of time they covered give them important scientific value and have aroused wide attention from seismologists throughout the world. Chinese seismologists meanwhile use this wealth of earthquake data in drawing seismographic charts and making various mathematical statistical studies. They thus provide important scientific basis for dividing the country into seismographic regions, for earthquake prediction and for rational distribution of national economic construction.

On the basis of continuous recording of earthquakes and accumulation of seismographic knowledge, Zhang Heng who had witnessed several earthquakes during his lifetime in the Eastern Han period (25-220) invented the world's first seismograph. According to statistical data, 20 earthquakes of which six caused damage occurred in the capital Luoyang and western Gansu Province between A.D. 92 and 139.

Zhang Heng, born in Nanyang of Henan Province, was working at the time in the capital Luoyang and was therefore able to gather much first-hand information about these earthquakes. Zhang Heng, who had held the post of Imperial Historian for many years, was the official astronomer and so charged with the duty of recording earth tremors in distant parts of the country. He felt the need of an instrument to register precisely earthquake phenomena throughout the country. Zhang Heng spent many years studying this problem and finally in the year 132 invented the world's first seismograph, an important device in man's fight against earthquake disaster.

Hou Han Shu (*History of the Later Han Dynasty*) describes this instrument as follows: "The seismograph is made of fine cast bronze. Its diameter is eight *chi*[1], with a protruding cover, and it is shaped like a wine jar." Inside was a sophisticated mechanism, consisting mainly of *duzhu* (similar to pendulums) in the middle and surrounded by *badao* (eight groups of lever mechanisms distributed in eight directions and connected to the body of the seismograph). Eight dragon figures were arranged on the outside of the seismograph. In the mouth of each dragon was a small bronze ball and underneath each dragon head was a toad with mouth opened upwards. If a strong earthquake occurred in any direction and seismic waves were transmitted to the seismograph, the *duzhu* inclined to one side and triggered the lever in the dragon head, opening the mouth of the dragon facing in the direction of the earthquake and releasing its bronze ball, which fell into the mouth of the toad with a clanging sound. This seismograph enabled monitoring personnel to know the time of the quake and the direction in which it occurred. Zhang Heng's seismograph was installed in Luoyang and detected an earthquake of magnitude six on the Richter scale that occurred in western Gansu Province in the year 138, initiating man's detection of earthquakes by scientific instruments.

---

1 A *chi* then equalled 0.237 metre.

The rulers of the feudal dynasties were never interested in scientific and technological inventions, and Zhang Heng's research and his invention of the seismograph received no support from them. Not only was this valuable instrument never widely used; the instrument itself was not preserved — a great loss in the history of science and technology. Still, Zhang Heng's seismograph could not be cast into oblivion, and his spectacular achievement has attracted man's attention ever since, later scientists having recorded for posterity the mechanism and principle involved. An invention 1,700 years in advance of similar apparatus in the world, Zhang Heng's seismograph caused a stir among scientific workers abroad. Some scholars believe that the pendulum device of Zhang Heng's seismograph was taken to Persia (Iran) and Japan in the Sui and Tang dynasties (581 -907). The description of the seismograph in *Hou Han Shu* was translated into foreign languages and spread to many parts of the world. Modern scholars both in China and abroad have made reproductions of Zhang Heng's seismograph, studied it and given it universal acclaim. Convinced that this seismograph was based on the principle of inertia, they see its structure as in full accord with principles of physics and as capable of detecting the major impact direction of seismic waves.

Besides this very early invention of the seismograph, the Chinese people have recorded and observed many foreboding phenomena of earthquakes, such as seismic sound, seismic light, forequake, strange activities of underground water, abnormal meteoric phenomena and unusual animal behaviour and so have abundant knowledge for earthquake prediction.

Seismic sound as an important fore-sign of earthquakes is described in chronicles. "Records of Forebodings" in *Wei Shu* (*History of the Wei Dynasty*) says that in Shanxi Province in A.D. 474 "there were sounds like thunder heard in Qicheng of Yanmen; they came from the west and sounded for a dozen times. When the sounds ended an earthquake occurred". "Records of the Five Elements" in *Jiu Tang Shu* (*Old History of the Tang Dynasty*), says that in A.D. 734 "in the 22nd year of the Kaiyuan period, on the 18th of the second month an earthquake occurred in Qinzhou. The people of Qinzhou first heard some underground sounds issuing from the northwest, then came the earthquake."

Cheng Hua Shi Lu (*Real Account of Chenghua*), Vol. 55, says that in Qiongzhou city of Guangdong Province on the fourth of April in 1468 "at the fourth watch in the night an earthquake occurred. Before the earthquake, sounds came from the southwest, then the strong earthquake. Another quake followed and lasted a long time before it stopped."

There are also descriptions in historical chronicles of seismic light. For instance, *Wan Li Shi Lu* (*True Account of the Wanli Period*), Vol. 55, says that in Hubei Province on May 26, 1509 "in Wuchang blue lights like lightning flashed in the sky about six or seven times, accompanied by sounds like thunder, then an earthquake occurred".

Zheng De Shi Lu (*True Account of the Zhengde Period*), Vol. 107, says that on 30 December, 1513 in Yuejun County of Sichuan Province "a fire ball was seen in the sky that gave out a sound like thunder. The next day there was an earthquake."

In the two cases cited above, it was seismic sound and seismic light that appeared before the earthquake.

Frequently a series of tremors called forequakes occurred before a strong earthquake, and many descriptions of these appear in China's earthquake records. *Er Shen Ye Lu* (*Accounts of Ershen*) written by Sun Zhilu for example says that in 1512 "in the fifth month earth tremors continued in Yunnan Province for 13 days, then in the eighth month a strong earthquake occurred in Yunnan".

Kang Xi Zhen Jiang Fu Zhi (*Chronicle of Zhenjiang Prefecture in the Kangxi Period*) says that on 25 July, 1668 in Zhenjiang City of Jiangsu Province "at about 8 o'clock in the evening an earthquake occurred. A few days before this there had been a slight quake, but on that day the earthquake was terrible, the mountains shook and the water in the rivers trembled so that most of the boats on them were overturned and sunk. Countless houses both within the city and outside were destroyed."

Frequently the underground water level shows unusual changes before a strong earthquake. Records of Tancheng in Shandong

Province from 1668 tell of a river that suddenly dried up in several places before an earthquake of magnitude 8.5 on the Richter scale occurred. *Min Guo Shou Guang Xian Zhi* (*Chronicle of Shouguang County in the Republic Period*) quotes *Qing She Yi Wen* (*Recollections of the Qingshe*) as saying that in Shouguang County of Shandong Province "the day before the earthquake occurred, I heard the river giving rumbling sounds. When I sent my son to see what was happening, he returned and told me that he had seen nothing unusual. Some said that the day before, the Midan and other rivers had become suddenly dry." *Kang Xi Hai Zhou Zhi* (*Chronicle of Haizhou Prefecture in the Kangxi Period*) quotes *Di Zhen Ji* (*Record of Earthquakes*) written by Ni Changxi as saying that in Ganyu of Jiangsu Province "bitter-tasting rain poured down for one month, then the canal south of the city suddenly overflowed only to dry up as suddenly within 12 hours in the daytime. Everyone who saw it was greatly surprised". Apart from these observations on the remarkable changes of water, some ancient books also contain descriptions of changes in water composition and taste before an earthquake, e.g., "the water in wells became suddenly muddy", "the water became as red as flowing red paint", "the taste of well water changed, sweet water became salty, salty water became sweet", etc.

Frequent also in historical chronicles are descriptions of abnormal meteorological phenomena before an earthquake, such as exceptionally high temperature, heavy rain and thunder, hurricanes, cloud and mist, droughts and floods. Before an earthquake of magnitude 8 on the Richter scale occurred in Sanhe and Pinggu counties in 1679 for example, "the weather was terribly hot, affecting many men and animals", while *Jia Qing She Hong Xian Zhi* (*Chronicle of Shehong County in the Jiaqing Period*) says of Sichuan that on 10 July, 1819, when an earthquake occurred, "it rained continuously for 10 days in the fifth month, then on the 19th torrential rain poured down at night; in the same night there was an earthquake and the rivers rose and flooded their banks". There were also other ominous signs preceding earthquakes, such as "clouds pervaded the whole atmosphere", "the sun looked dim and yellow. At noon the wind blew and everywhere was misty, at night one could not see the moon", and so forth. Such remarks are numerous.

Unusual animal behaviour before an earthquake has been recorded since the Tang Dynasty (618-907). One of these chronicles, *Kai Yuan Zhan Jing* (*Kai Yuan Classic on Astrology*) says that before an earthquake "mice gathered on the streets and alleys squeaking, and the ground broke open". *Shun Zhi Deng Zhou Zhi* (*Chronicle of Dengzhou Prefecture in the Shunzhi Period*) says that on 23 January, 1556 in Dengxian and Neixiang of Henan Province "sounds of wind and shower came from the northwestern direction, birds piped and animals cried. Soon after, an earthquake occurred with thundering noise."

After a strong earthquake in Pinglu County of Shanxi Province in the year 1815, a summary of earthquake experience was made. *Yu Xiang Xian Zhi* (*Chronicle of Yuxiang County*) explicitly stated that "when cows and horses lift their heads upwards, hens crow and dogs bark, these are forebodings of earthquake". After many prolonged observations of unusual animal behaviour before earthquakes, inhabitants of certain earthquake regions concluded that before a quake "animals on land and in the water exhibit strange behaviour".

It is obvious from the above that China's history is replete with records of macroscopic forewarning phenomena of earthquakes, and further that these were used successfully in earthquake prediction and precaution. A destructive earthquake that occurred in 1855 in Jinxian County of Liaoning Province was forecast on the the basis of seismic sounds, and a record preserved in the Beijing Palace Museum says:

> Before the earthquake occurred, sounds like thunder were heard and the people in that area took precautionary measures. They all went outside their houses, and so not many persons died from their crumbling — only seven, including men, women and children, were killed or wounded.

Another description in *Dao Guang Zun Yi Fu Zhi* (*Chronicle of Zunyi Prefecture in the Daoguang Period*) says that on 11 August, 1809 there were definite forequake activities: "In Xiaoxili and

Luoqianxi the mountains suddenly moved and stones fell down." Local inhabitants seeing these signs took the precautionary measures of transferring oxen, sheep and utensils to safe places. "Afterwards the earth shook, houses tumbled down and fields turned up their soil." In 1815 a strong earthquake that occurred in Pinglu of Shanxi Province was forecast by abnormal weather conditions. *Chronicle of Yuxiang County* says:

> Beginning from the sixth day of the eighth month torren-
> tial rain continued for 40 days. After the ninth day of the
> ninth month the weather cleared slightly, clearing up completely
> on the 13th. Some of the wise elders in the county warned
> that when hot weather followed continuous rain, precautions
> should be taken against earthquake.

Of the quake that actually occurred, the *Chronicle* comments that "houses crumbled".

Earthquake prediction in ancient China was based not only on single signs but on a composite of multiple forewarnings which were also summarized. *Yin Chuan Xiao Zhi* (*Brief Chronicle of Yinchuan*) written in 1755 tells of a cook who worked in a government agency consulting peasants and then summing up earthquake signs according to their experience. This book says that in Ningxia earthquakes "occur mostly in spring and winter. When the water in wells becomes suddenly muddy, seismic sounds are interspersed and prolonged, and dogs congregate and bark, then precautions should be taken against earthquake." This account indicates an awareness of a composite pre-earthquake picture 200 years ago, which is quite scientific in the light of modern knowledge. Various approaches and observations should be analysed, not just one unusual sign taken into account, in predicting a phenomenon as complex as an earthquake.

The observations of the Chinese people in former times, summed up, led to their being able to take various effective measures to prevent or minimize casualties and destruction caused by earthquakes.

Architecture is one field in which anti-seismic measures are manifest. In Taiwan, an earthquake-prone province of China, it was noted in building cities long ago that "it is very rare in Taiwan when no earthquake occurs throughout one whole year". Among the anti-seismic measures they took was building city walls incorporating bamboo and wood, as for example in Danshui. These materials are abundant in this region, making the structures economical as well as quake-resistant due to the resilience, light weight and flexibility of the materials. In Yunnan Province also, where earthquakes often occur, for the same reason walls are often woven of bramble twigs and other materials.

To make houses, bridges and temples durable and safe, especially in seismic regions, attention was paid to solid foundations, strong architecture and good integration. Ancient architectures indicate a wealth of experience and knowledge in designing and building structures that would withstand strong quakes. Examples are the Guanyin Pavilion of Dule Temple in Jixian County and the 60 metre-high wooden tower in Yingxian County of Shanxi Province, both built in the Liao period (916-1125) and Zhaozhou Bridge across the Xiaohe River in Zhaoxian County of Hebei Province which was built in the Sui Dynasty, all of which have stood for 1,000 years or more. All are situated in seismic regions of northern China where they have withstood frequent earthquakes of various intensity.

Strong quakes that toppled houses were followed by after-shocks, continuing the threat to human life and property. Ancient books record precautionary and anti-seismic measures taken by the populace to prevent injuries and losses — putting up shelters of wooden .boards, straw, or reed-and-straw matting in open places. From the Song Dynasty (960-1279) on, all chronicles contain such words as "the inhabitants, for fear of being buried by crumbling houses, built dwellings by weaving reeds", "people erected wooden framework covered with straw in open fields to serve as sleeping quarters", "in open fields near their houses the people built canopies of wood and covered them with straw to serve as shelter". Among these explicit historical chronicles is an account in the files of the Beijing Palace Museum which records a strong earthquake of magnitude 7.5 on the Richter scale that occurred in Cixian County of Hebei

Province on the 22nd day of the intercalary fourth lunar month in 1830:

> Slighter after-shocks continued until a strong after-shock occurred on the seventh day of the fifth month. All remaining houses crumbled. Fortunately the inhabitants were living in open places or in shelters, so there were no injuries.

These simple, safe and effective precautionary and anti-seismic measures are even now applied.

In addition to knowledge about precautionary and anti-seismic measures before and after earthquakes, the Chinese people learned also from experience certain emergency measures for preventing injuries when people were caught in their houses by sudden strong earthquakes. Qin Keda, a survivor of a strong earthquake of magnitude 8 on the Richter scale that occurred on 23 January, 1556 in Huaxian County of Shaanxi Province wrote *Di Zhen Ji* (*Account of the Earthquake*) in which he proposed emergency safety measures. He said:

> ... the inhabitants should use floor boards as cover, strengthened with a strong wooden bed; if a sudden earthquake occurs one should not run out at once but should lie still and wait. Even if the house falls down one would be safe. If one is not able to take such measures, then one should choose an open space beforehand in order to escape quickly.

Today, when earthquake forecasting is not yet wholly satisfactory, such measures as sheltering under sturdy furniture to prevent injury from sudden earthquake when immediate escape is impossible remain important precautionary means.

Historical chronicles also list places where one should not be during an earthquake. One of these in the archives of Beijing Palace Museum says that "one should not stand near a house corner or at the foot of a wall". In *Account of the Earthquake*, Qin Keda comments that during an earthquake disaster in central China in 1556 "whole families that lived in caves and canyons died; very few survived". On 2 August, 1733 an earthquake of magnitude 6.8

on the Richter scale occurred in Dongchuan of Yunnan Province. *Yong Zheng Dong Chuan Fu Zhi* (*Chronicle of Dongchuan Prefecture in the Yongzheng Period*) records that "three brothers and sisters who were living in houses facing a stone cliff were afraid and ran out when the earthquake struck. They sat down under the cliff and were killed by falling rocks". The chronicle points out further that minority nationals who lived in straw huts with wooden frames were safe even though their dwellings collapsed, but those who lived "beside mountains during an earthquake, with stones rolling down and mountains shaking, suffered much heavier casualties than those in other places". Such accounts are of value even now in preparing precautionary and anti-seismic measures.

Researches in Heredity and Breeding

Zhang Binglun

China is a large area in which animals and plants have originated. Many domesticated animals and cultivated plants were developed by the ancient Chinese labouring people from wild animals and plants. Further, the people of ancient China had for many centuries also creatively bred a large number of fine varieties of animals and plants and greatly enriched their knowledge of heredity and breeding through practice and research in improvement of nutrition, artificial selection and hybrid breeding.

Knowledge of Heredity and Variability in Ancient Times

Heredity and variability are universal in the biological kingdom. The process of evolution is the process of the unity of opposites between heredity and variability. It is precisely on the basis of heredity and variability that hereditary breeding is carried out.

We are referring to hereditary phenomena when we say, for example, "plant melons and you get melons, sow beans and you get beans", and "like begets like". The hereditary features of living things were known quite early in China from the universality of hereditary phenomena. Records concerning the nature of living things are to be found in ancient writings. These express the opinion that different species display different features. Many of the writings dealing with heredity take into consideration the close relationship between heredity and environment. To get good productive results

281

in cultivation and domestication, therefore, one should "adapt oneself to the nature of things and act according to the appropriateness of the season" to satisfy the living conditions of living things rather than "relying on one's own concepts and not on the orders of Heaven".

However, heredity is not unalterable, and people in ancient China noticed long ago a variability in living things. "From the same tree, some fruit tastes sour while some tastes sweet" refers to variability, a universal phenomenon. This may be illustrated by the ancient Chinese understanding of different results due to variation. As early as two millennia ago, *Zhou Li* (*Rites of the Zhou Dynasty*) records millet of a longer maturing period and of a shorter maturing period. Records concerning the variation of living things appear even more frequently in *Er Ya* (*Literary Expositor*). This work points out that in the case of horses there were as many as 36 varieties. It also discusses such differences as hair colour, some horses being dappled black and white or red and white. In *Qi Min Yao Shu* (*Important Arts for the People's Welfare*), Jia Sixie writes:

> As for cereals, some mature early while others mature late, some stalks are tall while others are short, some yield more seeds while others yield less, some plants are strong while others are weak, some of the husked seeds taste good while others taste bad, and some seeds are full while others are empty.

This sums up the varying features of different cereal varieties in respect to maturing period, shape, quality, output and yield of husked seed. Reflecting the ancient Chinese recognition of the universality of variation, Song Yingxing of the Ming Dynasty (1368-1644) writes in his *Tian Gong Kai Wu* (*Exploitation of the Works of Nature*):

> There are many kinds of cereal which are used as grain but not as flour. A distance of a few hundred *li*[1] will bring about

---

1 A *li* was about 578 metres in the Ming Dynasty.

differences in colour, taste, shape and properties of the plants. Though they differ according to the local environment, and are known by an infinite variety of names, yet they are basically the same species of grain.

Speaking of the causes of variation, we now know that nutrition and physical, chemical and biological factors can all bring about variation. Although this could not have been as fully understood as it is now, the ancient Chinese did recognize the tremendous influence that environmental changes have on biological variation. During the Song Dynasty (960-1279), Wang Guan in his *Yang Zhou Shao Yao Pu* (*The Peonies of Yangzhou*) observed:

> Herbaceous peonies grow with the breath of heaven and earth. Their size and colour can be controlled by changing their nature born of heaven and earth so that rare shapes and colours are produced in our human world.

Wang Guan added,

> The shades of colour of the flowers and the luxuriance of the leaves and stamens or pistils are all determined by the way they are manured and cultivated.

A similar observation was made by Li Shizhen in his *Ben Cao Gang Mu* (*Compendium of Materia Medica*):

> In order to obtain unusual flowers, the plants are grafted in autumn or winter and heaped with manured soil over and around their roots. In spring they put out blossoms with a hundred different shapes.

This shows that variation can be brought about by nutrition and by grafting.

Most remarkable is the realization by the ancient Chinese that formation of new varieties of living things is due to variability. In describing the 35 varieties of chrysanthemum in his *Ju Pu* (*Book on the Chrysanthemum*), Liu Meng of the Song Dynasty brilliantly observed:

Although the ancients wrote songs praising the chrysanthemum, how is it they never mentioned rare specimens like those I describe in my book? Perhaps it is because there were not so many varieties in ancient times as there are today. We now have 35 kinds. Also, I hear from flower growers that the blossoms of such plants as the peony change in colour and shape and that such changes are utilized each year to produce new varieties. Thus I am inclined to think that this is also the case with chrysanthemums. Though I might safely say that I have given a rather long list, yet I do not pretend that it is exhaustive, and the discovery of unexpected variations will be a task of all those who are interested in this field.

This remark is worthy of our attention as it faithfully records the valuable experience of flower growers. Variation in shape and colour is common in flowers like the peony, and new botanical types can be formed as long as the variants are selected and their variability retained. From this Liu Meng infers that the wealth of chrysanthemum types are formed also by selection of variations. The idea of using variations as material to create many varieties from a few species serves not only as a direct guide to production practice, but also reflects the concept the ancient Chinese had of the evolution of living things.

Application and Study of Artificial Selection

Not only did the ancient Chinese people understand the universality of variation and its relationship with environment, they also realized that man can use variation to breed new varieties through artificial selection. The essence of artificial selection is to discard variants that do not accord with people's demands and wishes and reserve those that do, and pass these variants on to succeeding generations so that new types that meet people's requirements are established by selection over a long period of time. By fully utilizing variations as material and extensively adopting the seed-selecting technique of retaining the good and discarding the bad in their production practice, the industrious and inventive Chinese labouring people contributed much for mankind by creating uncounted fine varieties since time immemorial.

The methods of selection are also many and varied. Ear selection was first noted in *Fan Sheng Zhi Shu* (*The Book of Fan Shengzhi*) of the 1st century B.C. in which appears: "Seed for wheat is selected when it is ripe." Directions are given for picking out the large and solid ears for cutting. These are tied inside a sheath, and when they have dried sufficiently in the sunshine, they are stored away. "Compared with ordinary seeds, they will crop far better if sown in the appropriate season." More records about artificial selection appeared during the Northern Wei Dynasty (386-534) in *Qi Min Yao Shu*. Among domesticated animals and cultivated crops such as pigs, sheep and goats, chickens and silkworms and glutinous panicled millet, spiked millet, non-glutinous panicled millet and glutinous millet, artificial selection was extensively used to select and breed new varieties. As for sporting selection, or selecting to breed types that deviate strikingly from the normal, a Song Dynasty work, *Luo Yang Mu Dan Ji* (*Peonies of Luoyang*) describes the Qianxi Red, a variety of peony, thus:

> Formerly the colour of its blossoms was purple, when suddenly a few red blooms appeared among them. Their scions were grafted on another peony the following year and people in Luoyang call this Transferred Twig Red.

This is a practical application of sporting selection in breeding. Now it is known that a sport is a cellular mutation in the body of plants and is hereditary. In selecting goldfish by artificial means, not only is the method of eliminating the bad and retaining the good applied, but isolation breeding is also made use of to determine the appropriate goldfish for copulation so as to select and breed new fascinating varieties. *Zhu Sha Yu Pu* (*Book on Cinnabar Fish*) written in 1596 says:

> The important thing is to keep as many kinds of goldfish

as possible while their selection should be very strict. Buy several thousand at the fair each summer and raise them in different jars. Get rid of undesirable ones until only one or two per cent remain. These are placed in two or three jars and bred with special care. In this way, you will be in possession of all sorts of marvelous fish.

Jin Yu Tu Pu (*Illustrated Book on Gold Fish*) in 1848 further records a creative method of artificial selection and breeding: "In choosing fish for mating, select a male of excellent variety that complements the female in colour, type and size." It means careful and conscious breeding. With elaborate selection, isolation breeding and the choosing of male and female individuals of similar variability for copulation, variations that accord with popular demand are accumulated to form new varieties. It shows that selection and breeding of fine varieties in China had already reached a rather high level long before Darwin's time. That is why when Darwin gives a systematic description of the processes and principles of artificial selection of goldfish in China in his *The Variation of Animals and Plants Under Domestication*, he adds, "The Chinese applied the same principles to plants and fruit trees." In *The Origin of Species*, Darwin also remarks:

> But it is very far from true that the principle is a modern discovery. . . . The principle of selection I find distinctly in an ancient Chinese encyclopaedia.

The two commonly used methods of selection in modern times, mixed selection and single plant selection, were also to be observed in ancient China.

Mixed selection. According to the aim of breeding and with a definite economic purpose in mind, select a fixed amount of fine plants in the field or seed plot each year, eliminate those found to be undesirable, thresh the remaining fine plants and store and sow the seed. After several successive years of such treatment, fine types of uniform properties and shape can be selected from the mixed group. The Chinese working people began to use ear selection in the Western Han Dynasty (206 B.C-A.D. 24) and by the Northern Wei Dynasty mixed selection had already reached a high level. *Qi Min Yao Shu* gives a detailed account of the latter method as follows:

> Seed for spiked millet, ordinary and glutinous panicled millets, and ordinary and glutinous millet is always to be collected separately every year. Choose plump ears uniform in colour; cut and hang up to dry. Next spring, thresh those choice ears and sow in a separate parcel of land for reproducing the following year. These special parcels ought to be hoed with particular care and diligence. When ripe, reap and thresh prior to any other parcel, and keep the kernels apart in a newly dug burrow. Stop the burrow with the straw just obtained from threshing.

This method is very similar to the mixed selection of modern times.

Single plant selection. This means selecting and breeding new varieties with a single plant or a single ear of excellent properties and shape. The method was widely used in China as early as in the Qing Emperor Kang Xi's reign (1662-1722). An account in *Kang Xi Ji Xia Ge Wu Bian* (*Kang Xi's Study of the Principles of Things*) (Part 1), tells of the chance discovery by labouring people in Wula of "a white millet plant growing in the hole of a tree" which was quite different from ordinary millet. Its seed was afterwards sown and "propagated until at last the fields were simply overflowing with such millet" so that a fine variety which was delicious in taste, mild in nature as well as high-yielding was eventually selected and bred. When Kang Xi got hold of this fine variety, he ordered his subjects to experiment with it in his mountain villa. It turned out that "the stalk, leaf and ear" of this fine variety "were really much bigger and matured earlier than other varieties". Besides, when it was cooked "it was as white as glutinous rice and even smoother and more savoury". This is a fairly early record and one known to workers in single plant selection. It is probable, however, that single plant selection was practised long before Kang Xi's time. *Kang Xi Ji Xia Ge Wu Bian* also states:

I am inclined to think that it must have been the same with various fine cereals in primordial times. They also did not exist at first and it was only in a later period that they came into being. They will fill the gaps in agricultural books.

Kang Xi also made use of the labouring people's development of single plant selection and succeeded in selecting, breeding and "causing to mature early" a high-yield, "fragrant and fatty" fine variety called imperial rice. *Kang Xi Ji Xia Ge Wu Bian* tells the story as follows:

In Fengze Gardens are a few plots of wet field where the rice is harvested and taken to the threshing ground in the 9th lunar month each year. Walking along a footpath between the fields one day late in the 6th lunar month, I noticed a rice plant standing out higher than the rest. When I found its grains full and mature, I stored them away. I wanted to see if the grain produced from them would ripen early the following year. It turned out that the seeds in fact matured again in the 6th lunar month. Since then, they have been doing very well, and for over 40 years my imperial kitchen has been using this rice as food. It is semi-red, long-grained, fragrant and fatty. Because it is grown on palace grounds it is called imperial rice.

Successfully bred by single plant selection this imperial rice has a short growing period and so can produce a crop in northeast China where the frost-free period is short. In southern China it gives two crops a year. In the 54th year of Kang Xi's reign (1715), therefore, it was popularized and grown in Jiangsu and Zhejiang provinces, where "the people were ordered to grow it". In the Suzhou district two crops were gained the first year. In the second year, experiments with imperial rice on the basis of past experience resulted in a marked increase in output, totalling 520 litres per *mu*[1] in two seasons, or 120 litres more than the yield of an ordinary variety. After four years the yearly output of two crops reached 680 or 690 litres per *mu*. This was 2.7 times the output of an ordinary rice field. Imperial

---

[1] 1 *mu* = 1/15 hectare or roughly 1/6 acre.

rice was later introduced into Anhui and Jiangxi provinces where it also gave high yields. Single plant selection in the case of imperial rice proved a great success.

The Application of Cross Breeding and Hybrid Vigour

Artificial cross breeding is the method of creating new biological types. The new types formed through cross breeding can combine the fine qualities and forms of two or more than two parents and become new biological types of higher productivity which are also more resistant to unfavourable conditions. The application of both methods of hybridization — sexual and asexual (or vegetable) — in ancient China is remarkable.

The mule is a typical example of the application of hybrid vigour. The Spring and Autumn Period (770-476 B.C.) records state that a certain Zhao Jianzi had two mules. Mules are mentioned also in *Chu Ci* (*Elegies of Chu*) written by Qu Yuan, indicating that the hybrids of horses and asses existed at least as early as the Spring and Autumn Period and that this new species which can be cross bred was used long ago in production by the Chinese labouring people.

Qi Min Yao Shu gives a more specific account of crossing horses and asses and of hybrid vigour:

"Since it is difficult to get mules by crossing stallions with female asses, mares are generally used to cross with jackasses to beget mules that look very strong and are superior to horses.

Ben Cao Gang Mu points out that "mules are bigger than asses and stronger than horses". Fang Yizhi remarks in *Wu Li Xiao Shi* (*Small Encyclopaedia of the Principles of Things*) that "the mule can make long journeys and rarely falls ill". The mule as a hybrid between the horse and the ass combines the fine properties and form of both and is better than either. It inherits its large size, swiftness, strength and liveliness from the horse, and its sure-footedness, gentle disposition, endurance and ability to cope with coarse fodder from the ass. It is an ideal draught and pack animal on steep roads as well as in rural areas. Even at present, racial hybrids like the mule are very rare, yet the Chinese labouring people were cross breeding 2,000 years ago and had produced the mule with marked hybrid vigour. Cross breeding has since been widely used in production.

Experimentation in cross breeding domesticated silkworms in the Ming Dynasty is described thus in *Tian Gong Kai Wu*:

> The cocoons are of two colours only, yellow and white. . . . Only the yellow kind is produced in Sichuan, Shaanxi, Shanxi, and Henan provinces. When a white male is crossed with a yellow female, the offspring will spin light brown cocoons. Recently some silkworm raisers have crossed an Early male with a Late female with the hope of producing an excellent breed. This is an unusual occurrence worth noting.

The grafting technique for asexual crossing also has its origin in China, though just when this began is not conclusively known. *Fan Sheng Zhi Shu* contains this account of grafting by approach. It says:

> Plant 10 gourds and wait until they have grown about two *chi*[1] high. Then bind the 10 vines together, wrapping them in cloth for about five *cun*[2] and smear them with mud. In a few days the 10 vines will become one at the joint. Retain the sturdiest vine and pinch off the rest. Train the retained one that is to bear gourds.

Because the one plant above ground is provided with the nutriment absorbed by the roots of 10, the gourds that are borne are exceptionally large and full.

After the Han Dynasty (206 B.C.-A.D. 220) grafting technique was further developed and used for plants of different species. *Qi Min Yao Shu* records the use of trees of different species in graft-

---

[1] A *chi* was then equivalent to 0.23 metre.
[2] 1 *cun* = 1/10 *chi*.

ing to obtain fruit of superior quality earlier. Pear trees were grafted, using crabapple as stocks and pear-tree shoots as scions. Pears from grafts on such stocks are large and fine, confirming that propagation by grafting is quicker and better than propagation by seedlings. It also studies different positions of scions as directly affecting fruiting time and the shape of the tree. *Wu Ben Xin Shu* (*New Book on the Essentials*) lists six methods of grafting in use on mulberry trees during the Kin and Yuan dynasties (1115-1368) as follows: trunk, root, bark, twig, dimple or bud, and splice grafting. *Wu Ben Xin Shu* states concerning the function and principle of grafting that the Chinese in ancient times also did some research and knew that "once grafted and bound, the two breaths communicate with each other so that the bad is replaced by the good while still retaining desirable qualities. The advantages are simply too many to enumerate." In *Hua Jing* (*Mirror of Flowers*), Chen Fuyao writes:

> It is indeed a universal truth that plants should be grafted. Small flowers may become big ones, single petals may become double, red may change into purple, fruits that are small may become large, sour and bitter may become sweet, and bad smell may be replaced by fragrance. Human efforts can change the very nature of things. It all depends on the proper use of grafting technique.

The best time for grafting is the period between the Spring and Autumn Equinoxes, because at this time the cortex is easier to separate and the sap flows in abundance.

Some Outstanding Works on Agriculture

Fan Chuyu

Agriculture has flourished in China since time immemorial. Excavations among ruins dating from primitive society show that the Chinese people were tilling paddy fields in the fertile Changjiang (Yangzi) River valley and growing such crops as Setaria in the Huanghe (Yellow) River valley 6,000 to 7,000 years ago. Three millennia-old inscriptions on bones or tortoise shells of the Shang Dynasty (c. 16th-11th century B.C.) bear the names of such crops as *dao* (rice), *he* (standing grain, especially rice), *ji* (millet), *su* (Setaria), *mai* (wheat) and *lai* (barley) as well as words related with land renovation in agricultural production such as *chou* (farmland), *jiang* (border), *zhen* (irrigation ditch), *jing* (well) and *pu* (garden or plot). Agriculture in China had obviously reached a fairly high level of development in that period.

In the course of China's long history of agriculture, there has accumulated a wealth of knowledge and theories on agriculture and agricultural technique. Quite a number of very early monographs on agronomy reflected the great achievements of the ancient Chinese in their struggle against Nature. The earliest agricultural treatise or book that has been handed down to the present day was written over 2,000 years ago. According to incomplete figures, 376 books on agriculture appeared in those 2,000 years. Some are extant while others have either been scattered or lost. These ancient books can be divided roughly into two main groups. One consists of comprehensive works on agriculture with crop cultivation, gardening, animal husbandry and sericulture as their main themes, and with emphasis on field production. Some deal, among their contents, with aquatic products as well as farm tools, water conservancy, disaster relief and the processing of agricultural products. The other group consists of what is called specialized books on agriculture including treatises on seasonableness in farming, various special guides, treatises on sericulture, veterinary books, monographs on edible wild herbs, and books on the elimination of locusts.

Comprehensive agricultural books can again be divided into three types: 1. Books on monthly family arrangements of agricultural activities, which originated as Cui Shi's *Si Min Yue Ling* (*Monthly Ordinances for the Four Classes of People*[1]) of the second century, to be followed later by a stream of other works including *Si Shi Zuan Yao* (*Outline of the Four Seasons*), *Nong Sang Yi Shi Cuo Yao* (*Essentials of Agriculture, Sericulture, Clothing and Food*), *Jing Shi Min Shi Lu* (*Practical Administration and People's Livelihood*), and *Nong Pu Bian Lan* (*Farming and Gardening Manual*). These books have the unique form of arranging agrarian operations in each of the twelve months according to urgency and precedence. 2. Books as represented by Jia Sixie's *Qi Min Yao Shu* (*Important Arts for the People's Welfare*) of the 6th century, with emphasis on systematic technical knowledge relating to farming, forestry, animal husbandry, side-line production and fishery. 3. Popular agricultural books of the nature of almanacs, or popular encyclopaedias for daily use. And *Ju Jia Bi Yong Shi Lei Quan Shu* (*Guide to Domestic Occupations*) of the Yuan Dynasty (1271-1368), *Bian Min Tu Zuan* (*Illustrated Compilation for the People's Welfare*) and *Duo Neng Bi Shi* (*Various Arts in Ordinary Life*) of the Ming Dynasty (1368-1644) all belong to this type.

The following introduces some agricultural works which are important in the history of Chinese agriculture and even had some influence on the world agricultural scene.

Earliest Extant Agricultural Treatises

According to "Yi Wen Zhi" ("Literary Records") of *Han Shu*

---

[1] The "four classes of people" at that time were scholars, farmers, artisans and merchants.

(*History of the Han Dynasty*), two monographs existed on agriculture during the Warring States Period (475-221 B.C.), *Shen Nong* and *Ye Lao*. Unfortunately these were lost long ago. Only the four essays in *Lü Shi Chun Qiu* (*Master Lü's Spring and Autumn Annals*), namely, "Shang Nong" ("Lay Stress on Agriculture"), "Ren Di" ("Capacity of the Soil"), "Bian Tu" ("Work the Ground") and "Shen Shi" ("Fitness of the Season"), which deal solely with agriculture can be claimed as China's earliest extant agricultural treatises. *Lü Shi Chun Qiu* was written collectively by scholars who were guests at the house of Lü Buwei (?-235 B.C.), the first Prime Minister of the Qin Dynasty (221-207 B.C.). According to textual research it was completed in 239 B.C. The above-said four essays are not independent specialized writings on agriculture, yet together they form a system and a complete set of treatises.

During the Warring States Period it was the landlord class that embodied the new relations of production. Being fully aware of the importance of promoting agricultural production, they gave primacy to agriculture and pursued the policy of *chong ben yi mo* (encourage the root and restrain the branch) with agriculture as the root and industry and commerce as the branch. Such theory and policy which attach great importance to agriculture are basically identical with the views of laying stress on agriculture held by Shang Yang (390-338 B.C.), Wu Qi (?-381 B.C.) and Han Fei (280-233 B.C.). The three essays, "Ren Di", "Bian Tu" and "Shen Shi", devote themselves entirely to agronomy. "Ren Di", which deals with the principles of land exploitation, discusses the 10 important problems with regard to enabling crops to grow well and obtain high yields. It begins with land preparation, soil utilization and improvement, and goes on to tillage, preservation of soil moisture, weeding, ventilation, and so on. Next, it puts forward the contradictions of soil such as between *li* (hard) and *rou* (soft), *xi* (to lie fallow) and *lao* (continuous cropping), *ji* (lean) and *fei* (fertile), *ji* (tight) and *huan* (loose), and *shi* (wet) and *zao* (dry), and points out that these contradictions can be transformed under given conditions and that land exploitation means improving the properties of soil through labour to make it fit for cultivation. "Bian Tu" discusses land operation, that is, changing the existing soil conditions by human effort, and answers in concrete terms the demands put forward in "Ren Di". In the first place, it makes different arrangements in the time of cultivation for soils having different properties. Next, it discusses the three evils caused by improper cultivation, too sparse sowing, lack of seedlings and weed encroachment. It also discusses the evils of cultivation which is not done on time, of improper land preparation, the most rational layouts for crops at a time when they have sealed the ridges and their impact on plant growth. "Shen Shi" discusses the effects of timely and untimely cultivation of crops in every respect, particularly on the quality of seed.

The three essays including "Ren Di" are highly dialectical in their treatment of agricultural production. They sum up the experiences of agricultural production of the labouring people before the Qin Dynasty and are a reflection of the scientific and technological level of agriculture during the Spring and Autumn Period (770-476 B.C.) and the succeeding Warring States Period.

Fan Sheng Zhi Shu (The Book of Fan Shengzhi)

"Yi Wen Zhi" of *Han Shu* states there were nine books on agriculture before the Western Han Dynasty (206 B.C.-A.D. 24). Except for the monographs *Shen Nong* and *Ye Lao*, four of the seven remaining books were entirely lost long ago. The other three books are *Dong An Guo* (12 essays), *Cai Kui* (1 essay) and *Fan Sheng Zhi Shu* (18 essays). Liu Xiang (77-6 B.C.) and Ban Gu (A.D. 32-92) who lived at the time junction of the Western and Eastern (25-220) Han dynasties affirmed that the three were written in the Western Han Dynasty. The first two of these books were also lost. All we know of Fan's book are the fragmentary remains of a little more than 3,000 words quoted and so preserved in several books including *Qi Min Yao Shu*. These surviving words, nonetheless, reflect the scientific and technological level of agriculture in the Western Han Dynasty.

Fan Sheng Zhi Shu sums up the basic principles of farming in northern China, particularly in the central Shaanxi plain: choose the right time, break up the soil, see to its fertility, hoe and harvest in good time. It lists more than 10 cultivated plants including such grain crops as spiked millet, glutinous millet, winter wheat, spring wheat, rice, lesser beans and soya, oil crops such as female hemp and common perilla, fibre plants such as male hemp and mulberry, and such non-staple foods as melon, gourd and taro. It gives an accurate description of each plant, seed selection, sowing, harvesting and seed storage. As regards seed selection, it proposes for the first time ear selection to guarantee the purity of wheat and glutinous millet. Ear selection, regulation of temperature of water by controlling its flow and the method of cutting down mulberry seedlings[1] are outstanding examples of the progress in agricultural technology during that period. Besides, there were two special measures in agronomy — "urination" in the treatment of seed and "shallow-pit cultivation" for the best use of water and manure — which are still worthy of our notice and study today.

Qi Min Yao Shu

Among the now existing Chinese classics solely devoted to agriculture, Jia Sixie's *Qi Min Yao Shu* is the best preserved and most comprehensive. It was written in the years 533-534. It can be seen from the records in *Qi Min Yao Shu* that in comparison with those of the Western Han Dynasty the farm tools in Jia's time are greater in number and show much improvement in their uses. For instance, there are as many as five kinds of farm tools for cultivation: hoe, harrow, rake, *feng* (a sharp tool resembling a ploughshare for turning up the soil) and drill (an animal-drawn seed plough).

Jia Sixie's attitude towards writing *Qi Min Yao Shu* is very serious. According to his own account, he gathered a great quantity

---

[1] Mulberry seedlings of the first year are cut close to the ground with a sharp sickle so that the next spring mulberry suckers appear, bearing more leaves and of better quality compared with uncut plants.

of his materials from classics and contemporary writings, collected many proverbs and folksongs, made inquiries from experts and drew from personal experience. Consisting of more than 110,000 words, the book is divided into 10 chapters, with 92 essays, and a "Preface" and "Miscellany" before the text itself.

The book embraces farming, forestry, animal husbandry, sideline production and fishery in the area comprising today's southeastern Shanxi, central-south Hebei, part of Henan north of the Huanghe River, and Shandong. As this area had been the site of general mixing of various nationalities ever since the Wei (220-265) and Jin (265-420) dynasties, the book can also be regarded as a summary of their experience in production. The merits of *Qi Min Yao Shu* may be summed up as follows:

1. It goes a step further than *Fan Sheng Zhi Shu* in realizing the importance of the proper choice of season and soil. Elucidating the advanced thinking in ancient China on regulating matters in accordance with season and place, it remarks:

> Follow the appropriateness of the season, consider well the nature and conditions of the soil, then and only then will the least labour bring the best results. Relying on one's own ideas and not on the orders of Nature will make every effort futile.

Basing on this fundamental principle, Jia Sixie gives the best, medium and latest time for producing various crops. For sowing spiked millet, the first 10 days of the second lunar month is best, the first 10 days of the third lunar month medium, and the first 10 days of the fourth lunar month the latest. He also divides the soil into three classes: best, medium, and least suitable. For spiked millet, it is best to sow in soil following mung bean or lesser bean, medium to sow in soil following hemp, panicled millet or sesame, and least suitable to sow in soil that has been planted to turnip. Besides, the sowing of the same plant must vary according to differences in place and time. All of this accords with science and reality.

2. In northern China, where it is "always very windy and dry

in spring-time", increased output depends much on the conservation of soil moisture at the time of spring sowing. Ploughing and hoeing are closely related to preservation of soil moisture, and *Qi Min Yao Shu* makes a rather deep inquiry into this matter. It further affirms the importance of autumn ploughing and points out that the depth of ploughing varies with circumstances, that "primary plough-ing should be deep; later, shallower", and that "in autumn, the drill of the plough ought to be deep; in spring and summer, shallower". When ploughing high or low fields, "no matter whether in spring or autumn, always look to the proper moisture; in years of unsuitable rainfall, it is best to plough when dry but never when wet". The experiences such as raking the soil level after ploughing, intertilling and weeding to prevent drought and preserve soil moisture and seiz-ing soil moisture for sowing are also summed up by Jia Sixie.

3.   Before Jia Sixie's time alternate fallowing was mainly used to restore soil fertility, but Jia Sixie summarized and studied the rotation system in addition. First, according to the characteristics of crops, he distinguished the ones that could be rotated from those that could not. He also summed up a set of rotation methods and pointed out that the bean family is the best forerunning crop. Next, affirm-ing the fertility of crops for green manure, he remarked, "To improve the soil, the best way is to plough down mung beans, next, lesser beans and sesame." Apparently, crops were already widely cultivat-ed and used for green manure. *Qi Min Yao Shu* also summarizes interplanting and thus provides a new direction in the maximum utilization of sunshine and cultivated land to raise the yield per unit area. Interplanting in *Fan Sheng Zhi Shu* is still limited to vegetables, but by the Later Wei Dynasty (386-534) field crops are also inter-planted.

4.   Crop varieties are specially discussed in *Qi Min Yao Shu*. Jia gives a total of 86 varieties of spiked millet and analyses their respective qualities and characteristics in his "Essay on Cultivation of Cereals". This is something that is not to be found among the agricultural books of former times. Furthermore, Jia Sixie related the varieties of plants to appropriate seasons and types of soil and rightly emphasized their regional differences,

Apart from the above, experience of the ancient Chinese in growing nursery seedlings for fruit trees and trees for timber, grafting and frost prevention by smudging is still used to this day. Of considerable scientific value is the knowledge of the plant organism and the theory of the development of fruit trees by stages. In animal breeding, ancient theory on livestock morphology and the practical summary of careful breeding and rational use have brought China's animal husbandry to a high level. Records also affirm fair development in the processing of farm produce and utilization of microbes at that time. In brief, *Qi Min Yao Shu* is not only rich in content, it is detailed and accurate. It sums up the vast amount of agricultural knowledge accumulated in China before the 6th century.

Chen Fu Nong Shu
(Agricultural Treatise of Chen Fu)

The natural conditions of the region south of the Changjiang River are quite favourable, and major efforts have been made for its development from the Wei (220-265), Jin (265-420) and Southern and Northern Dynasties (420-550) onwards. Agricultural production there was already fairly high after the Northern Song Dynasty (960-1127). *Chen Fu Nong Shu*, completed in 1149 in the Southern Song Dynasty (1127-1279), is China's first small-sized comprehensive agricultural treatise whose chief concern is to sum up paddy field cultivation. The book has only about 12,500 words including "Preface" and "Postcript", but it is substantial in content. It lays emphasis on recording and narrating what the author has learned in farming, and even if it quotes from ancient literature, that is done after the author has comprehensively studied the subject. It is different in style from *Qi Min Yao Shu*. The three-volume *Chen Fu Nong Shu* gives expression to many new developments in agriculture in ancient China and ought to be classified as one of China's first-rate comprehensive agricultural treatises. Volume 1 mainly discusses the cultivation of rice, but it also includes subsidiary crops such as hemp, spiked millet, sesame, radish and wheat. Volume 2 is devoted to the water buffalo, the sole draught animal for rice-paddy cultivation in the region south of the Changjiang River. Volume 3 focuses on sericulture, beginning with the planting of mulberry trees and ending with the cropping of cocoons.

Compared with the other agricultural writings of ancient China, *Chen Fu Nong Shu* has the following five features: 1. A special essay is employed for the first time to discuss land utilization systematically. 2. Two remarkable fundamental principles about views on soil are put forward explicitly for the first time: one is that although there are many kinds of soil that vary in quality, they can all be made suitable for the cultivation of plants through improvement; the other is that soil can always be kept fresh and vigorous if appropriately used. 3. Manure is not only dealt with in a separate essay, it is discussed in detail in various other essays as well. There are many innovations and developments as regards source of manure, preservation of soil fertility and methods of dressing. 4. It is the first extant monograph on agricultural technique in the rice-growing region south of the Changjiang River with a special essay on raising rice seedlings in beds. 5. It has a rather comprehensive theoretical system of its own.

Wang Zhen Nong Shu (Agricultural Treatise of Wang Zhen)

Wang Zhen Nong Shu dating from the Yuan Dynasty (1271-1368) is a large-sized agricultural treatise. Wang Zhen was a local administrative official in Anhui and Jiangxi provinces who often went to the rural areas for inspection and was very much interested in agricultural production. This treatise sums up the production practice of dry farmland cultivation in the Huanghe River valley and rice-paddy cultivation in the region south of the Changjiang River. Its present current edition has about 110,000 words and is divided into three parts: "Nong Sang Tong Jue" ("A General Survey of Agriculture and Sericulture"), "Bai Gu Pu" ("Guide to a

Hundred Cereals") and "Nong Qi Tu Pu" ("A Collection of Illustrative Plates for Agricultural Implements"), this third part taking up four-fifths of the whole book.

"Nong Sang Tong Jue" is meant to be a general introduction to agriculture. It is permeated with the concept that agriculture is the foundation and the thinking that agricultural production is jointly determined by weather conditions, terrestrial productivity and human effort. It explains in concrete terms the origins of agriculture and sericulture, discusses in generalities the various technologies and experiences in farming, forestry, animal husbandry, side-line production and fishery and describes the fitness of season and soil and other things.

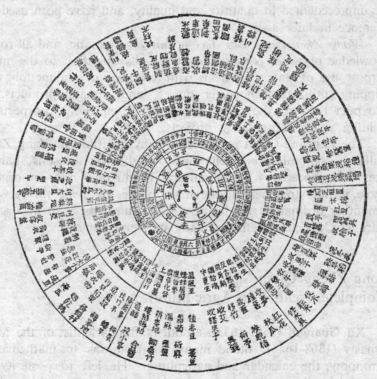

The Farming Chart in *Wang Zhen Nong Shu* (*Agricultural Treatise of Wang Zhen*).

"Bai Gu Pu" consists of various discussions on farm crops. It deals with technology and method in cultivation, protection, harvesting, storage and utilization of grain crops like millet, rice and wheat, as well as melons, gourds, vegetables and fruit trees. It also includes the growing and utilization of forest trees, and fibre and medicinal plants.

With its illustrations and descriptions, "Nong Qi Tu Pu" is a creative piece of writing. The majority of its 306 illustrations are portrayals of real objects of that time. Although specifications of the very few agricultural implements "of the remote past are not known", they are still included in it by Wang Zhen after searching and investigating in every way. The illustrations and descriptions are unprecedented in quantity and quality, and have been used for reference in later ages.

Wang Zhen Nong Shu has a fairly systematic and all-round knowledge of water conservation and calls attention to the multipurpose use of water. It deals with irrigation in conjunction with shipping, utilization of water power, and aquatic products. It also puts forward the conditions and the splendid prospects of building water conservancy projects.

The Farming Chart is a praiseworthy creation of Wang Zhen. Skilfully combining constellation periodicities, seasons, natural signs and procedures of agricultural production into an organic whole, it puts together all the main points of *nong jia yue ling* ("monthly ordinances for farmers") and sums them up in a small chart which is clear, definite, handy and very practical.

Nong Zheng Quan Shu
(Complete Treatise on Agriculture)

Xu Guangqi (1562-1633), an outstanding scientist of the Ming Dynasty (1368-1644), made many contributions to mathematics, astronomy, the calendar and agriculture. He left to posterity an unprecedentedly monumental work, *Nong Zheng Quan Shu*, which sums up and sets out traditional Chinese agriculture.

Xu's work contains more than 700,000 words and makes use of 229 written documents. Constantly busy with his activities, he was unable to finalize his manuscripts in his lifetime and these were afterwards compiled and block printed by Chen Zilong (1608-47) and others. Consisting of 60 volumes, the work is divided into 12 categories: 1. Agriculture as the foundation (classics, history, literary references, various schools of thought, miscellaneous essays, studies of various dynasties' policies stressing agriculture); 2. Farmland systems (studies of the *jingtian* or nine square system of land ownership and the various illustrations of farmland system in *Wang Zhen Quan Shu*); 3. Farming (management and administration, land reclamation, issuing almanacs for enforcement, divination of seasons, emphasis on opening up military colonies); 4. Water conservancy (hydraulic engineering, irrigation and water conservancy, *Tai Xi Shui Fa*[1]); 5. Agricultural implements; 6. Cultivation (cereals, vegetables, fruit trees); 7. Sericulture; 8. Extension of sericulture (silk cotton, ramie); 9. Planting (economic crops); 10. Animal husbandry; 11. Processing and building (foodstuffs, houses); 12. Disaster administration (preparation against natural disasters, with *Jiu Huang Ben Cao* [*Famine-Relief Herbs*] and *Ye Cai Pu* [*Manual on Wild Edible Herbs*] attached).

Nong Zheng Quan Shu differs from all preceding agricultural treatises in its systematic and centralized description of three items: military colonies, large-sized water conservancy projects (including irrigation works) and preparation against natural disasters. They are not general technological measures for agricultural production, but are indispensable to ensuring agricultural production and the safety of the lives of the peasants. *Qi Min Yao Shu* and *Wang Zhen Nong Shu* may be regarded as purely technological treatises on agriculture, while *Nong Zheng Quan Shu* emphasizes other measures for agricultural production such as farmland system, water

---

[1] Being very keen on irrigation and water conservancy, Xu Guangqi translated and edited *Tai Xi Shui Fa* (*Western Hydraulic Engineering*) and absorbed the advanced methods of the West on the basis of the knowledge of water conservancy China had already acquired.

conservancy and disaster administration. Guided by the thinking that "a workman must sharpen his tools if he is to do his work well", many illustrations and descriptions of agricultural implements in *Wang Zhen Nong Shu* are included along with supplements.

Although *Nong Zheng Quan Shu* attaches importance to politics, it does not neglect technology. From the Song Dynasty (960-1279) onwards, sericulture has been an important production activity in the region south of the Changjiang River and its newly acquired experiences were put down by Xu Guangqi in his work. His home-town Songjiang, which is in that region, was then the most advanced place in China for cotton textile industry. He recorded the new technology of cotton growing and cotton field management. He also particularly recommended the good method of stabilizing moisture for spinning and weaving. As for the sweet potato which had been introduced into China not long before, he wrote "Gan Shu Shu" ("Further Commentary on Sweet Potato") on the basis of personal experience. It was included in *Nong Zheng Quan Shu* in which he recommended growing the plant in large quantities to prepare against famine. He gave detailed descriptions of all newly introduced domesticated plants. In a word, *Nong Zheng Quan Shu* is the most comprehensive of all works on agriculture up to that time.

There are many valuable pre-Ming Dynasty books on agriculture. This article has listed only a few. All are summaries of farming knowledge in different periods from which may be seen in broad outline the situation in the science of agriculture in ancient China.

Sericulture

Wang Zichun

Silk is associated in the minds of a great many people with China, for it was the labouring people of ancient China who first raised silkworms, planted mulberry trees and wove this wonderful fabric. These processes of sericulture, developed and improved in production practice over a long period of time, were introduced from China into various other countries.

Legendary Origin

It is said that sericulture was invented by the legendary Yellow Emperor's wife Leizu in remote antiquity. This is certainly not to be believed, as the skills and art involved in sericulture could only be the result of the gradual build-up of experiences by the broad masses of people over a long period of time. Silk could in no way be the invention of any individual. The legend nevertheless indicates that sericulture in China does have a very long history.

After the founding of New China, Chinese archaeologists found among Neolithic ruins silk fabrics and thread in a bamboo basket. They included fragments of spun silk and silk ribbons. The site was Qianshanyang in Wuxing, Zhejiang Province. This find verified the claim that silk culture existed in China 4,000 years ago or before, for it was already well developed at that time. Even earlier evidence was found by archaeologists in 1926 in the form of a cut silkworm cocoon. This was among the Neolithic ruins at Xiyin Village in Xiaxian County, Shanxi Province, which pre-date the Qianshanyang relics. This cut cocoon is possibly linked with a phase of production by ancestors of the Chinese.

Silkworms were formerly wild, and because they fed exclusively on mulberry leaves, they were also called mulberry silkworms. Sericulture was already highly developed as early as the 14th century B.C., so that the Chinese must have begun rearing silkworms long before that time. On many bronze articles of the Shang Dynasty (c. 16th-11th century B.C.) are impressions of silk fabrics or fragments of spun silk. Silk-weaving technique was obviously already quite advanced at that time. A host of facts show that silk articles were becoming increasingly important in the social and economic life of the time, and that they had become media for the exchange of goods. The ensuing demand for silk fabrics led necessarily to the development of silkworm-raising in order to provide more and more raw material.

Archaeologists have found life-like carved jade silkworms in Shang Dynasty tombs in Anyang, Henan Province, and among the Yin ruins at Subutun in Shandong Province. Silkworms were also used for decoration on bronze objects of the Shang period, elevating them to an important place in people's minds.

Silkworm decorations on bronze objects of the Shang Dynasty (c. 16th-11th century B.C.).

We have also seen in ancient documents original records concerning silkworm rearing. *Xia Xiao Zheng* (*Lesser Annuary of the Xia Dynasty*), which reflects productive activity of the late Xia Dynasty (c. 21st-16th century B.C.) and the early Shang Dynasty, says, "In the third lunar month mulberry trees have to be pruned and women begin to rear silkworms." Bone and tortoise shell inscriptions of the Shang Dynasty include the characters for silkworm, mulberry tree, silk floss and silk fabrics, and there are also oracle inscriptions concerning sericulture preserved intact. Among these is the advice to consider raising silkworms only after as many as nine divinations. Sericulture had certainly become a very serious and important undertaking.

In the Zhou Dynasty (c. 11th century-221 B.C.) cultivation of mulberry trees and rearing of silkworms flourished widely in both north and south China. Silk was the main material used in clothing the ruling class. Silk production from worm to fabric was women's chief productive activity. Many sections and chapters of *Shi Jing* (*Book of Odes*) contain descriptions of sericulture. "Qi Yue" ("Seventh Month") in "You Feng" of *Shi Jing* says,

> *On sunny spring days*
> *Orioles sing in happiness.*
> *Baskets in hand, women*
> *Follow the footpath in endless stream*
> *To pluck tender mulberry leaves for silkworms.*

By the Western Zhou Dynasty (c. 11th century-770 B.C.) mulberry trees were already being cultivated on a large scale. "Within the Ten *Mu*" in "Wei Feng" of *Shi Jing* contains this description: "Within the ten *mu* of mulberry trees, pickers are coming and

Pictures of women picking mulberry leaves on a bronze vessel of the Warring States Period (475-221 B.C.).

going." According to archaeologists, two types of mulberry trees were grown in the Zhou Dynasty, the arbor type and the shrub type. Bronzes of the Warring States Period (475-221 B.C.) show women with baskets picking mulberry leaves and also the two types of mulberry.

According to accounts in the ancient books *Shi Jing*, *Zuo Zhuan* (*Zuoqiu Ming's Chronicles*) and *Yi Li* (*Rites*), silkworms were reared not only as a cottage occupation; there were also special mulberry and silkworm nurseries and utensils for silkworm-rearing. These utensils included the silkworm frame and silkworm mat. In other words, the Shang-Zhou period saw China already in command of a series of fairly mature techniques in sericulture.

Development of Silk Production

The book *Guan Zi* (*The Book of Master Guan*) of the Warring States Period says:

Anyone who is proficient in sericulture and can prevent disease in silkworms will be asked to introduce his experience and given one *jin* of gold and eight *shi*[1] of grain, and be exempted from military service as reward.

Indeed, sericulturists created and invented in their field and contributed their wide and valuable store of experience for China and the world.

Many books were written on sericulture in ancient China. In the records of the Han Dynasty (206 B.C.-A.D. 220) mention was made of such works on sericulture as *Can Fa* (*How to Raise Silkworms*), *Can Shu* (*Book on Silkworm Raising*) and *Zhong Shu Cang Guo Xiang Can* (*How to Plant Trees, Store Fruits and Judge Silkworms*). Unfortunately, none of these has come down to us. In the more than 2,000 years since the Han Dynasty, however, a number of ancient books discussing sericulture including *Fan Sheng Zhi Shu* (*The*

---

[1] 1 *shi* = 100 *jin* or 50 kilogrammes.

Book of Fan Shengzhi), *Can Shu* (*Book on Silkworm Raising*), *Bin Feng Guang Yi* (*Comments on the "Book of Odes"*), *Guang Can Sang Shuo* (*Treatise on Sericulture*), *Can Sang Ji Yao* (*Main Points of Sericulture*), *Ye Can Lu* (*Information on Wild Silkworms*) and *Chu Jian Pu* (*Information on Philosamia and Cynthia*) have been handed down. Some are monographs; others devote chapters to sericulture.

The leaves of the mulberry are the main food of the silkworm. As early as in the Western Zhou Dynasty the planting of mulberry trees from seedlings was already known. Layering, or using side-branches in the propagation of mulberry trees, was practised in the 5th century at the latest. *Qi Min Yao Shu* (*Important Arts for the People's Welfare*) describes this method, which greatly shortens the growth period as compared with planting with seedlings. From the Song (960-1279) and Yuan (1271-1368) dynasties onward, silkworm raisers in south China developed the technique of grafting mulberry trees, a further improvement in mulberry cultivation. Grafting was important in rejuvenating old mulberry trees, preserving desirable properties, accelerating the propagation of saplings and the breeding of fine varieties. It is a method that is still in use.

The quality of the mulberry leaves directly affects the health of silkworms and the quality of silk. Pruning, to improve the quality of the leaves, was used very early in China, *disang* (ground mulberry) or *lusang* as mentioned in later times being known in the Western Zhou Dynasty. *Fan Sheng Zhi Shu* tells how to grow ground mulberry trees. To quote: "In the first year, sow a mixture of mulberry seeds and glutinous millet. When the mulberry plants are as high as the millet, cut them down with a sickle close to the ground. Next spring, mulberry suckers will spring out. Such mulberry trees are convenient for plucking and management because of their shortness." Jia Sixie in his *Qi Min Yao Shu* speaks for farmers praising ground mulberry: "A hundred *lusang* will produce a large quantity of silk. This means that *lusang* is of high quality, labour-saving and much appreciated." With steady progress in pruning, the shapes of mulberry trees had also been changing rapidly from "the natural type" to types with tall, medium or short trunks, and *disang* from "fistless form" to "fist form". Leaves of fine quality can only come from new shoots, and the scissoring off of old shoots stimulates the growth of new, which are capable of absorbing large quantities of water and nutriment. The leaves of new shoots are dark green, succulent and big. Pruning thus not only increases leaf output; its skilful use also improves the quality of leaves. Again, this aid in sericulture was a contribution by the ancient Chinese labouring people.

The preparation of silkworm-moth eggs is important in silkworm rearing. There is an account of "bathing eggs in the stream" in "Ji Yi" of *Li Ji* (*Record of Rites*), indicating that the protection of silkworm-moth eggs by washing in clear water was known over 2,000 years ago. Cinnabar solution, brine and lime water were among the disinfectants subsequently used on the surface of eggs. As described in *Chen Fu Nong Shu* (*Agricultural Treatise of Chen Fu*) written in the Southern Song Dynasty (1127-1279): "In spring, bathe the eggs in warm water mixed with finely ground cinnabar when they are about to hatch." The washing of eggs with disinfectant solution just before hatching prevents silkworm diseases by removing bacteria which might invade and sicken the newly hatched silkworms.

Silkworm raisers at least 1,400 years ago paid attention to silkworm selection. As recorded in *Qi Min Yao Shu*:

> In collecting cocoons for reproduction, choose only those in the middle section of the straw cocks, as those above have thin silk envelopes while those near the ground lay less eggs.

Ancient raisers had two aims in mind. One was to eliminate weak or diseased silkworms, the other to ensure uniform growth of the next generation so as to facilitate their rearing and management. Silkworm selection consists of four items: selection of the worms themselves, selection of cocoons, selection of moths and selection of eggs. Only cocoon selection was done at first, as indicated by the *Qi Min Yao Shu* quotation. From the late Song Dynasty onward, selection went further to include the worms, moths and eggs. Selection of cocoons was for quality, time and position of their spinning, time of moth emergence, and the physical condition of the moths. In the Qing Dynasty (1644-1911) further attention was paid to the selection of silkworms, it being appreciated that "only healthy silkworms beget healthy offspring".

Ancient sericulturists were also aware of the close relationship between the silkworm and its environment. As early as during the Qin-Han period (221 B.C.-A.D. 220) it was already known that sufficiently high temperature and ample food promote growth and tend to shorten its life cycle. Hence silkworm raisers of past ages all paid great attention to controlling their living environment and conditions. *Qi Min Yao Shu* records the method of raising the temperature in the silkworm nursery by heating it all round with fire. *Shi Nong Bi Yong* (*An Indispensable Book of Agriculture*) also points out that more heat is required in the silkworm nursery when the worms are young because the weather is still cold at that time, and that after their final moulting less is required because the weather is already warm. *Wu Ben Xin Shu* (*New Book on the Essentials*) says:

> Whether it is windy or rainy, whether in the daytime or nighttime, one should judge changes in temperature with one's own body.

A silkworm raiser wears only a single garment.

> If he feels cold, the silkworms must also feel cold. In that case, fuel should be added to the fire. When he feels hot, the worms must also feel hot, so he should reduce the fire as he sees fit.

Ordinarily, an environmental temperature that is comfortable for the human body is suitable for silkworms, and so this advice is quite rational. Cui Shi (?-c. 170) writes in his *Si Min Yue Ling* (*Monthly Ordinances for the Four Classes of People*):

> On Pure Brightness Day in the third lunar month, women silkworm raisers are told to put the silkworm nursery in order, stop up cracks and holes and get the frames, mats and baskets ready for use.

This meant repairing and cleaning the silkworm nurseries and utensils before the breeding began. Fumigation was also used in ancient times to disinfect the silkworm nursery, and no doubt these measures did play a role in preventing silkworm diseases and eliminating pests. The prompt clearing away of silkworm droppings and sterilizing of silkworm utensils during the entire breeding course is prescribed in *Nong Sang Yao Zhi* (*Essentials of Agriculture and Sericulture*) which says that "two bottom mats should be laid" for the silkworm stand, and that each day, "after the silkworms are hatched, take out one of them and expose it to the sun until it is in the west. Then spread it again at the bottom. On the next day, take out the other bottom mat and expose it to the sun in the same way." The practice of changing alternate bottom mats is repeated. Also, it is economical and practical to make use of sunshine to disinfect silkworm utensils.

There are two medicinal methods for the prevention and cure of silkworm diseases. One is by adding medicine to the food, the other is medicinal fumigation. The former has been practised in China for 800 years. *Shi Nong Bi Yong* says that "silkworm diseases of a hot nature can be dispelled" by mulberry leaves which are about to fall and "have been pounded and ground to powder". Later on, *Yang Yu Yue Ling* (*Monthly Guide to Sericulture*) (A.D. 1633) and *Yang Can Mi Jue* (*Secrets of Silkworm Rearing*) also recorded the prevention and cure of various silkworm diseases by feeding silkworms with mulberry leaves sprinkled with licorice root solution, garlic sap and white spirit. This was followed by the use of specific prescriptions for the treatment of different silkworm diseases.

According to an account given in the Chapter "Nai Fu Pian" ("On Clothing Materials") of Song Yingxing's *Tian Gong Kai Wu* (*Exploitation of the Works of Nature*), since the Ming Dynasty (1368-1644) there was a fair knowledge of certain infectious silkworm diseases such as abscess, softening and rigidity, and also of preventing infection and spread by elimination and isolation.

As domesticated silkworms had been bred and selected over past ages, their characteristics had changed greatly and they had developed into breeds of different types in different districts during different historical periods. During the Song and Yuan periods, when sericulturists in north China were still raising mainly the Early Silkworm with three moulting stages, those in south China had already begun raising the Early or Late Silkworm with four moulting stages. Silkworms with three moulting stages are more disease-resistant than those with four, and hence are easier to rear. However, silk obtained from the cocoons of silkworms with four moulting stages is of better quality. This type made its appearance in the Northern Wei Dynasty (386-534). After a long period of breeding, large numbers in many excellent varieties were produced in the south China provinces of Jiangsu and Zhejiang. The successful propagation of the four-moulting-stage silkworm marked an advance in sericulture.

To promote silk production, people in ancient China also raised the Summer Silkworm, the Autumn Silkworm and even many crops a year besides the Spring Silkworm. Over 1,600 years ago the Chinese used multivoltine races for natural reproduction of many crops yearly, and also found a way to delay the incubation of silkworm eggs by using low temperatures. Hence, the same breed of eggs could be hatched in several successive batches in a year, giving many crops. This is considered another remarkable technical invention in sericulture in ancient China.

Particular mention should be made of silkworm raisers in the Ming Dynasty who discovered the hybrid vigour of domesticated silkworms while preparing the Summer Silkworm for propagation. Song Yingxing makes the earliest record of this in his *Tian Gong Kai Wu*:

> Recently some small silkworm raisers have crossed an Early (univoltine) male with a Late (bivoltine) female, hoping thus to produce an excellent breed. This is an unusual occurrence worth noting.

Spread of Sericulture to Other Countries

All original silkworm eggs and methods in sericulture in other countries were introduced directly or indirectly from China.

Ancient books record the introduction of Chinese silk-moth eggs and silkworm rearing into China's close neighbour Korea in the remote 11th century B.C. Legend has it that sericulture found its way from China to Japan during the reign of Shi Huang Di (First Emperor of the Qin Dynasty). Eager to promote silkworm rearing, the Japanese subsequently sent emissaries to China and Korea to learn the art and invited Chinese sericulturists to their country to pass on their experience.

A steady flow of the beautiful silks produced by the labouring people in ancient China were transported very early to Persia and Rome. In 138 B.C. the Wu Di Emperor of the Han Dynasty despatched Zhang Qian to open up the Western Regions. Zhang Qian went as far as Central Asia, his route being followed later in the transport of silk. It started at the northern foot of the Kunlun Mountains or the southern foot of the Tianshan Mountains and proceeded westward, crossing the Pamirs — the world-famous "Silk Road". Silk-moth eggs and sericulture also found their way at a later time from China's interior into Xinjiang where they were relayed to Arabia, Africa and Europe via this "Silk Road".

Sericulture spread to Arabia and Egypt in the 7th century, to Spain in the 10th and to Italy in the 11th. Silk-moth eggs and mulberry seeds were carried to France in the 15th century, when France's silk industry began. England received it from France.

In the Americas, silkworm rearing is said to have been carried on by the Mexicans in the mid-16th century. It was not promoted on a large scale in the Americas, however, until the 17th century when England thought it would be profitable to establish the silk industry in its American colonies where the climate was mild and the land fertile.

Horticulture

Dong Kaichen

Intensive and Careful Cultivation

Vegetables have been cultivated in China since very ancient times. Among the Neolithic Age ruins at Banpo near Xi'an, besides grains of cereal, seeds of mustard or cabbage have been discovered in an earthen pot. Tests place these seeds as belonging to a period about 6,000 years ago. By the Zhou Dynasty (c. 11th century-221 B.C.), the cultivation of vegetables was already well established. Accounts of vegetable production are to be found in *Shi Jing* (*Book of Odes*). The poem "Qi Yue" ("Seventh Month") in the chapter "Bin Feng" ("Odes of Bin") describes it thus:

> *In the seventh month, they eat the melons.*
> *In the eighth, they cut down the gourds.*
> *In the ninth month, they prepare the vegetable gardens,*
> *And in the tenth they bring in the crops.*

With the expansion of cities and towns there came to be a division of labour between the cultivation of field and vegetable crops so that vegetable growing became a specialized profession during the Spring and Autumn Period (770-476 B.C.) and the Warring States Period (475-221 B.C.).

The rational use of time and space to increase land utilization is of great importance in vegetable growing. It is mentioned in *Qi Min Yao Shu* (*Important Arts for the People's Welfare*) that mallow (*Malva verticillata L.*), which was called *kui* in ancient China, can be sown three times and chive harvested not more than five times a year, indicating that a vegetable was sown and harvested continually on one piece of land. As for sowing scallions or lesser beans among melon-vines and growing corianders and green onion together, it means that interplanting was practised in the cultivation of vegetables at that time. *Important Arts for the People's Welfare* also gives in its "Miscellany" notes on growing a large variety of vegetables:

> In places near cities and towns, more melons, fruits, eggplants, etc. should be grown, for what is not consumed by the family can be easily sold. If there are 10 *mu*[1] of land, pick out the five that are most fertile. Use 2.5 *mu* to grow green onion and the other half to sow sundry vegetables like gourds, radish, mallow, lettuce, turnip, white pea, lesser bean and eggplant in the second, fourth, sixth, seventh and eighth month respectively.

This frequent planting points to a rather high level in vegetable cultivation during that period. By the Song Dynasty (960-1279), while depicting the spectacular cultivation of vegetables on the outskirts of Kaifeng in Henan Province, Meng Yuanlao in his *Dong Jing Meng Hua Lu* (*Memories of the Eastern Capital*) writes, "There were gardens almost everywhere within 100 *li*[2] around the metropolis, and no land was lying idle." The imperial family had its special vegetable gardens, which supplied day-to-day needs. "The West Imperial Garden provided winter stores five days before the Begining of Winter."

In order to ensure high yield and balanced supply throughout the year, it is necessary to select the right kinds of vegetables and rotate them properly. Since intensive cultivation has long been a tradition for agricultural production in China, and the Chinese labouring people have succeeded in breeding a wealth of varieties, it is possible to organize many vegetables into a fixed cultivation system. The term "garden-style cultivation" or "gardenization" is sometimes used to describe agriculture in China because intensive cultivation characterizes its vegetable production.

---

1 A *mu* now equals 1/15 hectare or roughly 1/6 acre.
2 A *li* was about 540 metres in the Song Dynasty.

Numerous and Colourful Varieties

Vegetables in China are of many kinds, numbering about 160 in all, of which about 100 are commonly grown. Of these, native varieties and those grown from imported stock are roughly half and half. The earliest record of native vegetables of China appears in the *Book of Odes*. There are over 10 species including melon, gourd, chive, mallow, turnip, lotus, celery and vetch. It is uncertain at present however which of these were cultivated and which grew wild at that time. Jia Sixie in his *Important Arts* records 31 species of vegetables growing in the Huanghe (Yellow) River region, including melon, wax gourd, snake melon, cucumber, eggplant, mallow, turnip, radish, green onion, chive, mustard seed, coriander and even alfalfa. Of these, 21 are still grown today in kitchen gardens, while 10 have been withdrawn or turned to other uses. Among the existing species, cabbage, radish and mustard have been carefully bred through the ages so that cabbage and radish became staple vegetables; mustard was developed in many varieties to suit manifold uses.

Cabbage, or *baicai* (white vegetable), called *song* in ancient China, is widely cultivated and stored so that it is available at all seasons. Its adaptability in the Chinese menu and its delicate flavour make it a standard popular vegetable. The best-known *baicai* is the long Chinese cabbage of north China, with leaves folded into a compact head. In regions south of the lower reaches of the Changjiang (Yangzi) River such as Jiangsu and Zhejiang provinces, a small type of *baicai* with loose leaves was mainly cultivated in the past, and the large solid type was a rarity. Nor is it mentioned in horticultural accounts of north China before the Jin Dynasty (265-420). The large northern type does appear in the literature of the Southern and Northern Dynasties (420-589), however. For example, in *Nan Qi Shu* (*History of the Southern Qi Dynasty*) we find "early chive of early spring, late *song* of late autumn" described as tasting the best of all culinary vegetables. Discussing the cultivation of turnips, *Important Arts* mentions in passing that "the method of growing *song* and radish is the same as for growing turnip". Later, in the Song Dynasty, Su Song in his *Tu Jing Ben Cao* (*Illustrated Herbals*) remarks,

> According to ancient accounts, there was no *song* in the north. Today it is grown in the capital Luoyang the same as in the south, but it is not as big and fleshy.

In the Ming Dynasty (1368-1644) Li Shizhen in his *Ben Cao Gang Mu* (*Compendium of Materia Medica*) also says,

> We might suppose that there had been no *song* in the north before the Tang Dynasty [618-907], but today purple and white *song* are both quite common in the north as well as in the south.

Thus we see that although the cultivation of cabbage was highly developed in the south in the Southern and Northern Dynasties, it was not grown extensively in the north until Tang and Song. The loose-head and semi-loose-head types of *song* were intermediate stages in cabbage cultivation before the compact-head type was developed. Chinese cabbage is definitely one of the products listed in *Shun Tian Fu Zhi* (*Records of Shuntian Prefecture*) of the Qing Dynasty (1644-1911). Through meticulous breeding, there are at present more than 500 local varieties in north China. Some have been introduced into the south with good growing results. Since the small cabbage and the large, compact type both originate in China, they are called in Latin *Brassica sinensis* and *Brassica pekinensis* respectively.

China is also one of the homes of the radish. Its earliest account appears in *Er Ya* (*Literary Expositor*). Su Gong of the Tang Dynasty in his *Xin Xiu Ben Cao* (*Revised Materia Medica*) writes,

> A large portion of radishes are grown in the area north of the lower reaches of the Changjiang River and in Hebei, Shaanxi and Shanxi provinces, while in Denglai they are also extensively cultivated.

By the Song Dynasty,

> There are radishes on both sides of the Changjiang River....

In the region north of the Huanghe River big radishes are very common, and in districts south of the lower reaches of the Changjiang River such as Anzhou, Hongzhou and Xinyang they are also quite big, some weighing five or six *jin*[1] each or more when properly cultivated." (*Illustrated Herbals*)

Having been bred over a long period and grown in various habitats, there are far more varieties of the radish in China than in any other country in the world.

Mustard is a special product of China and grown in many varieties, some for the root, some for the stem and others for the leaf. Only the seeds of wild mustard were used at first as a condiment. Li Shizhen remarks in his *Compendium of Materia Medica* that besides pungent mustard, which can be used as medicine, there are also horse mustard, stone mustard, purple mustard and variegated mustard whose leaves are fit for human consumption as a vegetable. At present, varieties of mustard that are eaten for their leaves include potherb mustard and big-leaf mustard; among those with edible tubers there is the famous hot pickling mustard tuber (*zhacai*) of Sichuan Province. This variety is an achievement of the Chinese labouring people in transforming the habits of plants.

In addition to domesticating vegetables, China has been introducing new varieties since early times, careful cultivation gradually acclimatizing them to China. Many fine new types and varieties have been bred. The cucumber, for instance, which was of small size and had little flesh, has been improved in shape and quality, while new varieties that adapt to different seasons and climates have been developed, ensuring supply without the benefit of hot houses from spring to autumn. The eggplant, which originated in India, started out the size of a chicken egg. Introduced into China, it was cultivated long ago into an oval shape with a length of 20-30 centimetres; round eggplants weighed several *jin*. The big dark purple round eggplant of north China has in turn been introduced into many countries. The pepper, a native of America, was introduced into

---

[1] 1 *jin* = 0.5 kilogramme or 1.102 pounds.

China via Europe as recently as 300 or 400 years ago, yet China already has the greatest variety of peppers in the world. Besides the long hot pepper, the Chinese have bred many varieties of sweet pepper.

Various Methods of Vegetable Cultivation

The area in the middle and lower reaches of the Huanghe River is one of China's early agricultural regions. Here the winters are long and dry, and a year-round balanced supply of fresh vegetables is certainly very difficult to obtain. In their effort to wrest bigger and earlier harvests, the Chinese labouring people of past ages developed in the course of production different methods of [vegetable cultivation such as growing in protected ground, blanching and heeling-in, besides open-field cultivation. The ancients, windbreaks, cold frames, hotpits, hotbeds and greenhouses have continued in use to this day.

China was probably the first country to use protected ground for growing vegetables, the practice likely beginning in the Western Han Dynasty (206 B.C-A.D. 24) at the latest. The book *Yan Tie Lun* (*On Salt and Iron*) mentions "winter mallow and hothouse chive" as one of the luxuries enjoyed by the rich of that time. *Han Shu* (*History of the Han Dynasty*) says in more concrete terms, "Houses which are heated day and night with a slow fire have been built in Taiguan Garden for growing green onion, chive and other vegetables in winter." This describes vegetable growing in winter in the Imperial Palace during that period. Legend has it that in the reign of Qin Shi Huang (221-207 B.C.) melons ordinarily grown in a warm climate were cultivated in winter at Lishan, where there were hot springs. By the Tang Dynasty, detailed accounts of using hot springs for warmth in vegetable cultivation began to appear. Glimpses of such accounts are seen in these lines by Wang Jian:

> *Early in the middle of the second month*
> *Stocks of melons are already laid in,*

Simply because the Imperial Garden
Is supplied with the hot water from the spring.

In *Wang Zhen Nong Shu* (*Agricultural Treatise of Wang Zhen*) of the Yuan Dynasty (1271-1368) there is this specific account of using cold frames for growing chive:

> Seed beds with windbreaks covered with horse manure in winter are also used to cultivate chives which will be fit for eating by early spring when they have grown to a height of two or three *cun*[1].

As it is much cheaper to grow vegetables in cold frames than in greenhouses, cold-frame products "came into use as food for ordinary people in the cities and prefectures."

But vegetables grown in cold frames and greenhouses were limited in variety and quantity, and eating stored radish and cabbage day after day during the long winter months is none too pleasant. This situation gave rise to the simpler method of producing blanched vegetables. The earliest record of blanching culture appears in the book *Shan Jia Qing Gong* (*Fresh Food of a Mountain Dweller*) written by Lin Hong in the Song Dynasty. The soybean sprout, produced by blanching with the seed allowed to sprout in the absence of sunlight, is a unique culinary contribution of the Chinese labouring people. Mung beans and peas can also be sprouted by blanching. Tender, crisp and tasty as well as nutritious, blanched bean sprouts are a widely popular food item.

Blanched vegetables are not restricted to bean sprouts. The seedlings of Chinese chive, green onion, garlic and even celery can all be treated by blanching, while blanched garlic leaves have always been valued. During the Song Dynasty, Su Shi wrote in a poem about "tasting green crown daisy chrysanthemum and yellow blanched chive in spring". Meng Yuanlao in his *Memories of the Eastern Capital* also observes that during his time blanched chives were sold on the street in the capital Kaifeng in the 12th month. *Agricultural*

---

[1] A *cun* now equals 3.33 centimetres or 1.312 inch.

Treatise of Wang Zhen describes the cultivation of blanched chives in the following concrete terms:

> Nowadays roots of chives are transplanted in greenhouses and hilled up with horse manure. The heat causes the roots to sprout quickly, the sprouts reaching a height of about 1 *chi*[1]. They are called yellow chives because of the light yellow colour of their leaves produced by growing in the absence of wind and sunshine.

Besides the conventional methods of pitting and burying, fresh vegetables can also be stored through heeling-in. Volume 9 of *Important Arts* says:

> Dig a hole four or five *chi* deep at any time from the ninth to the middle of the 10th month on the sunny side of the wall where there is plenty of sunshine. Place into it the various vegetables separately, one layer of vegetable alternately with one layer of earth until the pit is filled to within one *chi* of the mouth. Cover with a thick layer of straw. After such treatment the vegetables are able to pass the winter. They can be taken out for use at any time and yet remain as fresh as vegetables in summertime.

This method utilizes facilities similar to cold frames for storing and preserving vegetables such as celery, rape and lettuce.

Fruit Cultivation in Ancient China

North and south China are the main native habitats of deciduous fruit trees of the temperate zone and evergreen fruit trees of the subtropics respectively. In *The Origin, Variation, Immunity and Breeding of Cultivated Plants* (*Chronica Botanica*, Vol. 13, No. 1/6, 1951), N. I. Vavilov remarks, "China occupies first place in the wealth of fruit species of *Pyrus, Malus, Prunus*. Many citrus fruits have

---

[1] 1 *chi* = 10 *cun*.

their origin in China." Among the fruit plants of Chinese origin, some have been cultivated for at least 3,000 years, whereas most have found their way to all parts of the world. Fruit trees grown in China include the peach, Chinese plum, apricot, Chinese pear, persimmon, jujube, chestnut, citrus, longan, litchi and loquat. Not only are some of Chinese origin but they have continued to be special products of China to this day. In recent years China has introduced *yangtao* (*Actinidia chinensis*, kiwi fruit or Chinese gooseberry) into some countries where it has been cultivated commercially in large areas. Now this and other such fruits are highly valued on the international market. Here we shall say something about the peach and citrus fruits.

The peach, a fruit cultivated since antiquity, has its origin in northwest China. In the past the West mistook it to be indigenous to Persia and called it *Persica*, from which its English name is derived. However, research and investigation in recent years have established it as a fruit tree of Chinese origin. Depicting the peach tree in an ode, *Book of Odes* has this to say: "What a beauty is the peach, and how brilliant are its blossoms!" In *Er Ya* is also mentioned "the *mao* which means winter peach" and "the *ce* which is another name for David peach". These are but two of the very early references to this tree. A detailed description of its characteristics, propagation and cultivation is found in *Important Arts*, which shows that peach cultivation at that time was already quite advanced. *Qun Fang Pu* (*Beautiful and Fragrant Flowers*) of the Ming Dynasty (1368-1644) gives a fairly detailed description of peach tree varieties.

Investigations into peach varieties in recent years also confirm that the peach is indigenous to China. Wild species and closely related species of the plant such as the David peach and the Gansu peach have been discovered in mountain areas of west and northwest China, while the different types of present-day cultivated species such as the clingstone and the freestone, the soft-fleshed and the firm-fleshed, the pointed and the flat-topped as well as the two mutations, the flat peach and the nectarine, are all among Chinese cultivated varieties.

The peach was probably introduced into Iran from northwest

China via Central Asia between the 2nd and 1st century B.C. From there it went to Greece before being reintroduced into European countries. The cultivation of peaches began to flourish after the 9th century. In the second half of the 19th century, China's honey peach and flat peach were introduced into Japan and the United States, where further new varieties were bred.

Citrus is a general term which includes many species, varieties or cultivars. The sweet orange, tangerine, shaddock and lemon are of economic importance, the first three plants being of Chinese origin. The mandarin orange and the tangerine are both termed *Citrus reticulata Blanco* in Latin, and the only distinction between them is a slight difference in the shape of their fruit and a greater resistance to low temperatures in the latter.

The duration of citrus cultivation in China may be seen from a reference in *Yu Gong* (*Tribute of Yu*) that "the tangerine and the shaddock have become articles of tribute", the time being the Zhou Dynasty (c. 11th century-221 B.C.). By the Han Dynasty (206 B.C.-A.D. 220) these were grown extensively. It is noted in *Shi Ji* (*Records of the Historian*) that the income from "a thousand tangerine trees in the district of Jiangling in the kingdom of Shu Han [221 -236]" is comparable to the revenue collected by "a marquis over a thousand families". Later, during the Song Dynasty, Han Yanzhi wrote in 1178 a monograph on citrus, *Yong Jia Ju Lu* (*Citrus Horticulture*), in which are discussed 27 citrus plants including the mandarin orange (8), tangerine (14) and orange (5). The last volume of the monograph gives a detailed account of citrus culture under nine headings: 1. seed treatment; 2. planting; 3. cultivation; 4. curing disease; 5. watering; 6. picking; 7. storage; 8. processing and 9. use as medicine. Ever since the appearance of "Ju Song" ("In Praise of the Tangerine") in *Chu Ci* (*Elegies of Chu*), literary works on the theme of citrus are uncountable, showing the Chinese people's admiration of the plant.

Today the citrus is extensively grown in all the 15 provinces and districts south of the Changjiang River valley. In spite of its liking a warm climate, the Chinese labouring people of past ages increased the citrus plant's hardiness through careful breeding and obtained varieties resistant to low temperatures. They also summarized such measures in cultivation as domesticating introduced seedlings, moving up the rest period and hilling up the roots. Because it was already known in ancient China that citrus groves must be located in places sheltered from wind and frost, citrus growing continued to develop in the Changjiang River valley despite periodic frosts. In his *Zhong Shu Shu* (*Book of Afforestation*), Yu Zongben of the Ming Dynasty writes, "Although in Dongting frost is quite common, no damage is done, and the tangerine grows extremely well and bears each year."

In Europe, only the citron was used as medicine in ancient times, and not until after the 10th century were the lime and sweet orange known. In 1545 the sweet orange was introduced from China into Lisbon by the Portuguese. After that Western countries began to grow it extensively and gradually spread its culture to other parts of the world.

The Wonders of Grafting

Grafting is of great importance in the propagation of fruit and other economic trees, because plants mature and bear much sooner when propagated by vegetative means than by the sexual means of seedage. Grafting, or asexual propagation, has the advantage of retaining all the original characteristics of cultivated varieties. Grafting apparently was practised in the latter Warring States Period (475-221B.C.). A detailed account of the principles and methods of grafting appeared in *Important Arts*.

Chapter 37, "Pear Growing", in *Important Arts* points out:

Grafts come into bearing sooner than seedlings. Use birchleaf pear as stocks. Slight bursting of leaf-buds indicates the optimal time for grafting. Be careful not to injure the green bark or it dies, and see to it that the bark and wood of the scion are in good contact with those of the stock.

As survival of a graft is determined by perfect union of cambiums of the stock and the scion, *Important Arts* is correct in the light of present-day knowledge. In order to stress the advantages of propagation by grafting, *Important Arts* also makes comparisons and introduces propagation of fruit trees by seedlings. It points out that wild pear trees and seedlings that have not been transplanted come into bearing very late, and seedlings are always subject to deterioration. Only two of the ten or so seeds in each pear will develop into cultivated pear trees, while all the remaining seeds will become birchleaf-pear trees. This observation shows that people at that time had already noticed the serious deterioration and degeneration of seedlings as well as genetic dissociation as a result of sexual propagation. In asexual propagation by grafting all daughter plants are practically identical to the parent plant from which they are taken so as better to retain the desirable characteristics of their parent.

Grafting methods were also improved with the passage of time. In *Important Arts* we find in Chapter 37 reference to twig grafting of pear trees by using one stock and one or several scions, and Chapter 40, "Cultivation of Persimmons", tells of root grafting by "taking a branch and grafting it onto dateplum persimmon". Chapter 13, "Essay on Cultivation", in *Agricultural Treatise of Wang Zhen* of the Yuan Dynasty sums up the methods thus: "There are six methods of grafting, namely, (1) trunk grafting; (2) root grafting; (3) bark grafting; (4) twig grafting; (5) dimple or bud grafting and (6) splice grafting." "Trunk grafting" resembles today's top grafting; "root grafting" is different from today's root grafting and resembles grafting at lower positions on the stock. Such division is defective because of its inconsistency in bases. Some are classified according to the methods by which they are grafted as is the case with splice grafting, while others are classified according to the grafting positions of the stock and the scion as in the case of trunk grafting, root grafting and twig grafting. However, being concise and well-organized in its presentation, it was still used in writing agricultural works in later ages. Today, some of the grafting terms continue to be used as nomenclature not only in China but also in Japan.

The proper use of key grafting technique to ensure the survival of a grafted plant is of crucial importance and may be regarded as a sign of progress in grafting. Xu Guangqi of the Ming Dynasty says in Volume 37, "Cultivation", of his *Nong Zheng Quan Shu* (*Complete Treatise on Agriculture*):

> There are three secrets of success in grafting a tree: (1) Do it when the bark is of a green colour, i.e., when it is still young and tender; (2) Choose the part where there is a node; (3) See to it that the scion and the stock are in good contact where they are joined. With these in mind, there will be no danger of anything going wrong.

The book also treats in concise and definite terms plant age, positions for grafting and points for attention. Since the meristematic cells are most well developed in places where there is a node, there is scientific basis for choosing this position for grafting.

Hua Jing (*Mirror of Flowers*), written by Chen Haozi in the Qing Dynasty (1644-1911), explores the physiology of grafting. While *Agricultural Treatise of Wang Zhen* contains only the words "once grafted and bound the two breaths communicate with each other" to illustrate the internal mechanism, *Mirror of Flowers* explains that "the sap of a tree flows through its bark, and when the union is formed the scion will continue to grow". This explanation of survival in grafting in terms of physiological mechanism follows the principle that survival is achieved by the transportation of nutrients through the xylem and the phloem of both scion and stock.

Floriculture also made remarkable progress in ancient China. The peony is a typical example. Discussing the function of artificial selection in his *Variation of Animals and Plants Under Domestication* (1866), Charles Darwin points out: "According to the Chinese tradition, the *P. moutan* has been cultivated for 1,400 years. At present there are about 200-300 varieties in China."

The peony has its origin in northwest China. Its large brightly coloured blooms are elegant and varied. The peony was first used for its medicinal qualities only, and it was not until the Sui-Tang period (581-907) that it was bred as an ornamental flower. Grafting as a method of propagation was largely responsible for accelerating changes in shape and colour of the flowers so that many new varieties were obtained. There were already detailed accounts of this in the Song Dynasty. *Luo Yang Hua Mu Ji* (*Flowers and Trees of Luoyang*), compiled by Zhou Shihou in 1082 and handed down to the present day, lists as many as 109 varieties of peony. Zhou also discusses the art of grafting the peony in his *Luo Yang Mu Dan Ji* (*Peonies of Luoyang*). The grafting of peonies became the means of livelihood of many families in Luoyang, and grafted peonies became a special commodity. Other graft flowers included the chrysanthemum, the Japanese apricot (*Armeniaca mume*) and the Chinese flowering crabapple (*Malus spectabilis*), to mention only a few of the floricultural wonders achieved by the Chinese labouring people in ancient times.

Tea

Zhang Binglun

Origin and Cultivation

Besides being the home of the tea tree, China is credited with its discovery and being the first country to use its leaves. Chinese tea has always been noted for its variety and fine quality, and tea-producing countries the world over have directly or indirectly received tea-tree plantings from China. *Cha Jing* (*Canon of Tea*) written by Lu Yu in about the year 780 during the Tang Dynasty (618-907) is the world's first book on tea.

The use of tea in China has a long history. In ancient times, tea was called *tu* and other names such as *jia*, *she*, *ming* and *chuan*. About 50 B.C. Wang Bao in his *Tong Yue* (*Contract with a Servant*) mentions boiling *tu* that was purchased from Wudu, a mountain in one of the tea districts of Sichuan Province. This fairly early record of buying and brewing tea has led to the belief that the drinking of tea originated in Sichuan.

The fresh leaves of wild tea trees were first directly used as a medicine or beverage, and only later were tea trees cultivated, though their cultivation in China has been of two millennia duration. By the Tang Dynasty tea was cultivated in the present provinces of Jiangsu, Anhui, Jiangxi, Sichuan, Hubei, Hunan, Zhejiang, Fujian, Guangdong, Yunnan, Shaanxi and Henan. In that era, tea farmers cultivated tea trees planted in odd corners and on the hillsides, while fair-sized government tea plantations also made their appearance. Detailed early records of the cultivation of tea trees are to be found in *Cha Jing* by Lu Yu and in *Si Shi Zuan Yao* (*Outline of the Four*

329

Seasons) written in the Tang Dynasty, and also in *Si Shi Lei Yao* (*Classified Outline of the Four Seasons*) of the Yuan Dynasty (1271-1368). The tea tree is a short-day plant that grows best in shady places. Song Zi'an comments in his *Shi Cha Lu* (*Tea Sampling*) that "it likes the shade of high mountains and welcomes the early morning sunshine". An appropriate amount of sunshine enables it to grow luxuriantly, but too intense sunshine ages the leaves too quickly and they will not make good tea. Tea produced where there are mountains subject to cloud and mist is tender of leaf and fragrant, and so it is said that "famous teas come from high mountains". The shady side of hillsides, or slopes shaded naturally or planted to trees have always been favoured as tea-plantation sites, as such locations afford good drainage, raise the survival rate of tea trees and improve the quality of the leaves.

Tea seed must rest for a certain period in definite conditions of temperature and humidity before budding is possible. Tea workers in Tang Dynasty times used "the budding method of burying

Page from Lu Yu's *Cha Jing* (*Canon of Tea*).

tea seeds in the sand". They mixed the ripe tea seed with moist pebbly soil and placed this in a basket covered with stalks to prevent frost damage. The following year the seeds were bunch planted in holes which had been dug and well supplied with base manure beforehand. The seedlings were given the best of care — watered when dry and provided with organic manure. Then,

> after three years each cluster will yield eight ounces of tea. Since there are 240 clusters to a mu^1, the total output of tea per *mu* will be 120 catties. Male hemp, millet and broomcorn are planted between the clusters when they are not yet fully grown. (*Si Shi Zuan Yao*)

This indicates a fairly high output of tea already in the Tang Dynasty, and that intercropping of tea with grain was practised.

Plucking tea leaves was a highly developed art in ancient times. Lu Yu notes in his *Cha Jing*: "One is likely to fall ill if he drinks tea which was plucked at the wrong season, improperly manufactured or has rank grasses mixed in." The time for plucking tea leaves is generally earlier in areas of higher temperature than where it is lower. Within the same plucking season, early mornings and cloudy days were considered in the Tang and Song dynasties as the right time for plucking. Further, fingernails and not fingers were to do the job. "No abrasion will be caused by snapping off with nails while harm will be done if plucking is done with fingers because of their warmth." In manufacturing high-grade teas, the leaves should be plucked separately according to their age whether they are "buds like the tongue of a sparrow or like grain", "a spear and a flag" (i.e., a bud and a leaf), "one spear and two flags" and so on. Handling the leaves and temperature control in processing will be facilitated so that uniformity in shape is assured.

Zhang Yi of the state of Wei (220-265) of the Three Kingdoms period bases himself on Lu Yu's account when he writes in his *Guang Ya* (*Enlarged Literary Expositor*) concerning tea processing:

---

[1] 1 *mu* = 1/15 hectare or roughly 1/6 acre.

In the district between the provinces of Hubei and Sichuan the leaves are plucked and made into cakes; those made of old leaves are mixed with rice. To make tea as a drink, bake the cake until reddish in colour, pound it into tiny pieces, put the pieces in a chinaware pot, pour boiling water over them and add onion, ginger and orange. The brew sobers one after drinking and keeps one awake.

During the Tang Dynasty changes were made in the processing which greatly improved the brew. Steaming out of the "green" colour was invented. The steaming was the first process when the fresh tea leaves were brought in, then they were pounded and made into cakes which were perforated, strung together and baked. Their former offensive "green" odour was removed and they were also in a form that was convenient for storage and shipping. This stimulated the transport and sale of large quantities of tea produced south of the Changjiang (Yangzi) River in north China and outside the Great Wall during the 7th to 10th centuries.

In the Song Dynasty (960-1279) the fresh tea leaves were washed before being steamed. Their juice was then removed by extraction before they were made into cakes. In order to simplify the processing of tea and preserve its true flavour, tea handlers in the 10th to 14th centuries gradually replaced solid cake and ball-shaped tea with the loose form, in which the leaves were neither rolled nor pounded but roasted directly after steaming out the "green". This ushered in whole-leaf tea, while the old processing method was largely discarded.

In the last years of the Yuan Dynasty (1271-1368) and early Ming (1368-1644), the labour- and time-saving simple method of *chaoqing* or "roasting out the green" was invented, greatly improving the colour, aroma, flavour and shape of the tea leaves. Scented, or flower, tea and black tea came into existence after the Ming Dynasty.

Scented teas are prepared by mixing fresh flowers of heavy fragrance with high-grade green tea. And in the Song Dynasty other aromatic substances were used. Cai Xiang writes in his *Cha Lu*

(*Treatise on Tea*): "Though tea has a natural fragrance, as an article of tribute to the emperor its fragrance is enhanced by mixing in a minute amount of borneol and other extracts." The use of flower petals with tea began with the Ming Dynasty. According to an account given by Gu Yunqing in his *Cha Pu* (*Manual on Tea*), there are two kinds of scented tea — lotus-flower tea, and tea scented with different sweet blossoms.

> Mignonette, sweet osmanthus, jasmine, rose, rambling rose, orchid, orange blossom, gardenia, banksia rose and plum blossom can all be used to scent tea. Blossoms which are half-opened and whose fragrance is at its best are plucked in a quantity according to the amount of tea to be scented, the ratio of tea and flowers being three to one. These are put in a porcelain pot, the tea and flowers in alternate layers till the pot is filled. It is then sealed with broad bamboo leaves and boiled in hot water, after which the contents are taken out and cooled. They are wrapped in paper and baked over a fire until dry enough to keep for use.

The heavy fragrance of fresh flowers mingled with the refreshing flavour of tea gives scented tea the distinctive feature of the mutual enhancing of fragrance and flavour.

There were already "a great number of famous teas" early in the Tang Dynasty, because "they were by then highly valued". These included the Shihua of Mengding, Shisun from Guzhu, Luya of Fangshan in Fuzhou, and Huangya of Huoshan, which were all renowned teas of that period. In the Song Dynasty there were as many as 41 "tribute teas" such as Silver Leaves of Ten Thousand Springs and "best-quality picked tea" in the single province of Fujian. Today, famous names in tea are too many to enumerate, a great number of these prized highly on the domestic and foreign markets. Among these are the Qimen black tea of Anhui Province, the Maofeng of Huangshan, Guapian of Qiyun, Houkui of Taiping and Longjing of Shifeng in Zhejiang Province, the Pingshui Gunpowder and Wuyi Rock Tea of Fujian Province, Biluochun of Jiangsu, the black and Pu'er tea of Yunnan Province, Sichuan's Erui,

Maojian from Guizhou's Douyun, the Yulucha of Hubei Province, the Silver Needle and Silver Sword Edge of Hunan Province, Lushan's Cloud and Mist Tea from Jiangxi Province, the Oolong of Taiwan Province and high-grade scented teas from various provinces.

Efficacy of Tea as Described in Ancient Books

Drinking tea has been popular throughout the world since ancient times because tea makes a delicious beverage that benefits the health. As soon as tea found its way into Europe, it joined coffee and cocoa as one of man's main beverages. As for the efficacy of tea, much was written in Chinese ancient books. *Shen Nong Ben Cao Jing* (*Shen Nong's Materia Medica*) states for example that when Shen Nong was poisoned trying various herbs he was "easily relieved by tea as an antidote", and that "tea makes one less desirous of sleep, strong and exhilarated". Hua Tuo, a celebrated physician and surgeon who died in A.D. 208, noted in his *Shi Lun* (*On Food*): "To drink bitter *tu* constantly makes one think better." Gu Yunqing of the Ming Dynasty writes in his *Cha Pu*:

> Drinking genuine tea aids in quenching thirst and in digestion, checks phlegm, wards off sleepiness, stimulates renal activity, improves eyesight and mental prowess, dispels boredom and dissolves greasy food. One cannot do without tea for a single day.

Li Shizhen writes in *Ben Cao Gang Mu* (*Compendium of Materia Medica*):

> Tea is bitter and cool . . . can best assuage internal heat which is the root of all diseases. One is in full possession of his faculties once internal heat is got rid of. . . . Internal heat is checked by the cooling effect of tea when it is drunk warm, and dispelled with hot tea.

Modern scientific research shows in fact that drinking tea dissipates internal heat, helps digestion, promotes the secretion of saliva, stimulates the kidneys, ameliorates certain diseases, refreshes one and calms the mind, counters fatigue and restores physical strength. Experience demonstrates the prompt exhilarating effect of a cup of hot tea after physical or mental labour. Its caffeine content stimulates the nerves, limbers up the muscles, facilitates muscular contraction and promotes metabolism. Even on a hot summer day a cup of hot tea is refreshing. After a feast, strong tea helps digestion. This is due to the essential oil in tea which dissolves fat. Experiments made on guinea pigs reveals two-thirds less fatty acid in the faeces of animals that have been given 10 ml. of tea after each meal than in the controls. It has also been shown that tea extract neutralizes acid poisoning due to its affinity to protein or fat. Heavy meat-eaters are known to say, "Better live a day without oil and salt than without tea." In short, modern science supports the views of ancient books concerning the efficacy of tea.

Tea also contains vitamins C and P, against scurvy and cerebral haemorrhage respectively, amino acids and minerals. The tannin in tea acts on protein and inhibits the growth of colon bacillus, streptococcus and pneumonia bacteria. It therefore has a beneficial effect in bacillary dysentery and even typhoid fever and cholera. Tea promotes the resilience of blood vessels and so counters arteriosclerosis. Tea as a beverage is considered by researchers in China as well as in other countries to ameliorate chronic nephritis, hepatitis and atomic radiation. Tea has often been used in China since ancient times in traditional medical prescriptions, while Songluocha, a tea produced in Anhui, is still used by practitioners of Chinese medicine in Jinan, Shandong Province. It is not without scientific basis that tea drinking has since ancient times been considered health-inducing and a remedy for disease.

Not only has tea been the traditional beverage of the Chinese people, it is very popular in other parts of the world as well, and long ago became China's main export product.

Chinese tea found its way to other Asian countries during the 5th century, and in the 17th it was transported to various countries in Europe and America where it was soon appreciated and became a popular beverage. The efficacy and use of tea spread from country to country, and the custom of drinking tea became increasingly prevalent throughout the world. Tea rose rapidly on the list of China's exports, reaching a peak of 134,000 tons in 1886 and ranking first among Chinese exports at that time.

Since liberation, Chinese tea has been exported to about 100 countries and regions of the five continents. In the interest of friendship with the Asian and African peoples, the Chinese government has been helped Mali, Guinea, Morocco and Afghanistan to plant Chinese tea on their soil. These tea bushes of friendship have grown and yielded good crops.

Two Celebrated Medical Works

Yu Yingao

Traditional Chinese medicine has a long and independent history, being the accumulated experience of the Chinese people in their millennia-long struggle against disease. China's treasury of medical works is not only voluminous and extensive. It is of valuable content, the works occupying an important place in the ancient and glorious civilization of the country. They have served as a media in spreading information collected by different schools of physicians for identifying, observing, analysing and treating a wide range of diseases. Statistics, unavoidably incomplete, give the number of these works as nearly 8,000. Most discuss clinical medicine.

Inscriptions on tortoise shells and bones excavated in the 13th-century B.C. ruins near Anyang, Henan Province, record parasitological diseases, tooth decay and various ailments. There is also a rudimentary classification of diseases according to location. Medical writings dating back to the 3rd and even the 8th century B.C. such as *Wu Shi Er Bing Fang* (*Prescriptions for Fifty-Two Diseases*), *Zu Bi Shi Yi Mai Jiu Jing* (*Eleven Channels for Moxibustion of the Arms and Feet*) and *Yin Yang Shi Yi Mai Jin Jing* (*Eleven Channels for Moxibustion in the Yin and Yang System*) were found among the books made of silk discovered in Tomb 3 excavated in 1973 near Mawangdui Village in Changsha, Hunan Province. Archaeological research has established the period from the 11th to the 3rd century B.C. as the time when writings on medicine first appeared. The partition of the country into feudal states with different political systems and calligraphic styles hampered the reproduction and propagation of these books. *Huang Di Nei Jing* (*The Yellow Emperor's Canon of Medicine*), also known simply as *Nei Jing* (*Canon of Medicine*),

337

and *Shang Han Za Bing Lun* (*Treatise on Febrile and Other Diseases*) are the two most important medical documents extant.

Canon of Medicine

Canon of Medicine, the earliest and most comprehensive medical classic from both the theoretical and clinical standpoints, was compiled around the 3rd century B.C. A compilation of treatises by different authors of different times, it consists of 18 volumes, half of which are entitled *Su Wen* (*Questions and Answers*) and the rest *Zhen Jing* (*Canon of Acupuncture*), renamed *Ling Shu* in the 7th century.

This collection, based on rudimentary materialism, stressed the fundamental theories of Chinese traditional medicine. It also discussed the maintenance of health, clinical symptoms, prescriptions, acupuncture and moxibustion. It was important in preparing for the future expansion and growth of the systematic theories of Chinese medical science.

The theory of *yin* and *yang*, a concept of the ancient Chinese spontaneous materialistic outlook and simple dialectics, appears in the *Canon of Medicine* in an attempt to explain the structure of the human body, physiological and pathological phenomena, and the laws governing the incidence and development of various diseases. It is the guiding principle in clinical diagnostics and therapeutics. The theory views the changes and developments of things in the light of the opposition, interdependence, mutual conflict and transposition and mutual status change observed in the relationships between opposites — the positive (*yang*) and the negative (*yin*). It affirms the relative balance and co-ordination of *yin* and *yang* in the human body as prerequisite to the maintenance of health, stating conversely that imbalance and inco-ordination of these result in disease. Fever, for example, may result from disproportionate increase in either *yin* or *yang*. In other words, its cause and pathology may be different. Overall analysis is necessary, taking into account the specific features of the fever and other clinical data. Overall analysis, a concept which arose in the era when *Canon of Medicine* was written and which has been developed since, remains a principal method in treating and studying different diseases.

In the unique traditional Chinese system of medical science, the visceral organs theory and that of the channels and collaterals are important branches dealing with physiological and pathological problems. *Canon of Medicine*, which provides extensive information on the visceral organs and the channels and collaterals, contains a highly significant discourse in a chapter entitled "Treatise on the Channels and Collaterals" in *Su Wen*, which states that food and drink ingested are absorbed by the stomach and other organs of the digestive system. The "lighter elements extracted from the food and water" are then distributed to the liver, while the heavier elements are sent up into the heart. The heart in turn delivers these heavier elements into the blood, which transports them to the lungs and to all parts of the body including the skin, hair and visceral organs. This narrative of the general and pulmonary circulations is mainly correct. *Su Wen* also contains a passage to the effect that "the heart is the main organ for blood circulation" and "blood flow is constant and circulative". It thus affirmed the relationship between the heart and the blood and its circulation.

In anatomy, a chapter on acupuncture titled "Jing Shui" says:

The method to use in examining a human body of average stature is to take measurements and feel the pulse if there is life. For a dead body, perform autopsy. The stiffness and size of every organ, contents of the digestive system, length of the blood vessels and the purity of the blood may be indicated in figures.

This passage suggests that even at that early time rudiments of pathological anatomy were known and anatomical data were valued in medical research. Another chapter entitled "Gastroenteric Study" gives the length of the digestive tract and every section of it from the pharynx to the rectum. The data largely agree with those provided by modern anatomy.

As for diagnostics, the contributors to *Canon of Medicine*

summarized diagnostic knowledge from the time of the celebrated physician Bian Que, (a pioneer in medical examination and pulse feeling) of the 5th century B.C., and further provided certain supplements and revisions. These contributors' discussions on pulse feeling mentioned feeling the temporal arteries on the head and anterior tibial arteries of the legs besides feeling the radial artery on the wrist, which is mainly practised today. The *Canon of Medicine* also summarizes experience in examination, which had been largely perfected. The necessity of combining pulse feeling with examination is stressed in order to avoid one-sidedness in diagnosis.

This ancient medical work deals with 311 diseases in 44 categories, including various fevers, sunstroke, malaria, cough, asthma, diarrhea, parasitic diseases, nephritis, icteric hepatitis, diabetes, mumps, various gastroenteric ailments, epistaxis and blood in the faeces, haematuria, haematemesis and other hemorrhages as well as pectoral angina, rheumatic arthritis, neurasthenia, mental diseases, epilepsy, leprosy, boils, hemorrhoids, thromboendarteritis, cervical tuberculous lymphadenitis, oesophageal tumours and a number of gynopathies as well as otolaryngological, ophthalmological and dental diseases. There are informative and penetrating discourses on causes, symptoms and treatment of various diseases. One is laryngeal disease, including oesophageal tumour in which the main symptom is that "eating and drinking are hardly possible" and that "whatever has been eaten will be brought up". Dealing with tuberculous lymphadenitis it points out that "the disease, apparent in the cervical region or the armpit, has its root in the viscera". This shows, quite correctly, the relationship between this condition and tuberculosis of the visceral organs. *Canon of Medicine*'s pathological discourses have provided valuable information for researchers of later generations.

In therapeutics, this brilliant medical text attaches great importance to the "treatment of potential diseases", or in modern terms, preventive medicine. In one chapter of *Su Wen* is an interesting comparison between treating a disease after its onset to digging a well only after one is thirsty. Another aspect of "treating a potential disease" is the prevention of aftereffects. An experienced doctor, it says, should effectively treat a disease at its first stage. "The best remedy is one applied before the sickness grows." Another feature is *Canon of Medicine*'s refutation of the superstitious. "It is meaningless to talk about medical science to those who adhere to their belief in the supernatural," declares that ancient book.

There are deep-going analyses based on the proposition that "treatment should be aimed at the root of the disease" as well as expositions of the relationship between radical and palliative treatment. *Canon of Medicine* discusses specifically a variety of treatments, including oral administration of drugs, diet, surgical intervention, moxibustion, acupuncture, massage and tapping of fluid. A detailed account of treating ascites by abdominal paracentesis, especially at so early a time, is noteworthy. This was performed by inserting a needle into the abdomen at a point called Guanyuan (Ren 4) 10 centimetres below the umbilicus, inducing the fluid with a stylet until a certain amount was released. The abdominal region was then tightly bandaged to avoid the patient's feeling suffocated due to the change in abdominal pressure. The surgery involved and the post-operative measures outlined in this paragraph shed light on the ingenuity and skill of Chinese physicians in ancient times.

Another chapter titled "Carbuncles" teaches that advanced thromboendarteritis can be treated by amputation of the blood vessel affected as an emergency measure to prevent spread to the upper arm or leg. It is thus evident that *Canon of Medicine* offers scientific and dialectical views on the prevention and treatment of various diseases and also a rich store of practical clinical experience. These certainly contributed to the subsequent development of medical science in China.

Treatise on Febrile and Other Diseases

Treatise on Febrile and Other Diseases, written by Zhang Zhongjing early in the 3rd century, was later re-edited and divided into two separate books, *Shang Han Lun* (*Treatise on Febrile Diseases*) and *Jin Gui Yao Lue* (*Jingui Collection of Prescriptions*). This work systematically summarized contemporary knowledge in diagnosing and treating the acute fevers and a number of other disorders. Zhang was an advocate of "elaborate study on the prescriptions available" and an enemy of witchcraft. In the preface of his book he advocates "diligently studying the ancient instructions and extensively collecting known prescriptions". And he followed his own advice. Zhang's seriousness, his respect for the achievements of his predecessors and his comprehensive summarization of medical prescriptions in folklore, plus his own experiments, made his book important in clinical practice.

The main value of this *Treatise* to the science of healing is as a dialectical diagnosis and compendium of tested treatments and prescriptions.

In diagnostics, Zhang proposed "four methods" — examination, auscultation, enquiry and pulse feeling. He classified the acute fevers into six groups named after channels and took the typical symptom of each group as basis for dialectical diagnosis. He monitored the development of disease and proceeded to grasp its variations and essence. This approach was later named "knowing diseases by the six channels". A rudimentary concept of the "eight keys" in diagnostics is apparent in Zhang's *Treatise*. Those "keys", or the different aspects in analysing diseases, were summed up as *yin* and *yang*, external and internal factors, fever and cold, deficiency and excess. Such dialectical principles and methods help to better understand the properties of disease, its penetrating power, various symptoms and the patient's resistance to it. They thus give a doctor systematic knowledge of a specific case.

Combining a variety of theories that appeared after the 3rd century B.C., when the *Canon of Medicine* was compiled, and the pathogenetic theories of the "channels and collaterals" on the one hand with the "four methods" and "eight keys" on the other, the author produced a general system of therapeutical methods now known as "the eight therapies". They are induction of perspiration for disorders over the surface of muscles, induction of vomiting for sluggish digestion, hiccups, oversecretion in the trachea and similar illnesses, purgative measures for ascites due to intestinal obstruction, mild treatment for ailments at a medium depth of the organism, warming measures for the effects of cold, purification for fevers, tonics for ailments arising from weakness, and dissolution for stagnation, swelling and similar disorders. These specific, comprehensive and systematic general laws of therapeutics constitute practical rules which can be applied separately or in combination to suit different symptoms and diseases. Zhang Zhongjing also proposed the approach of combining diagnostic theory, therapeutic laws and the prescription and administration of drugs in treating a specific condition. His principles and theories laid the foundation for the clinical therapeutics of the Chinese traditional school of medicine.

Treatise on Febrile and Other Diseases contains more than 300 prescriptions which are well-arranged and clearly drawn up for specific diseases. These prescriptions have been called "the forerunners of all prescriptions" or "classical prescriptions". It is agreed that some are of high clinical effectiveness and have served as basis for advances made in Chinese pharmacology.

Besides drugs for oral administration, this *Treatise* deals with acupuncture, moxibustion, scalding, local warming, administration of drugs by rubbing over the affected region, use of suppositories, baths, foot-baths, injecting water or air into the ears, and the sublingual administration of drugs. Here we may quote the book's directions for rescuing a person who has just hanged himself.

> The suicide should be cut down and laid out, then the noose removed. After the patient has been covered with a quilt to keep him warm, the rescuer should stand with a foot on each of the suicide's shoulders and straighten his neck by pulling him by the hair. Another man should then press the patient's chest repeatedly with his hands. A third person should massage his arms and legs, bending and stretching them alternately while gradually increasing the amplitude of the movements. Pressure should also be applied to the patient's abdomen. . . . Cinnamon decoction or porridge may be offered and observation made to see whether the patient can swallow. If so, air should be injected

into the suicide's ears.

The above is indeed a concise and specific introduction to a co-ordinated first-aid measure which conforms to modern scientific principles.

Treatise on Febrile and Other Diseases was the first medical work providing an overall expatiation on the whole process of making combined analysis of the symptoms by every available diagnostic means and proceeding to find the correct treatment in accordance with the therapeutic principles of traditional Chinese medicine. The process, accepted by later generations as a guiding principle in therapeutics, exemplifies the dialectical approach characteristic of traditional Chinese medicine.

Canon of Medicine was one of a number of classic medical works in existence before the 3rd century. *Shen Nong Ben Cao Jing* (*Shen Nong's Materia Medica*), which is China's earliest systematic pharmacological discourse, and *Nan Jing* (*Classic of Difficulty*), which discusses the medical theories and a number of diseases besides acupuncture and acupuncture points, are two other influential medical books that have come down to us today. After the 3rd century more medical writings appeared covering a wide range of subjects — general medical theory, physiology, pathology, diagnostics, pharmacology, moxibustion and acupuncture. There were also collections of prescriptions and clinical treatises on such particular diseases as leprosy and tuberculosis. To these were added medical encyclopaedias and works on hygiene, nursing, massage, surgical treatment, forensic medicine, veterinary medicine etc. All contributed to improving the health of mankind.

Acupuncture and Moxibustion

Ma   Jixing

Acupuncture and moxibustion are singular therapies developed by the ancient Chinese. No drugs are needed in these treatments. The curative effect is obtained simply by puncturing or applying heat to certain "points" of the human body.

Acupuncture originated in the New Stone Age out of the practice of giving treatment by stimulating various parts of the body with stone slivers called *bian*. Moxibustion, the burning of moxa near the skin, came into use at roughly the same time. Metallic pins began to be employed in acupuncture in the 8th century B.C. Gold pins designed for this therapy were recently excavated in a tomb attributed to the 3rd century B.C. in Mancheng County, Hebei Province. It is thus evident that acupuncture and moxibustion have occupied important places among therapeutic techniques in traditional Chinese medicine for several millennia.

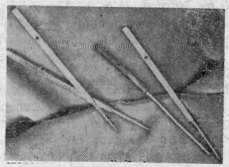

Gold acupuncture needles of the 3rd century B.C. excavated in Mancheng County, Hebei Province.

The advantages of acupuncture and moxibustion may be summed up as follows:

1. They are applicable in a wide variety of diseases and suit therapeutic needs in internal medicine, surgery, gynaecology, paediatrics, ophthalmology and otolaryngology, as well as in preventive medicine.

2. They produce prompt and appreciable cures, increase resistance to disease by providing stimulation at certain points, and alleviate or stop pain.

3. They are easily practised and mastered.

4. They involve little expense.

5. They produce no or slight side-effects, are generally safe, and can be employed together with other therapies.

History of Acupuncture and Moxibustion

Archaeological findings from a 3rd-century B.C. tomb near Mawangdui Village in Hunan Province provide convincing evidence that Chinese physicians had systematized clinical experience in both acupuncture and moxibustion more than 20 centuries ago. Excavated in 1973, this tomb revealed a series of medical works. Two of these entitled *Zu Bi Shi Yi Mai Jiu Jing* (*Eleven Channels for Moxibustion of the Arms and Feet*) and *Yin Yang Shi Yi Mai Jin Jing* (*Eleven Channels for Moxibustion in the Yin and Yang System*) discourse respectively on pain, spasm, numbness and swelling that may occur along the channels, mouth and sense-organ symptoms as well as symptoms of vexation, cold and drowiness, which are all amenable to treatment by moxibustion.

Later, in the 3rd century B.C., *Nei Jing* (*Canon of Medicine*) dealt in various ways with diseases curable by acupuncture and moxibustion and exemplified the application of these therapies to various visceral diseases, fevers, malaria and carbuncle. The *Canon* also offered detailed discourses on certain techniques in these treatments. Examples are: re-enforcing and reducing by manipulation, and puncturing the corresponding points on the left and right sides of the body alternately.

Among the physicians who specialized in acupuncture and moxibustion was Bian Que whose biography appears in *Shi Ji* (*Records of the Historian*). This doctor is said to have arrived in the state of Guo in what is now Baoji in Shaanxi Province at a time when the crown prince of that state had just expired. Bian Que rushed with his apprentices to the royal palace, enquired concerning the prince's symptoms and pronounced his "death" reversible. The king was informed and lost no time in requesting Bian Que's help in saving his son. The physician examined and felt the pulse of the patient, enquired as to further symptoms and concluded that the prince was in a coma or a state of shock, and not dead at all. Acupuncture and moxibustion were included in the rescue, which resulted in the coming round and full recovery of the patient. News of Bian Que's remarkable "resurrecting of the dead" spread, the report indicating not only wide application of acupuncture and moxibustion between the 5th and 3rd centuries B.C. but that the methods were considerably improved.

Two medical works, *Huang Di Ming Tang Jing* (*The Yellow Emperor's Classic on Acupuncture and Moxibustion*) written towards the end of the 3rd century B.C. and *Zhen Jiu Jia Yi Jing* (*A Classic of Acupuncture and Moxibustion*) compiled in the 3rd century A.D., gave more reliable and comprehensive information on the experiences gained in treating patients by these therapies. These works provided more systematic clarification on selection of "points" in treating different diseases as well as the therapeutic properties of the points. Besides these books, on which the development of these therapies in succeeding centuries was based, there were a number of other important medical writings illustrated to show the points for acupuncture or moxibustion.

During the period from the 4th to the 10th century, works dealing with these therapeutic techniques grew not only in number but also in variety. These centuries also saw the publication of coloured charts and diagrams for acupuncture and moxibustion, special books on moxibustion, and writings on the veterinary application of these techniques. Sun Simiao (581-682) and Wang Tao (702-

772) were celebrated physicians whose works *Qian Jin Yao Fang* (*The Thousand Golden Formulae*) and *Wai Tai Mi Yao* (*Medical Secrets Held by an Official*) especially emphasized acupuncture and moxibustion. Sun drew three large-size coloured charts showing the anterior, posterior and lateral views of the body. The 12 channels were marked out in coloured lines, the eight extra channels in green. Wang Tao's charts numbered 12, the lines representing the regular and extra channels also in different colours. At that time acupuncture and moxibustion were officially recognized as courses in the curriculum of the imperial medical college, while *Canon of Medicine* and *The Yellow Emperor's Classic on Acupuncture and Moxibustion* were among the textbooks selected. Such titles as master, assistant master, lecturer, technician and apprentice of acupuncture were accorded physicians of the Imperial Medical Bureau.

An even greater number of writings on acupuncture and moxibustion were published from the 7th century to modern times. Best known among these are *Tong Ren Yu Xue Zhen Jiu Tu Jing* (*Illustrated Manual on the Points for Acupuncture and Moxibustion on the Bronze Figure*) compiled under the supervision of the imperial physician Wang Weiyi of the 11th century, and *Zhen Jiu Da Cheng* (*Compendium of Acupuncture and Moxibustion*) by Yang Jizhou of the 16th century.

When Wang Weiyi was preparing his *Illustrated Manual* in 1027 he headed a group that cast two bronze figures inscribed with the acupuncture channels and points. Besides serving as teaching materials, the figures were used in examinations given to students of acupuncture. The examinee was asked to puncture the figure, which was filled with mercury, coated over with wax, and clothed. The accuracy of the student's puncturing was easily determined by whether or not the mercury leaked.

Such bronze figures proliferated, some cast by imperial hospitals, others by civilian doctors or apothecaries. Many of these valuable figures were destroyed in wars, while some were seized by imperialist powers in their aggressive military campaigns against China. The one cast by the Imperial Hospital towards the middle of the 15th century was among looted items taken out of the country by the

Russian army that invaded China in 1900 along with seven other imperialist powers. This invaluable bronze figure is still being held by the Leningrad Museum in the USSR.

Improvements in the materials and technique used in acupuncture and moxibustion over the centuries include red-hot needling, warm needling and plum-blossom needling. Moxibustion was enriched and refined by applying heat from burning cones of drug or rush and the use of "moxa rolls".

The Theory of Channels and Collaterals

The effectiveness of acupuncture and moxibustion in treating a wide range of diseases depends among other factors on the nature and intensity of the stimulation produced by puncturing or applying heat. It also depends on the point selected, and the organism's power to transmit the stimulation resulting from the acupuncture or moxibustion. The distinctly Chinese traditional theory of channels and collaterals elucidates this. Long experience in treating disease lies behind and substantiates the theory.

Named in the 3rd-century B.C. medical books found in the tomb near Mawangdui Village are the "tooth channel", "ear channel" and "shoulder channel", all taken from the main transmission route of the stimulation produced by acupuncture or moxibustion at a certain point on one of the channels. This nomenclature was the forerunner of the theory of channels and collaterals. The authors of such ancient medical works as *Eleven Channels for Moxibustion of the Arms and Feet* pioneered in dealing systematically with the channels and numbering them as 11. They went further to rename and classify the channels, basing themselves on the theory of the upper and lower extremities of the body and also the theory of *yin* and *yang*.

Canon of Medicine carried these researches to a new level by establishing the number of channels as 12. It also modified the discourses on the specific route of each channel and its diagnostic and therapeutic significance. The *Canon* thus laid down the scientific principles of acupuncture and moxibustion by developing the theory of channels and collaterals.

The main features of the theory of channels and collaterals are the view that these routes exist over the entire human body, and their function as transmitters of vital energy and co-ordinators of the different parts of the body. The whole network of channels and collaterals reaching every part of the body extend deep into the visceral organs. There is circulation along these routes. Channels are the trunks, while the collaterals and the sub-collaterals are their branches and sub-branches. Besides this system of 12 channels there are eight extra channels.

Physicians named the acupuncture points over the body on the basis of the theory of channels and collaterals. The authors of *Canon of Medicine* located all these points as along the channels. *A Classic of Acupuncture and Moxibustion* set the total of these points at 654.

Chinese physicians of later generations steadily clarified the therapeutic value of each point and the relationship between points and the visceral organs. Improvement continued in therapeutic effectiveness of acupuncture and moxibustion with the identification of additional points.

The Spread of Acupuncture and Moxibustion from China

Playing an important role in the advance of Chinese medical science, acupuncture and moxibustion have also served to promote medical and health work world-wide.

China's friendly trade relations and cultural exchanges with Korea, Japan and the southeastern and central Asian countries date from the 3rd century B.C. Chinese medical science, essentially acupuncture and moxibustion, were introduced into those countries at that time and won recognition by both rulers and people. The Chinese doctor Yang Er went to Japan as a professor of medicine in A.D. 513, while Zhi Cong took medical writings and acupuncture and moxibustion diagrams to Japan when he went as a doctor in 550. In 552 the Emperor Wen Di of the Liang Dynasty presented *Zhen Jing* (*Canon of Acupuncture*) to the Japanese court. This was followed by visits to China by Japanese students of medical science including acupuncture and moxibustion. The Taihō Code promulgated by the Imperial Government of Japan in 701 stipulated that medical institutes include compulsory courses based on *The Yellow Emperor's Classic on Acupuncture and Moxibustion* and *A Classic of Acupuncture and Moxibustion*. Measures were appended to ensure enforcement of this stipulation. There thus grew up in Japan a circle of Japanese physicians and writers specializing in these methods, and institutes of acupuncture and moxibustion were founded.

In what is now Korea the ancient kingdoms of Silla, Paekche and Koguryo adopted a civil examination system comparable to China's between the 7th and 10th centuries, making *Canon of Acupuncture*, *The Yellow Emperor's Classic on Acupuncture and Moxibustion* and *A Classic of Acupuncture and Moxibustion* compulsory reading for medical students.

The development of navigation after the 10th century favoured China's trade and other exchanges with Africa and Europe. Acupuncture and moxibustion were among the Chinese techniques taken to those parts of the world. English, French, German, Dutch and Austrian physicians took up these techniques in their clinical practice and research. Textbooks on these branches of medical science were translated from Chinese into a number of other languages.

Achievements in Ancient Chinese Pharmacology

Cai Jingfeng

Chinese pharmacology, created by generations of physicians through their experiences, is an important branch of traditional Chinese medicine.

History

There is a legend that tells of an emperor, Shen Nong, tasting herbs in his quest for medicines. The reign of that emperor, believed to be a specific period in the New Stone Age, saw the beginnings of primitive farming. People began to study and understand the properties of crops and other plants and came to know the pharmacological functions of some of them. The "tasting" of herbs by that legendary emperor should be taken as an ancient version of the story of research workers personally investigating the therapeutic effects of herbal medicines. A number of herbs, Asiatic plantain, fritillary, motherwort and others, are mentioned in *Shi Jing* (*Book of Odes*). *Shan Hai Jing* (*Classic of Mountains and Rivers*) written over 2,000 years ago named a total of 120 drugs of vegetable, animal and mineral origin and described their effects in treating and preventing diseases as well as their processing and administration. Though further archaeological research has yet to be done to identify many of the drugs named, such ancient works revealed the progress of pharmacological studies at that time.

Around the 2nd century B.C. *Shen Nong Ben Cao Jing* (*Shen Nong's Materia Medica*) appeared as China's earliest pharmacological work. It dealt with 365 drugs divided into three categories. This *Materia Medica* discussed at length the geographical origin, properties, collection and therapeutic value of each medicinal herb. It also told how to prescribe, administer and process each. Special mention should be made of the fact that the ancient Chinese found through long clinical experience specific medicines for certain diseases, such as ephedra for cough and asthma, rhubarb as a purgative and dichroa root for malaria. The effectiveness of these specific medicines has been confirmed by modern pharmacologists.

By the 6th century, however, *Shen Nong's Materia Medica* had fallen behind clinical needs. To fill this gap, the renowned physician of that time, Tao Hongjing, comprehensively reviewed all the known drugs, relying on the empirical knowledge so far gained. He produced his *Ben Cao Jing Ji Zhu* (*Commentaries on Materia Medica*) in which he described 730 medicines, or twice the number dealt with in the original work. He also improved on the classification system of the drugs, dividing them into seven groups: herbs, arboreal plants, cereals, animal products, minerals, garden products and those with unidentified effects. His classification, which went far beyond that of his predecessors who considered only the toxicity of the drugs, was the standard for physicians and research workers during the following 10 centuries. Tao also pioneered in listing the drugs according to their respective therapeutic action, placing medicines that relieve rheumatic pains and colds as *fangfeng* (*Saposhnikovia divaricata*), large-leaved gentian, *fangji* (*Stephania tetrandra*) and *duhuo* (*Angelica grosseserrata*) in one category. Such classification better suited therapeutic needs and facilitated the advance of medical science and pharmacology.

The Tang Dynasty (618-907) was the golden period of Chinese feudal society, with feudal culture developing tremendously. A pharmacological compilation sponsored by the imperial court was published in A.D. 659. It listed 844 medicines in nine categories with illustrations based on specimens collected in different parts of China. This great work, *Xin Xiu Ben Cao* (*Revised Materia Medica*), summed up the pharmacological knowledge accumulated through more than 10 centuries and was China's first pharmacopoeia. Its publication stimulated the later development of pharmacological science by standardizing the nomenclature of the medicines and revising their descriptions.

The 16th century saw an unprecedented upsurge in Chinese pharmacology. Capitalism in embryo had already emerged in the national economy. Transactions with foreign countries were facilitated by improved means of transportation and development in mining, and the science of farming made it both necessary and possible to sum up anew pharmaceutical knowledge. The celebrated naturalist Li Shizhen (1518-1593) was the one to fulfil this task. Through long clinical practice and learning from the common people, Li completed his *Ben Cao Gang Mu* (*Compendium of Materia*

Ben Cao Gang Mu (*Compendium of Materia Medica*) by **Li Shizhen in different languages.**

Medica), a masterpiece known the world over. Completed in 1578, the 52-volume encyclopaedia contained not only descriptions of 1,892 medicines with illustrations but also 11,000 prescriptions in 16 different parts. It reviewed the discourses by earlier pharmacologists, making corrections in them, and gave serious materialistic refutation to the Taoist alchemists who preached the attainment of immortality by means of magic pills. This voluminous work touching upon many branches of natural science, including zoology, botany, chemistry, mineralogy, geology, agronomy, astronomy, geography, etc., gave strong stimulus to later generations. Having been translated, wholly or partly, into Japanese, English, German, French, Latin, Russian and some other languages, it enjoys recognition by medical men of many countries. It has been confirmed that when the English scientist Charles Darwin quoted from what he called "the ancient Chinese encyclopaedia" he referred chiefly to this great work. Its author, Li Shizhen, was indeed a brilliant world scientist of his time.

We have so far dealt with only a few typical works in the extensive treasure of ancient Chinese literature of *Materia Medica*, but even these suffice to reveal the long history and world significance of Chinese pharmacology.

Unique Content

Chinese pharmacology is unique in its theory. As an independent system, the theory is based on the knowledge of the diseases and of the natural properties and therapeutic effects of the medicines. Medicines are classified according to their "temperature" (cold, cool, warm and hot) and "tastes" (hot, bitter, salty, sour and sweet) as well as their nature of being "ascending" or "descending". The "cold" or "cool" medicines are for fevers, the "hot" or "warm" type to be applied in diseases with the symptom of weakness or low temperature, especially cold extremities. A medicine is referred to as "ascending" if it induces perspiration (like ephedra) or relieves the feeling of pressure at the rectum (like dahurian bugbane). Such herbs as lily magnolia and perilla whose flowers or leaves are used as medicine are generally "ascending" in the direction of their action, while seeds like those of trifoliate orange or minerals like red ochre are mostly "descending". Modern experiments and analysis show that such classifications are founded on clinical experience. The "bitter" and "cold" medicines coptis and the root of large-flowered skullcap, for example, have been confirmed as containing bacteriocidal and bacteriostatic substances and are thus antipyretic. And betelnut palm, a medicine often administered as a vermicide, is found to contain arecain which is an effective paralyser of parasites, especially tapeworms.

Chinese pharmacology is unique also in the process of preparing the medicines. There are the techniques of fluid treatment, heat treatment and fluid-heat treatment. Fluid treatment means soaking in wine or vinegar or washing in water. Frying, baking and singeing are various methods of heat treatment. Among the techniques of combined fluid-heat treatment are steaming and boiling. Medicinal herbs or other substances are processed for the purpose of detoxication, enhancing their action, changing their properties or easy administration. Thus the herbs monkshood and the tuber of pinellia, which are poisonous, are soaked in water with ginger and vitriol to be detoxicated without harming their therapeutic potency. Glutinous Rehmannia which is "cold" in nature and cures febrile diseases in the unprocessed state, becomes "warm" and hematinic after steaming and drying several times. Other herbs and substances are processed to be purged of extraneous matter or to be reformed into pills or tablets for preservation.

Another significant respect in which Chinese pharmacology is unique is its compound prescription and the use of different parts of herbs for different therapeutic purposes. Generally speaking, doctors specializing in traditional Chinese medicine prescribe several or even dozens of different items for a dose, all of which are in balance for co-ordinated action. Different ratios of ingredients in a prescription and different dosages yield different effects. Compound prescriptions were first found in Nei Jing (Canon of Medicine) compiled in the 3rd century B.C. Zhang Zhongjing of the 3rd century

A.D. offered a number of compound prescriptions in his *Shang Han Za Bing Lun* (*Treatise on Febrile and Other Diseases*). He prescribed cassia twigs with ephedra to induce perspiration in treating influenza or cold, and he used ephedra with almond and plaster for asthma and cough with high fever, while the same ephedra supported by the rhizome of large-headed atractylodes and ginger was given to patients suffering from swelling. Chinese angelica, as correctly indicated in Zhang's book, is a hematinic when the whole plant is used but its tips activate blood circulation. All these were discoveries made in long practice.

Different forms of processed medicine have been developed. The decoctions, pills and powders which we now use originated more than 2,000 years ago, when there were more than a dozen forms of medicine for internal and external administration.

Many specific medicines were found in ancient China. To those listed in *Shen Nong's Materia Medica* which we mentioned above may be added Java brucea for amoebic dysentery, chinaberry and stone-like omphalia for parasitic diseases, alga for goiter, animal liver for night-blindness, etc. The effectiveness of these specific substances has been confirmed by modern researchers.

What deserves special attention is the application in the 3rd century by the great surgeon Hua Tuo of an effervescing powder called *mafeisan* as an anaesthetic. Hua's initiative was of inestimable world significance.

Alchemy in ancient times also contributed to the advance in pharmacology, though that was not the aim set by the alchemists for themselves. Such alchemical products as mercury bichloride and mercuric oxide remain useful in surgery today.

Diagnosis by Pulse Feeling in Chinese Traditional Medicine

Ma Kanwen

To cure a disease a correct diagnosis must be made. Many diagnostic means based on advanced technology exist today, but in ancient times physicians had to depend on examining, inquiring, listening, smelling and feeling. Feeling the patient's pulse is an original diagnostic measure of Chinese physicians. Together with other diagnostic methods, pulse feeling is indispensable in obtaining the information necessary to differentiating the patient's symptoms and signs.

A Long History

Pulse feeling in China is based on centuries of clinical experience. A chapter in *Shi Ji* (*Records of the Historian*) written in 104-91 B.C. tells how the celebrated physician Bian Que, who lived in the 5th century B.C., employed especially the diagnostic method of pulse feeling. Sima Qian, author of the book, wrote: "Bian Que was the first to adopt pulse feeling in this country." This was challenged, however, by other evidence that pulse feeling was practised by physicians before Bian Que. The legendary doctors of prehistoric times Jiudai Ji and Guiyu Qu gave discourses on pulse feeling. *Nei Jing* (*Canon of Medicine*) compiled during the Warring States Period (475-221 B.C.) and *Nan Jing* (*Classic of Difficulty*) written later carry comprehensive discussions on this method. The books on silk entitled *Mai Fa* (*Methods of Pulse Feeling*) and *Yin Yang Mai Zheng Hou* (*Symptoms of the Yin and Yang Pulse Patterns*) found in the 2,100-year-old tomb excavated in 1973 near the village of Mawangdui in a suburb of Changsha, Hunan Province, also offer valuable information about the ancient diagnostic technique of pulse feeling. From these evidences we see that pulse feeling as a means of clinical examination was already well systematized into the content of Chinese medicine before the 5th century B.C.

Chunyu Yi (205-? B.C.), also called Cang Gong, another physician whose biography is included in *Records of the Historian*, studied pulse feeling for three years under Gongcheng Yangqing, who presented *Bian Que Mai Shu* (*Bian Que's Book on Pules*) to him. Case records of Chunyu Yi included in *Records of the Historian* show that he began his examination of a patient by feeling his pulse.

Zhang Zhongjing (150-219), an outstanding physician in the Eastern Han Dynasty, revealed in his *Shang Han Za Bing Lun* (*Treatise on Febrile and Other Diseases*) that pulse feeling was already fairly well developed in clinical examinations by his time. China's first book specially devoted to pulse feeling, however, did not appear until Wang Shuhe (180-270) compiled his *Mai Jing* (*Classic on Pulse*). This great physician described in detail 24 different patterns of pulse beat and their diagnostic significance. He also discussed various methods of taking the pulse, making it a more practical diagnostic measure. Further works on pulse followed, and physicians began to specialize in this branch of medicine. Notable among them was Li Shizhen (1518-93), who wrote among other books one entitled *Bin Hu Mai Xue* (*Bin Hu's Study on Pulse*). It is estimated that by the beginning of the present century more than 100 works had been written on the pulse. These chronicled the successive stages in the development of this aspect of medicine.

Factors to Be Considered in Pulse Feeling

A knowledge of anatomy and physiology are indispensable in pulse feeling. And in fact valuable references in these fields are found in ancient Chinese medical literature.

First to be considered in pulse feeling is the relation between the heart, blood and blood vessels. The pulse directly reflects the condition of the blood and the heart function. *Canon of Medicine* describes the blood and blood vessels as controlled by the heart, and the vessels as canals through which the blood flows. It also describes the heart and pulse as co-related. The pulse is described as keeping pace with the heart beat and stopping with cessation of blood circulation when the heart dies. It also points out that nourishment absorbed through the digestive system is transmitted to the liver from which it goes to the lungs by way of the heart and then returns to the heart, to be carried to all organs and tissues. *Canon of Medicine* likens the circulation of blood through the vessels to its coursing through an endless ring. Rudimentary as the theory was, it was highly significant at that early time, serving as a fairly scientific basis for the study of the pulse and its feeling.

The second consideration in pulse feeling is the rate of blood circulation, and physicians in ancient China paid due attention to this factor. *Classic of Difficulty* noted that blood travels three *cun*[1] on inhalation and another three *cun* on exhalation, a total of six *cun* in one breath. Inaccurate as it was, this observation recognized that the rate of blood flow is an important indicator of the condition of blood circulation. Modern clinical examination of patients with cardiovascular disease also usually includes measuring the time taken for a round in the systemic and pulmonary circulations.

The third consideration is the relation between breathing frequency and pulse rate. A passage in *Canon of Medicine* states that two pulse beats are felt when one inhales and another two follow when one exhales. Or, when an average of 18 breaths are taken per minute, there are 72 pulsations, a 1:4 ratio. The book also gives data on the normal frequency of pulse rate that conforms with modern physiological observations. *Canon* mentions too that disproportion between the frequency of breathing and that of circulation suggests abnormality. According to the ancient physicians, the ratio of 1:2 or, at the other extreme 1:6 pulsations for a breath

---

1 A *cun* was then about 2.29 centimetres.

indicate disease. As is known, the ratio between these frequencies reflects the co-ordination between heart and lung functions. In clinical practice we often see patients with fluctuations in pulse revealing arrhythmia arising from abnormal pulmonary circulation and lack of oxygen in arterial blood.

The ideal site for pulse feeling is the fourth consideration. There was a very long period of probing to find this. The earliest practice was feeling at all arteries where a pulse could be felt, including those on the head, at the neck and cheeks, as well as the radial arteries in the arms and the posterior tibial, dorsal pedal, popliteal and femoral arteries of the legs. These are the hypodermic arteries and those over the bones. It is mainly this type of pulse feeling that was dealt with in *Canon*. There is also the "three-site pulse feeling," which is done at the temporal, radial and dorsal pedal arteries. The latter is practised in modern medicine and is generally termed radial-artery pulse feeling. It involves the radial artery at the radial styloid process only. Discourses on this particular type of pulse feeling can be found in such medical works as *Classic of Difficulty* and *Classic on Pulse* as well as *Canon of Medicine*.

Though rarely used, general pulse feeling has not been discarded. Because the pulse wave or blood pressure arising from the contraction of the heart is transmitted through the arteries to all parts of the body, fluctuation occurs in this pressure corresponding to alterations in the condition of the circulatory system and indicates heart function and the elasticity of the artery walls. Taking the pulse at more than one site therefore enables the doctor to obtain a more reliable and extensive basis for diagnosis. Local anaemia in the arms or legs, for example, resulting from defects at such main arteries as the aortic arch will make pulse feeling in the extremities difficult. Pathology of the blood vessels can therefore be affirmed by general pulse feeling, which is also useful in diagnosing cardiopathy or thrombotic arteritis.

Other factors involved in pulse feeling are listed in *Canon of Medicine* as time of day, recommending morning as best, when vital energies function in balance; seasonal changes; disposition, constitution, weight and mental state, pointing out that fear, worry and physical exertion cause fluctuation in pulse. Later medical writings mention such factors as sex, age and build. Modern physiology attributes various factors as affecting heart function, i.e. pulse rate and blood supply. Blood supply, for example, is usually stable after a night's rest. Affecting the volume of blood delivered by the heart are also sex, age, posture, activity and mental state. The conclusion can therefore be drawn that the Chinese physicians in ancient times made elaborate observations necessary in differentiating normal from abnormal pulses.

Clinical Significance of Pulse Feeling

Ancient Chinese medical practitioners were enabled by pulse feeling to know whether a disease was "cold" or "warm" in nature and whether the patient's vital energy was growing or declining. They were able also to determine the cause of a disease, the part of the body affected, and prognostic signs. *Canon of Medicine* points out that every physician should be skilled in pulse feeling as an aid in judging whether a disease is fatal, and if it is not, in drawing up his therapeutic plan so as to regulate the body energy and restore the patient to health. It also states that the pulse pattern indicates not only the visceral origin of a disease but also its basic cause. This formula stems from the concept that the human body should be viewed as a whole, based on a theory in Chinese traditional medicine that channels and collaterals in the body are routes along which vital energy circulates and which connect the visceral organs with the extremities, muscles, skin and joints, into an organic whole. The pulse, then, as part of the whole reflects physiological changes within the whole. *Canon*'s affirmation that "what exists inside will inevitably be detected from outside" summarizes this concept concisely.

Pulse is a significant indicator of the functioning of the circulatory system. It indicates whether or not the aortic valve of the heart is functioning normally and the beat is in proper rhythm, also the elasticity of the arteries. Moreover, the circulatory system is so closely related to the various visceral organs that any change in the metabolism of the tissues and any major abnormalities in the organism will affect blood circulation. The pulse pattern represents therefore not only changes in the circulatory system itself but also those in the other systems and individual organs. Fever and inflammation will produce an increased number of white blood corpuscles; hepatocarcinoma, diabetes and certain other diseases will cause structural change in the blood which in turn induces alteration in the pulse pattern. The nervous system is the most closely related to the circulatory. This is manifest in the blood vessels being subject to functional changes in the sympathetic and accessory sympathetic nerves, which control the blood vessel walls — changes that occur in a disease. The ancient Chinese physicians were on the right track to rely heavily on pulse feeling in diagnosis, as we can see, though they understandably did it quite empirically.

Pulse pattern, or characteristics of pulse as felt by the doctor, includes depth, rate, force, rhythm, etc. A healthy person's pulse is normal in these respects and is termed by Chinese physicians mean pulse. A sick person has a symptomatic pulse, varying in different diseases. Ancient Chinese practitioners did elaborate research into pulse patterns, more than a dozen different patterns being described in *Canon of Medicine*, 24 in *Classic on Pulse* and 30 or more in later medical works. Of special interest is the book entitled *Cha Bing Zhi Nan* (*Guide to Diagnosis*) written in 1241 by Shi Fa, which contains 33 diagrams symbolizing different pulse patterns. Modern diagrams of pulse patterns date only from 1860 when the French physician Etienne Jules Marey invented the sphygmograph, while centuries before that invention the Chinese doctors had drawn equally elaborate sphygmographic diagrams simply by feeling the pulse of the patients.

Of the two dozen or so pulse patterns generally listed in ancient medical works, we shall discuss the following.

The "floating" and "sinking" pulses, first mentioned in *Canon of Medicine* and elaborated on in *Classic of Difficulty* and *Treatise on Febrile and Other Diseases*, differ in depth — that is, the specific depth of the optimal site for feeling the pulse. A "floating" pulse is one easily felt when the doctor presses the patient's artery lightly but weakens when he presses it harder. This pattern usually suggests "external" disease, that is, conflict between the organism and external pathological factors, a phenomenon characteristic of certain diseases in their early stage. In physio-pathological language, such a pulse is in most cases due to weakening of the blood vessels' elasticity and resistance, as well as to relaxation of the radial artery resulting from increased blood supply from the heart and acceleration of blood circulation. A "sinking" pulse is one felt only when the artery is pressed fairly forcibly and indicates "internal" feebleness. Resulting from decreased blood supply, low blood pressure, little blood passing through the artery endings and stronger resistance to circulation from the blood vessel walls, this pattern may be taken as a symptom of heart disease.

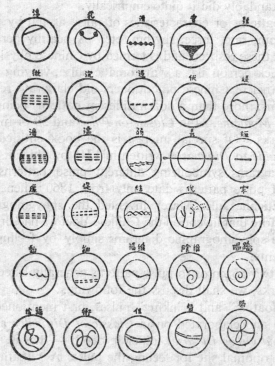

Diagrams of various pulse patterns.

Pulse may be felt as "retarded" or "rapid". First mentioned in *Canon of Medicine* and then in other classics, the concept refers to the pulse rate. A pulse rate of less than 60 beats per minute is referred to as retarded and points to a disease "cold" in nature. Modern researchers attribute such abnormality in blood circulation to over-excitement of the vagus nerve and obstruction in atrium-ventricle transmission. A disorder "warm" in nature, or hyper-function of the organism, is indicated by rapid pulse, over 90 beats per minute.

What *Canon of Medicine* calls "intermittent" pulse is referred to in modern diagnostics as bigeminal, trigeminal or quadrigeminal pulse. This is characterized by several slow beats followed by an interruption of equal length and suggests obstruction in metabolism (resulting from heart disease), heart failure or crisis in the function of the organism. *Canon* notes that such a pulse pattern indicates feebleness of visceral organs, an observation not remote from modern theories.

The concept of "slipped" or "rough" pulse refers to abnormal fluctuation in rhythm. A "slippery" pulse, first mentioned in *Canon of Medicine*, was described in *Classic on Pulse* as "fluid", while the celebrated physician Sun Simiao (581-682) compared it to "a stream of pearls". "Slippery" pulse indicates hyperthyroidism, arteriosclerosis or other diseases with such symptoms as accelerated metabolism and vascular expansion-contraction, as well as over-fluency of the blood. This occurs normally after the third month of pregnancy, suggesting increased volume and supply of blood. Chinese doctors long ago felt the pulse as a pregnancy test.

A "rough" pulse is described in *Canon of Medicine* as "staggering along a rugged road" and "symptom of heart pain". This may indicate either a lack of vital energy, or circulatory obstruction arising from excitement, indigestion or excess secretion in the respiratory tract. Anaemia, hemorrhage, or heart dysfunction which decreases the supply and retards the flow of blood usually cause this "rough" pulse pattern.

Ancient Chinese physicians summarized their experiences in pulse feeling:

A case of intoxication is easily curable when the patient's pulse is full, critical when it is weak; a case of obstruction or stasis in blood circulation in the abdomen is easily curable when the patient's pulse is smooth, critical when it is feeble; a case of apoplexy is easily curable when the patient's pulse is floating and retarded, critical when it is full and rapid.

They also listed various patterns of pulse indicating various diseases. All their records remain valuable in modern medicine, for ascertaining the cause, pattern, trend and aftermath of a disease.

Pulse feeling is not to be taken as a total diagnostic means, however. The ancient medical works were correct in stressing that pulse feeling should be practised in co-ordination with examination, auscultation, smelling and enquiring.

Pulse Feeling Across China's Borders

There has been a very long history of the Chinese skill of pulse feeling introduced into various foreign countries. Such medical classics as *Canon of Medicine* and *Classic on Pulse* containing discourses on pulse feeling were taken to Japan and other neighbour countries of China as early as the beginning of the 7th century. According to sources quoted in *History of Japanese Medicine* by Fukuji Fujikawa, the Japanese Buddhist monks E-Nichi and Fukuyin who had come to China as students carried Chinese medical works with them when they returned in 632. In 702 the Japanese government in its Decree on Medicine listed *Classic on Pulse* among medical textbooks.

Researchers claim that Chinese impact is evident in the voluminous Arabic work *Canon* by the celebrated physician Abu Ali al-Husain ibn Abdula ibn Sina (Avicenna) with respect to knowledge on pulse.

The Chinese skill of pulse feeling was introduced into Persia in the 14th century when an encyclopaedia published there included discourses on Chinese medicine and particularly pulse feeling. The

List of names of various pulse patterns in Arabic and Chinese.

work also mentioned the Chinese *Classic on Pulse* and its author Wang Shu Khu (Wang Shuhe). The 17th-century Polish physician Michael Boym, who had visited China, had published in 1686 his

Illustration on Chinese pulse feeling in a Persian medical book.

illustrated book entitled *Clavis medica ad Chinarum doctrinam de pulsibus* — a book which had first appeared in 1682 under the title *Andreas Cleyer*. Special mention should be made of the work of the noted English physician John Floyer (1649-1734), who did research on pulse, quite evidently using his knowledge of Chinese traditional medical science. He also invented an instrument with which the doctor could measure a patient's pulse, and wrote *The Physician's Pulse-Watch*, an essay explaining the old art of feeling the pulse and improving it with the aid of his pulse-watch. This was published in London in 1707. Floyer's invention and book are considered by Western scientists as of great historical significance.

The spread of Chinese knowledge on pulse continued on from the 17th century, with more than a dozen works in this field translated into a number of Western languages.

Surgery in Ancient China

Li Jingwei

Zhou Li (Rites of the Zhou Dynasty), written during the Warring States Period (475-221 B.C.), contains an entry on medical science classifying "the healing of wounds" as one branch, indicating the importance attached to surgery as a science even at that time. Extensive discussions on surgery are found in China's earliest medical work, *Nei Jing (Canon of Medicine)*, written in the 3rd century B.C. from which time China has contributed no small amount in surgery research, literature and practice.

Abdominal Operation and Anaesthesia

In the 2nd century A.D. there was a doctor in China, Hua Tuo, known for his surgical operations. This is described in *Hou Han Shu (History of the Later Han Dynasty)* as follows: When a disease had been confirmed as affecting a visceral organ and neither orally administered drugs nor acupuncture had proved effective, Hua Tuo gave the patient a dose of the herbal preparation *mafeisan* dissolved in wine and the patient became unconscious. This early surgeon then opened the patient's abdominal cavity and removed the "accumulated mass", which might very well have been a tumour. Hua treated a patient suffering from gastroenteric disease by opening the stomach or intestine, taking out the "dirty substance", cleaning and suturing the wound and applying a plaster. The wound would close in a few days and the patient return to normal life within a month. These accounts indicate that Hua successfully removed abdominal tumours and also resected the stomach and intestines. Present-day surgeons will appreciate that this was no small feat in those days.

Anaesthesia is a leading factor in the safety and success of a major operation. Hua's success 1,700 years ago was due to his ability and skill in inducing narcosis. The *mafeisan* dissolved in wine was a great discovery. We know that alcohol is in itself a narcotic to which surgeons in modern times may have to resort. Unfortunately, the composition of *mafeisan* has long been lost. Modern research workers have deduced that it probably was similar to *shuishengsan, caowusan* or *menghanyao* described by Dou Cai of the 12th century, Wei Yilin of the early 14th century and Li Shizhen of the 16th century, who were all surgeons. Dou Cai in his book *Bian Que Xin Shu* (*My Understanding of Bian Que*) published in 1146 mentions the narcotic *shuishengsan* and advises its application before moxibustion, which might cause a local burn. The active component of this narcotic is believed to be the flower of datura (*Datura innoxia*). The same substance was used by Wei Yilin as the active ingredient in his narcotic, which he called *caowusan*, for bone-setting operations. The Japanese surgeon Hanaoka Seishu is credited with being the first to use the datura flower as the active ingredient of a narcotic in 1805. He was in fact anticipated by the Chinese surgeons mentioned above.

Satisfactory results were achieved in recent experiments made by Chinese surgeons who tested the effectiveness of datura flowers as a narcotic. The experiment showed this narcotic to be not only reliable and safe. It also proved valuable in reducing the danger of shock and infection. These properties are significant, particularly the former, as shock is a known menace in many cases.

Instances of surgical operation under anaesthesia as performed by Hua Tuo were to be found elsewhere in Chinese history. *Zhu Bing Yuan Hou Lun* (*Causes and Symptoms of Diseases*) written by Chao Yuanfang in the 7th century and *Shi Yi De Xiao Fang* (*Tested Prescriptions of Veteran Physicians*) compiled by Wei Yilin in the 14th century report anastomosing ruptured intestines. Prominent surgeons like Fang Gan who lived between the 8th and 9th centuries and Gu Shicheng of the 19th century were known for their success in the surgical repair of harelip. Wang Kentang (1549-1613) and Chen Shigong (1555-1636) were reported to have successfully rejoined severed ears and mended the trachea.

Treating Fractures and Dislocations

Ancient documents report as part of the pre-3rd century B.C. medical system the treatment by surgeons of fractures and dislocations. And in a "list of the wounded" in military camps of the 2nd century are descriptions of fracture cases, evidence that the art of dealing with fractures was maturing at that early date.

Xian Shou Li Shang Xu Duan Mi Fang (*Secret Healing of Wounds, Fractures and Dislocations*) published in 841 is the earliest work in China on these subjects. It lists as necessary 10 measures to be taken in treating a fracture or dislocation, which include methods of washing and examining the wound, applying skeletal traction and reduction, using splints, and administering drugs. Followed by later surgeons as guiding principles, these measures continue useful in combined treatment of fractures or dislocations by doctors of the Western and traditional Chinese schools. A passage in *Secret Healing* on complicated fracture points out:

When a broken bone has pierced the muscle and skin, or simple reduction operations have failed, the surgeon may scrape off the point of the broken bone with a sharp knife or excise the muscle in order to expose the fracture before reducing it.

This technique is still used today.

The treatment in the 14th century of spinal fracture by Wei Yilin is revealing on the level of surgical experience at that time. The principles outlined by this surgeon agree largely with modern theory. His use of large pieces of mulberry bark for fixation compares favourably with the Western technique of using plaster.

Sun Simiao, celebrated doctor of the 7th century, was a pioneer

An illustration in the ancient medical work
Yi Zong Jin Jian (*Golden Mirror of Medicine*),
showing the treatment for spinal fracture.

in reducing mandible dislocation by a method quite in line with modern rules. More comprehensive study on such reductions was made in the 17th century by the surgeon Chen Shigong. Later, in the 18th century, Chinese surgeons produced a series of illustrations depicting various reduction operations, the most notable of which are those on treating dislocation of the shoulder and hip joints.

Early in the 19th century the Chinese orthopaedist Jiang Kaoqing performed bone grafting. His specific operation was to transplant a piece from one of the patient's healthy bones to the fissure in a comminuted fracture. This provision of a bridge for quicker healing is a technique frequently used in present-day military surgery. Doctor Luo Tianpeng of the same period invented a "rocking bed" resembling the modern swing bed for facilitating the nursing of pa-

An illustration in the 18th-century medical work *Shang Ke Hui Zuan* (*Collected Treatises on Surgery*) showing the treatment for dislocation of the mandible.

tients with spasm of the extremities or serious fractures and also for preventing secondary disorders.

Extraction of Cataract

Cataract is an ocular disease with comparatively high incidence in China in the area of Guangxi and Fujian Province. Long experience in treating cataract led to its extraction with a needle as the most notable cure, a technique initiated in the 5th century. Wang Tao (702-772) in his book *Wai Tai Mi Yao* (*Medical Secrets Held by an Official*) published in 752 presented the earliest clinical description of cataract and distinguished between the congenital and traumatic types. He also gave the first systematic discourse on the operation of extracting cataract with a needle, an operation that had become so popular by the 16th and 17th centuries that older women practising healing arts were performing it successfully. Simple and not requiring complicated equipment, the method has been improved in the recent two decades to become a new technique of couching performed with only a needle and a tube. The method enables the doctor to separate the opaque lens and also to remove it. It is superior to other techniques in involving a simpler process, less time of operation, smaller incision and less pain.

Minor Operations

Described below are a few interesting examples of minor operations performed in ancient China.

1. Excision of nasal polyp. Though not fatal, nasal polyps cause obstruction and distortion in pronunciation and even deformation of the nose. Dr. Chen Shigong in his *Wai Ke Zheng Zong* (*Orthodox Manual of Surgery*) published in 1617 describes an instrument invented by himself for the excision of such new growths. It was a pair of brass wires with a hole at one end of each through which fine silk thread was drawn. After applying a few drops of narcotic to the patient's nostril, he inserted the wires to the base of the polyp, leaving a small space between them. He then tightly encircled the polyp with the silk thread attached to the end of the wires. A single quick pull removed the new growth.

2. A simple catheter. Obstruction to urination resulting from certain diseases causes pain and even acidosis due to accumulation of wastes in the organism, a condition which could be fatal. In the 7th century Dr. Sun Simiao used the tubular leaf of scallion as a catheter, truncating the leaf and inserting it gently into the patient's urethra to a depth of 10 centimetres. He blew into the tube at the other end to inflate it, whereupon urine flowed out through his "catheter". Sun recorded his successful experiment in a book entitled *Qian Jin Yao Fang* (*The Thousand Golden Formulae*).

3. Tapping. *Canon of Medicine* (3rd century B.C.) contains a passage on tapping the abdomen with a tubular needle, but the description is vague. Dr. Zhang Zihe (c. 1156-1228) invented a "leaking needle" to tap the scrotum, thoracic cavity and abdomen. The "tubular needle" or "leaking needle" was in fact the ancient form of the stylet for puncture.

4. Removal of foreign body in the larynx. Foreign body in the larynx, trachea or aesophagus is a common occurrence especially among children. Death may result when a foreign body in the trachea causes asphyxia. Various instruments were devised and different operations performed by Chinese doctors long ago to relieve such suffering. A 4th-century alchemist named Ge Hong initiated the removal of a needle in the throat by using a magnet in the patient's mouth. Surgeons stepped in and improved the method by reducing the size of the magnet to that of a date kernel, polishing it, attaching a string and having the patient swallow it. Thus a needle or nail that had been accidentally swallowed could be brought out along with the magnet when the doctor pulled the string even if the foreign body had found its way into the oesophagus. In the 13th century Dr. Zhang Zihe pioneered in a technique to remove a holed coin from the larynx or upper oesophagus. He made two paper tubes as thick as chopsticks, cut one tube at one end to give it a soft tuft and attached a tiny hook to the end of the other. He inserted the tubes down the patient's throat until he was sure he had caught the coin with the hook attached to the first tube. Then he wrapped the coin in the tuft at the end of the other tube. This way he brought out the foreign body without injuring the wall of the larynx or upper oesophagus. Though primitive and simple, this experiment made more than 700 years ago followed the same principle on which the modern aesophagoscope is based.

Concepts of Surgery

Chinese surgeons in ancient times stressed examining the patient as a whole, observing the changes innate in the entire patient's organism rather than focusing attention on the pathological site alone. Greater importance was attached to fortifying organic resistance to disease or infection than to surgical intervention. Discourses in *Canon of Medicine* point out the interactions between faulty blood circulation, drunkenness or overeating on the one hand and ulcer on the other. These also emphasize improving circulatory function and preventing the aggravation of inflammation in treatment. The author of *Secret Healing of Wounds, Fractures and Dislocations* dealt in detail with the importance of firmness in splinting and bandaging while at the same time allowing for movement of the part splinted or bandaged. Chen Ziming of the 12th century, Qi Dezhi of the 14th century and Wang Ji of the 16th century discussed at length in their respective books the necessity of medical measures in surgery. They noted that awareness of the internal condition is prerequisite to knowing an external disease, and that latent disease inevitably becomes apparent. These surgeons rejected the approach of taking into account only local and superficial pathological changes, an approach which they described as attending to trifles while neglecting essentials. Improving digestion and nutrition was also stressed by the 17th-century doctor Chen Shigong.

Ancient Chinese experts in treating fractures and dislocations agreed on proceeding from the whole-body approach. One wrote: "When there is an external disorder, one's vital energy and the normal circulation are harmed." "Failure of nutrition causes trouble in the visceral organs." Listed by various early surgeons as necessary in the healing of fractures and dislocations are movement to optimal extent and functional training accompanied by administration of medicine to improve blood circulation and organic function.

The World's Earliest Works on Forensic Medicine

Gao Mingxuan and Song Zhiqi

In the very early *Li Ji* (*Book of Rites*) appears a royal order to court judges that in examining criminal cases involving death they must base their judgement on careful inspection of the victim's wounds. Such examinations were the earliest form of medical jurisprudence. Data on an autopsy are contained in *Nei Jing* (*Canon of Medicine*), China's first systematic medical work compiled in the 3rd century B.C. In the 1st century, an imperial surgeon is reported to have performed an autopsy on a murder victim. The surgeon Wu Pu who lived in the early 3rd century examined the exhumed body of a sing-song girl's husband and drew the conclusion that echthotoxin in an old well had caused his death. There were no books devoted to forensic medicine, however, until the 10th century. China's earliest extant medico-legal work, *Yi Yu Ji* (*Difficult Cases*), was written by He Ning and his son He Meng in A.D. 951. *Nei Shu Lu* (*Clear Conscience*) by an unknown author, *Ping Yuan Lu* (*Wrongs Righted*) by Zhao Yizhai, *Jian Yan Ge Mu* (*Forensic Investigations*) by Zheng Xingyi, *Zhe Yu Gui Jian* (*Examples of Misjudged Cases*) by Zheng Ke and *Tang Yin Bi Shi* (*In the Shade of a Crabapple Tree*) by Gui Wanrong, all published between the 10th and 13th centuries, though scientific like *Difficult Cases*, lack necessary elaboration and systematization. It is *Xi Yuan Ji Lu* (*Manual of Forensic Medicine*) written by Song Ci in 1247 that merits the distinction of being called China's first systematic work on forensic medicine. It anticipated by more than 350 years the work by the Italian author Gortunato Fedeli in 1602, the earliest known Western publication on the subject.

Song Ci (1186-1249) who four times served as a high-ranking judge in criminal cases, has been described by historians as an investigative, serious and prudent examiner for his elaborate research and repeated examinations of major cases. He attended examinations by coroners, often helping to make tests. In one alleged suicide case he noticed that the knife with which the man, a peasant, was said to have killed himself was lying loosely in his hand, while the stab wound suggested that more force had been used in extracting the knife than in thrusting it in. These clues and careful investigation and analysis led to the verdict that murder had been committed by a gentryman who coveted the victim's comely wife. The murderer rearranged the murder scene and bribed some officials to declare the death a suicide, but the plot was exposed with ironclad evidence by Song Ci.

This judge, Song Ci, gained a wealth of experience and knowledge of forensic medicine through many examinations of corpses of victims killed in various ways. He also read widely on the subject besides seeking information from veteran judicial officials. His own work, begun in 1245, summarized the achievements of his predecessors. He wrote in the preface:

> It has often occurred to me that a wrong judgement usually results from erroneous investigation which is due to lack of experience. This belief led me to make extensive study of books on this subject among which *Nei Shu Lu* was the earliest written. I made revisions and supplements and thus produced the present book which I have named *Xi Yuan Ji Lu*. It is my hope that this collection will help in preventing wrong judgement and benefit innocent defendants who might otherwise be unjustly sentenced to death.

The book was published on imperial decree two years later and became a manual for judicial officials of succeeding generations.

Song Ci's *Manual*, which survives in editions of four and five volumes, is rich in content and extensive in scope. It contains elaborate discourses on anatomy, physiology, pathology, pharmacology, diagnostics and therapeutics. It explores emergency treatment and also goes deeply into medicine, surgery, gynaecology, paediatrics and orthopaedics. Song's researches largely coincide with modern forensic medicine not only in topics covered and approach, but in methodology as well. An example of the sound basis in physiology, pathology and general medicine of the *Manual* is its affirmation that red spots over the corpse are formed by the settling of blood due to the sudden stoppage of circulation and the argument posed on the relation between the cadaveric spots and the posture of the body. It is established today by forensic physicians that cadaveric spots do serve as an indicator of the cause of death. The book also noted that before posthumous examination the body must be bathed and the wounds cleansed with a decoction of vinegar, brewer's dregs and unripe gallnut to prevent infection and deformation of the wounds. This 13th-century medico-legal work points out that when typhus was the cause of death the corpse bears crimson spots, and when tetanus is the cause there are such signs as shrinkage of muscles at the mouth, eyes, hands and feet. First-aid measures similar to the modern technique of artificial respiration are outlined for a suicide victim by hanging.

The structure of the skeleton and function of its various members are described in detail, as well as methods of treating fresh wounds, including debridement. On first-aid for a snakebite victim, the author says, "It is necessary quickly to excise the part bitten with a sharp knife." The healing of fractures and fixation with the aid of splints are also discussed.

Also reported is his examination under a yellow umbrella of invisible bone wounds in a murder victim. The wounds became clearly visible. This finding by Song followed the optical principle of ultraviolet examination used today.

In the field of toxicology, Song's *Manual* lists poisonous elements including *Illicium lanceolatum*, croton, white arsenic, mercury, poisonous mushroom and the globefish. It goes into symptoms of poisoning and criteria for differentiating murder from suicide.

A passage from the book states: "The body of one killed by gas produced by coal in a stove is soft and bears no wound." Ob-

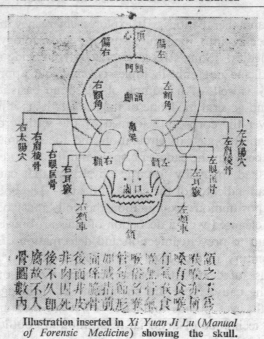

心頂
偏右   偏左
顖門
腦顱
右額角   左額角
顬顳
鼻梁
右太陽穴   左太陽穴
右顳後骨   左顳後骨
右眼匡骨   顴右   顴左   左眼匡骨
右耳竅   左耳竅
齒口
右頰車   左頰車
頜

骨腐後非後頄後有喉頜
圖故不肉而係或每俗各氣有咽之
數不久因井臨折面名官致食亦下
肉人即死皮肝前危整絕食喉頜骨

Illustration inserted in *Xi Yuan Ji Lu* (*Manual of Forensic Medicine*) **showing the skull.**

viously, the Chinese knew about carbon monoxide poisoning at this early date.

Another paragraph deals with detoxification:

> The venom of the viper or other poisonous snakes will penetrate and cause death within a short time. Quickly bind with string or a handkerchief the part bitten to prevent the venom's penetration. Let another person, holding vinegar or wine in his mouth, suck out the venom and spit it out. This is to be repeated until there are no longer signs of inflammation and swelling at the wound. The person who does the sucking must be cautioned against swallowing any venom.

Written more than 700 years ago, this paragraph is remarkable. Ligation of the artery leading to the part bitten is still done to prevent the poison from spreading. Sucking the venom out is of course dangerous, though less so given the precautions suggested. Against arsenic poisoning, Song wrote:

> The way to save one who has just taken white arsenic is to spoon-feed him with more than a dozen beaten eggs mixed with three *qian*[1] of alum. This feeding, which induces vomiting, should be repeated until the patient has brought up all the arsenic he has taken. This must be done before the arsenic has time to penetrate.

White arsenic, or arsenic trioxide, which passes easily into the blood by absorption through the stomach wall, causes quick death. The albumen in the eggs however, combines chemically with the arsenic forming a product that is solid and insoluble in water, while alum is emetic. Song's 13th-century prescription is therefore quite affirmed by modern medicine.

Also included in the *Manual* is the following test for poisoning by mouth:

> Take a silver hairpin, wash it clean in an infusion of honey-locust and place the silver end in the victim's throat while covering his mouth with paper. After some time remove the hairpin. If it is a dark blue that withstands washing, poison has been taken; otherwise not.

This method is effective in testing for poisoning by sulphides especially.

The Manual of Forensic Medicine remained useful from the 13th to the 19th century, and served as the chief source for later medico-legal works. These, with revisions or supplements to Song's book or with added material on further theoretical research, include *Wu Yuan Lu* (*Judicial Infallibility*) (1308) by Wang Yu, *Xi Yuan Lu Jian Shi* (*Comments on "Manual of Forensic Medicine"*) (mid-17th century) by Wang Kentang, *Xi Yuan Hui Bian* (*Examples of Redressed Cases*) (mid-17th century) by Zeng Shenzhai, *Xi Yuan Wu Ji Zheng* (*In Support of "Manual of Forensic Medicine"*) (1796)

---

[1] 1 *qian* = 0.05 gramme.

by Wang Youhuai, *Xi Yuan Lu Bian Zheng* (*Amendments to "Manual of Forensic Medicine"*) (1827) by Qu Zhongrong and *Bu Xi Yuan Lu* (*Supplement to "Manual of Forensic Medicine"*) (early 20th century) by Shen Jiaben.

Beyond the borders of China, Song's work has its Dutch version of 1862 and French version of 1908. The book has also been translated into Korean, Japanese, English, and Russian among other languages.

The Invention and Development of Printing and Its Dissemination Abroad

Xing Runchuan

Printing, an art which has had an important bearing on the development of human civilization, is known to have begun in China. The ingenuity of the ancient Chinese people is shown through this invention.

Origins of the Art of Printing

Before the printing press, the dissemination of culture depended mainly on hand-copied books. As the economy and culture developed, however, this meagre supply of hand-copied books could by no means satisfy the growing demand. A method of printing was urgently called for.

The inspiration for printing came from the very early use of seals and incised stone for making inscriptions. The seal, from which a positive impression is produced by engraving it in the negative, did print a limited number of words. Incised stone was an enlargement of the seal. The 10 stone drums dating from the state of Qin of the Warring States Period (475-221 B.C.) are the earliest inscription stones still existing. The earliest book printing was done from many of these stone engravings and provided textbooks for ancient scholars.

By the 4th century, rubbings were produced using paper on incised stone tablets. The paper, which was thin and strong, was moistened and spread on the stone. It was patted gently and repeatedly till it conformed to the carving. Ink was applied when the paper was dry. When removed from the tablet, the incised characters showed up in white against the black background produced by the ink. Thus positive copies were obtained from incised positive characters, and this was what led to the invention of block printing.

Wood-Block Printing

Sui Shu (History of the Sui Dynasty) and Bei Shi (History of the Northern Dynasties) contain references indicating the strong possibility that the art of wood-block printing was invented during the Sui Dynasty (581-618), or about 1,300 years ago.

The plate was generally carved from the wood of the date or pear tree, which was suitable for engraving. The text was first written on a sheet of thin, translucent paper, which was glued face downwards on the plate. The characters were then carved out in relief with a knife. For printing, the engraved plate was inked, a sheet of paper spread on it and this brushed over gently and evenly, imprinting the characters in the positive on the paper. Block printing was linked closely with the people's life and production from its beginning, when most of the books dealt with agriculture or medicine, or were almanacs or rubbings of calligraphy. After 762 these were printed by merchants and sold in the Tang Dynasty capital of Changan (now Xi'an in Shaanxi Province) at its East Market, the commercial centre of the city.

About 20 years later, "printed papers" appeared on the market. These served as bills of business transaction or tax receipts. In the preface written in the year 824 to the collected poetical works of Bai Juyi, another noted poet of the time, Yuan Zhen mentioned that printed copies of the book were exchanged for wine and tea. This marks the copying of popular poems by the printing process.

The almanac, used by the peasants to time farming operations and therefore in popular demand, was printed by wood blocks and available on the market in the areas of Sichuan and northern Jiangsu about the year 835. This came to the attention of Feng Su, the Military Governor of Dongchuan, who noted that privately printed almanacs "were widespread under Heaven". Now Feng Su considered this detrimental to the integrity and prestige of the emperor as "director of the people's activities in connection with the farming seasons", and he suggested to the emperor that the printing be prohibited.

This in itself points to the popularity of block printing already at that time, and though the Tang rulers decreed that printing be stopped, the craft continued. And when the peasant uprisings led by Huang Chao toppled the Tang Dynasty, the number of block-printed books increased greatly. Sichuan was then the centre of the block-printing trade. Buddhists were quick to utilize the invention of block printing to reproduce large numbers of Buddhist scriptures, Buddha portraits and other religious pictures.

Records say that Xuanzang, well-known Buddhist monk of the Tang Dynasty, used large amounts of paper for printing portraits of Buddha. In the year 1900 an exquisitely printed volume of *Diamond Sutra* was found in Thousand Buddha Grotto at Dunhuang, Gansu Province, bearing the date: "Fifteenth Day of the Fourth Moon in the Ninth Year of Xiantong" (A.D. 868). This is the world's earliest printed work known which bears a specific date of production. *Diamond Sutra* is in the form of a scroll more than five metres long, being made up of seven printed sheets glued together. Its frontispiece is a picture of Sakyamuni preaching Buddhist doctrine, followed by the full text. The printing is exquisite, indicating great mastery of the art of plate engraving. Both printing and calligraphy are highly refined though restrained. The clarity of the printing could only have been achieved with high-quality printing ink and expertise in its application. This relic of such perfection and immense value was, however, to the Chinese people's great indignation stolen and carried away to London in 1907 by Aurel Stein, an agent of the old imperialism.

During the period of the Five Dynasties and Ten Kingdoms (907-979) the cultural departments of the feudalist government block-printed large numbers of old classics. Private printing of books was likewise popular. Kaifeng in Henan Province became a printing centre, and also western Gansu, eastern Shandong, Nanjing in Jiangsu Province, and Fujian. The greatest numbers of printed books were turned out in the Sichuan and Zhejiang areas. The Song Dynasty (960-1127) saw further development in the craft of block printing and a zenith in its technique. The engraver Jiang Hui was outstanding among thousands of accomplished block printers. The products of Hangzhou in Zhejiang Province, Fujian and Sichuan were best in quality. The books printed during the Song Dynasty were both many and of excellent workmanship, making the Song editions of classics the most highly valued.

In the fourth year of the Kaibao period during the reign of the Emperor Tai Zu of Song (A.D. 971), Zhang Tuxin had the complete text of *Tripitaka* printed in Chengdu, Sichuan. This work of 1,076 volumes, 5,048 chapters, took 12 years to complete. Altogether 130,000 plates were engraved for this most voluminous of the early block-printed books.

Copper-plate printing, which arose after the Song Dynasty, was used mainly in the manufacture of banknotes. The highly complicated designs made by the very fine strokes on the copper plates made possible the printing of banknotes that were not easily counterfeited.

The most remarkable accomplishment in the subsequent development of block printing was the colour plate, which made its appearance in China in the 14th century at the latest. The world's earliest colour-printed book, *Notes on the Diamond Sutra*, produced at Zhongxinglu (now Jiangling County, Hubei Province) in the 14th century was done in red and black. The colour-block technique did not, however, come into popular use until the end of the 16th century. In the years of Wanli during the reign of the Emperor Shen Zong of the Ming Dynasty (1368-1644) a number of experts in colour-block printing attained fame, among them Min Qiji, Shan Zhaoming, Ling Ruxiang, Ling Mengchu and Ling Yingchu. The technique saw further development in the subsequent Qing Dynasty (1644-1911).

In co-ordination with woodcut art, colour-block printing brought forth exquisite reproductions of paintings. The best were included in *Jian Pu* (*Ornamental Letter-Paper*) and *Shi Zhu Zhai Hua Pu* (*Ten-Bamboo Studio Painter's Manual*) done towards the end of the Ming Dynasty. With perfection in the use of dark and light tones, these colour-prints brought their themes to life, and some of these ancient reproductions are indeed art treasures in themselves.

The Beginning and Development of Movable-Type Printing

Block printing had the enormous advantage over hand-copying of producing hundreds or even thousands of copies at a time. The carving of the wood plates was, however, a time-consuming task. Each page of a book required the carving of a plate, or block, and for a book of any length this took several years, while special store-rooms had to be provided for the preservation of the blocks. If another book was to be printed at the same time, another whole set-up and procedure would have to begin. The problems involved in time, skilled engravers and material may be imagined.

In the Song Dynasty, there lived in an area where block printing was in its prime a commoner by the name of Bi Sheng who for years engaged in block printing and introduced the world's first block-letter printing. Dispensing with the traditional process of plate engraving and so reducing the time required to print a book, the new method was economical as well as convenient. Revolutionizing the printing press and having far-reaching impact, Bi Sheng's invention is essentially the same as the contemporary block-letter printing with lead type as still widely used in today's world.

Bi Sheng's feat is described in *Meng Xi Bi Tan* (*Dream Stream Essays*) by Shen Kuo, an eminent scientist of the Song Dynasty. In the years 1041-48, according to Shen Kuo, Bi Sheng started making clay types, one for each character. These were fired for hardness. For typesetting, a square sheet of iron was prepared with a layer of resin, wax and paper ashes mixed and spread on it. The mixture was circumscribed with an iron frame. A plate was complete when the frame was full. This was heated over a fire until the mixture melted. The types meanwhile were pressed down to the height of the frame with a wooden board and the plate was ready for printing. For higher efficiency two iron sheets were used, one for fresh typesetting and the other for printing, so that a new plate was ready before the specified number of copies had been made from the previous one. Several duplicate types were made for each character, the number depending on the frequency of its use. As for rarely used characters, they were carved and fired when necessary and used on the spot. Bi Sheng's method had great merit, with its notable speed, when hundreds or thousands of copies were made.

In the reign of the Emperor Dao Guang (1821-1850) of the Qing Dynasty there lived in Jingxian County, Anhui Province a schoolmaster named Zhai Jinsheng who made over 100,000 clay types after reading *Dream Stream Essays*. The work took him many years. With these clay types he printed *Ni Ban Shi Yin Chu Bian* (*Initial Notes on Printing with Clay Types*) and other books. Additional books printed later by the same method have been located in Beijing Library in recent years, demonstrating the accuracy of the records in *Dream Stream Essays* concerning Bi Sheng's clay-type printing.

By the Yuan Dynasty (1271-1368) Wang Zhen, an agronomist, succeeded in trial-producing wooden type. He also invented the rotating apparatus for type-holding, a simple mechanism that raised typesetting efficiency. Detailed descriptions of his method of making wooden type and its use in printing are found in his *Wang Zhen Nong Shu* (*Agricultural Treatise of Wang Zhen*). Wang Zhen's method was as follows:

The characters were chosen from officially sanctioned dictionaries, classified according to sound and hand-written on paper. These hand-written sheets were glued onto wooden plates, leaving space between the characters, which were engraved and each character sawed off with a very fine-toothed blade into small cubes. The finishing touches were then done with a cutting tool to ensure that all the types were of unified size and height.

The typesetting was done in a wooden tray, the types being picked out by hand with bamboo tweezers. When the tray was

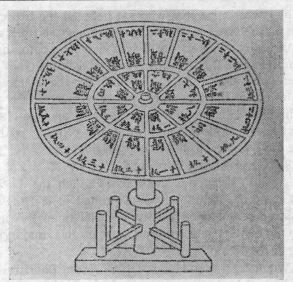

Revolving type frame invented by Wang
Zhen in the Yuan Dynasty (1271-1368).

full, the type was held in place by inserting wooden wedges and then
secured by a wooden bolt so that none got loose in printing. Any
difference in the height of the type was corrected by placing tiny
slices of bamboo beneath low types to build them up. The plate
was now ready for use. The ink was applied with bristle brushes
in lengthwise strokes parallel to the lines of types. The brush for
mopping the paper while printing was wielded in the same direction.

A large wooden rotating frame was used for storing the types.
This looked like a round table about 2.3 metres in diameter, with an
axle about one metre long. Both were made of light-weight wood.
The types were placed in bamboo vessels fastened to the circular
frame, all the characters in any one of the many vessels rhyming.
The characters were also serial-numbered and indexed. Two rotating frames were used, one for characters selected for use in the work
to be printed, the other for the characters most often used, these classifications being the basis for two index books. The typesetting itself
was done with one person calling out the numbers of the characters

wanted while another person, who sat between the two circular rotators, picked the types by their serial numbers and placed them in a tray. As both frames revolved freely, the operator had only to sit between them and "push either wheel to pick the types", to quote Wang Zhen, who also wrote: "Looking for the characters is very strenuous work for the typesetter, but it is much easier when they are made mobile. The rotating frames enable the typesetter to reach the characters sitting down, and when the printing is through, he can return the types as easily to their original places."

In 1298 Wang Zhen test-printed *Jing De Xian Zhi* (*Chronicle of Jingde County*) using this method. Though the book ran to some 60,000 words, 100 copies were printed in less than a month — high efficiency and good quality for the time. Wang Zhen's method of typesetting and printing was another major innovation in printing history.

After Wang Zhen, wood block-letter printing flourished in China, reaching its acme in the dynasties of Ming and Qing. In the 38th year of the Emperor Qian Long's reign (1773) in the Qing Dynasty, the government had 253,500 type pieces of various sizes made from datetree wood. In the course of years, *Wu Ying Dian Ju Zhen Ban Cong Shu* (*Editio Princeps Books of Wuying Hall*), a compilation of 138 books by different authors running to a total of 2,300 volumes, appeared in print as the largest-scale wood block-letter printed opus ever done in China.

Further development of block-letter printing came with the manufacture of metal types. Even before Wang Zhen, in the 13th century, attempts had been made to produce tin types, the world's earliest metal printing type. These attempts failed, however, because the tin types did not catch the ink, and it was not until the reign of the Emperor Xiao Zong of Ming (1488-1505) that copper types came into wide use in the copper-rich areas of Wuxi, Suzhou and Nanjing in Jiangsu Province. The lengthiest work printed with copper type in China was *Gu Jin Tu Shu Ji Cheng* (*Collection of Books, Ancient and Modern*), an encyclopaedia published in the Qing Dynasty.

China's Printing Goes Abroad

China's art of printing inspired many countries of the world, either through direct or indirect contact, to develop their own.

Block-printed books of the Tang Dynasty were taken to Japan, where late in the 8th century *Invocation Strata* was block-printed with success. In about the 12th century, or even earlier, the art of block printing was carried to Egypt. In the 13th century most Europeans travelling to China traversed Persia, or present-day Iran, where the Chinese printing technique had been mastered and was used for the manufacture of paper currency. Persia was virtually a relay station for the spread of the Chinese art of printing to the West. By the end of the 14th century there appeared in Europe block-printed playing cards, portraits of Christ and books in Latin for school pupils.

About the 14th century also, China's early block-letter printing was passed on to Korea and Japan. On the basis of this, the ingenious Korean people created the world's earliest copper types, making a major contribution to the world's art of printing. From the 15th century, the Korean copper types in their turn effected further development in China's printing technique.

In the Yuan Dynasty, block-letter printing was being used by China's minority nationalities. Based on the Uyghur rules of spelling, the Uyghurs made printing type of whole words rather than individual letters. These were perhaps the world's earliest block-letter printing for a language using an alphabet. In time the Chinese block-letter printing spread from Xinjiang to Europe through Persia and Egypt.

The dissemination of printing to Europe terminated the monopoly by clergymen of the right to learning and higher education. It provided important conditions preparatory to the whirlwind advance of science following a long period of medieval darkness and to the Renaissance movement. In his letter to F. Engels in January 1863, Karl Marx referred to the discovery of gunpowder, the compass and printing as "prerequisites of bourgeois development", a remark that places the art of printing in its properly significant role.

Metallurgy

He Tangkun

Iron Smelting

Invention of Cast Iron

In early times, iron smelting was done at temperatures of 800-1,000°C, the ores being directly reduced with charcoal. The wrought iron obtained by this method was called "block iron" and was a spongy solid when it came from the furnace. This "block iron" was inferior to cast iron in the following ways: 1. It did not flow out of the furnace but had to be removed after the fire was extinguished, which damaged the furnace lining. Production was low, as the smelting was interrupted, and the furnace was small. 2. Being very hard to forge, only simple-shaped utensils could be made of it. 3. It usually contained large amounts of non-metallic impurities which could be removed only by forging. 4. "Block iron" had a very low carbon content, rendering it very soft. Cast iron was smelted at the higher temperatures of 1,150-1,300°C and flowed from the furnace in a liquid state; its production could be continuous, and the product could be cast in moulds. It was harder and easier to smelt and cast, and it was superior in both quantity and quality. The change-over from the spongy "block iron" to cast iron was a great step forward in iron smelting.

Iron smelting in China developed differently from other countries. Although wrought iron was produced abroad from early times, it was not until the 14th century that Europe produced its first cast iron, while in China excavated iron artifacts indicate that iron metallurgy had its beginning not later than the middle of the Spring and Autumn Period (770-476 B.C.), the invention of iron smelting being followed closely by the appearance of cast iron. This date is not comparatively early, but iron smelting developed at a faster rate in China than in other countries.

A block of white cast iron from the late Spring and Autumn Period which was excavated in Luhe County's Chengqiao Middle School in Jiangsu Province is analysed as China's (and also the world's) earliest cast iron article found to date. Most iron articles of the Warring States Period (475-221 B.C.) that have been excavated are of cast iron. During the later mid-Warring States Period, iron utensils were widely used in China, more than 16 kinds of production implements alone, most of them of cast iron, being among the excavated iron objects of this period. Wrought iron had already taken second place to cast iron, indicating great progress in the production of cast iron during this period.

Malleable Cast Iron

The earliest malleable cast iron discovered in China, dating from the early Warring States Period, is in the form of an iron spade excavated in Luoyang. Analysis reveals it to be partially constituted of black cored malleable cast iron, though its main constituent is ferrite with flakes of annealed graphite. Malleable cast-iron making developed further after the later mid-period of the Warring States, many castings being annealed for malleability. Examples are an iron spade of the middle Warring States Period which was excavated in Changsha, and iron butt-ends and hoes of the later mid-period of the Warring States excavated in Yixian County, Hebei Province. This process reached maturity during the Han Dynasty (206 B.C.-A.D. 220). An iron axe of the Western Han Dynasty (206 B.C.-A.D. 24) that was excavated in Xuecheng and some iron pickaxes, spades and ploughshares found stored in Han and Wei Dynasty cellars in Mianchi had all undergone proper heat treatment. Some of these are made of white-heart malleable cast iron, others of black-heart malleable cast iron, both being distributed over wide areas.

The conversion of cast iron into malleable cast iron is most significant. At normal temperatures carbon is present in cast iron in two forms: cementite and free graphite (as tissues, flakes or nodules). These different forms of carbon give cast iron of the same composition different characteristics. In white cast iron all the carbon is present in the form of cementite, a very hard compound of poor ductility. White cast iron is therefore hard and brittle. In malleable cast iron most of the carbon is in the free graphite state, giving it relatively great hardness and better ductility. After the middle of the Warring States Period, the invention and development of malleable cast iron promoted the wide use of iron implements in agriculture, handicraft and everyday life.

Nodular Cast Iron

In 1974 a cellar of the Northern Wei Dynasty (386-581) was opened in Mianchi of Henan Province and some 4,000 iron implements of the Han up to the Northern Wei Dynasty came to light. These included weapons, tools, domestic utensils, founding moulds and iron materials. Among them was an axe which had undergone decarbonized annealing and whose composition was largely steel containing 0.4 per cent carbon with no flake-shaped graphite. The part of the axe holding the handle contained graphite nodules (resembling those in modern nodular cast iron) 20 microns in diameter and distributed over a U-shaped cross-section of 3.2 mm mean thickness and 50 mm in length. There were 30 nodules, all of regular shape.

Some scholars believe that these graphite nodules probably resulted from the white cast iron being annealed.

For many years iron workers had tried to obtain nodular graphite rather than the flaked but had met with difficulties. For a long time in ancient China cast iron had a very low silicon content, yet the ancient Chinese were able to produce large quantities of malleable cast iron with flake-shaped graphite by annealing. They also produced malleable cast iron containing some nodular graphite, Han Dynasty implements of this type having been found in Tieshenggou of Gongxian County and Wafangzhuang of Nanyang County — rare items in the history of metallurgy.

Steel Smelting

Steel of a Hundred Forgings

"Forging steel a hundred times" is an ancient Chinese steelmaking process. In 1961, a steel knife of the Zhongping reign in Eastern Han (184-189) was dug up in an old cemetery at the Todai Temple of Ichinomoto in Nara, Japan. On this knife was inscribed in gold: "Blue steel of a hundred forgings." In 1974 another steel knife, made in the 6th reign year of Yongchu of Eastern Han (A.D. 112), was excavated in Cangshan of Shandong Province bearing the gold-inlaid inscription: "This knife has been forged thirty times." The Beijing Institute of Iron and Steel analysed the steel in this knife and found it to consist of puddling steel with 0.6-0.7 per cent carbon content. It had undergone repeated folding and forging. Compared with the swords and knives found in the tomb of Prince Liu Sheng of the mid-Western Han Dynasty, this knife has more folding layers and less impurities. The impurities that do occur are smaller in size and more evenly distributed. There is also a more even distribution of carbon (i.e., there is no layer of different carbon content). This demonstrates that repeated smelting improves the quality of steel. *Meng Xi Bi Tan* (*Dream Stream Essays*) by Shen Kuo of the Song Dynasty describes in detail "steel that has undergone a hundred forgings". It tells of "fine iron" being forged a hundred times and weighed after each forging. It was said to be pure steel when there was no further decrease in weight. It also says that there is steel in iron just as there is gluten in flour, and that after washing many times the gluten will appear. The so-called "fine iron" is apparently neither cast nor wrought iron in the modern sense, but was likely a malleable iron consisting of steel and wrought iron. It would contain many non-metallic impurities, and the decrease in weight after each forging would be due to impurities being removed. The cessation of weight reduction would be only relative, for with repeated heating and forging iron oxide layers appear and are removed, so the decreases in weight will not cease. Carburization depends on the carbon content and the property of the iron product. No carburization is necessary for high-carbon steel, but low-carbon steel or wrought iron requires carburization to prevent its being too soft. Most steel, however, has a high carbon content.

The chief feature of "steel of a hundred forgings" is repeated heating and forging, the latter having special significance in ancient steel making. Except for the decarburization of cast iron, forging was always a major step in the ancient Chinese steel-making process. Forging removed impurities, decreased the granular size of residue impurities, made the composition homogeneous and fine in crystalline structure, and increased strength. Such is the secret and technical significance of "steel that has undergone forgings".

The production of this type of steel flourished on the basis of repeated forging of steel made of carburized iron blocks and accompanied by the development of steel puddling. Its main raw material is puddling steel. Steel made of carburized iron blocks was invented in China between the 5th and 4th centuries B.C. Forging 10 and 30 times began in the 1st century B.C. when it was first applied to copper smelting and then to steel. The earliest demonstration of the hundred forging process has so far been fixed as the Zhongping Knife, but the process became quite common in the first half of the 3rd century. Cao Cao, king of the state of Wei, ordered five "weapons of a hundred forgings" made. The Emperor Sun Quan of the state of Wu had three valuable knives called Bailian (Hundred Forgings), Qingdu (Blue Hue) and Louying (Cloud Pattern) respectively. Liu Bei, emperor of the state of Shu, ordered Pu Yuan to make 5,000 such knives with the words "72 forgings" inscribed on them. "Steel of a hundred forgings" reached its zenith in the Wei and Jin dynasties but declined with the development of other steel-making pro-

cesses. This steel was used mainly to make valuable knives and
swords and epitomized the wisdom and labour of the ancient Chinese
toiling people as well as reflecting the advanced steel fabrication
technique of the time.

Puddling Steel

Puddling got its name from the continuous stirring in the smelting process. The raw material of puddling steel is cast iron, which
is first heated to a liquid or semi-liquid state. The silicon, manganese and carbon ore are then oxidized by wind blowers or the addition of fine ore powder so that the carbon content is lowered to
within the limits of steel composition. Puddling steel is mostly
low-carbon steel. When the process is well controlled it could also
be medium-carbon or high-carbon steel, or it might be wrought
iron. The earliest Chinese object made of puddling steel discover-

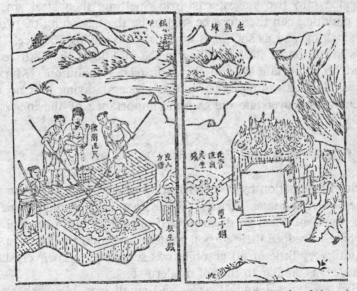

Illustrations in *Tian Gong Kai Wu* (*Exploitation of the Works of Nature*)
written in 1637 showing the smelting processes of cast and wrought iron.

ed to date is the "big knife of thirty forgings" excavated in Cang-shan of Shandong in 1974. This knife was forged directly from puddling steel and its blade, which is of even structure, is composed mainly of pearlite with a small amount of martensite. It is high-carbon steel. Puddling steel objects were also found in the Han Dynasty iron-smelting site in Tieshenggou of Gongxian County. They have some non-metallic impurities, mainly silicate.

The earliest account of steel puddling is found in Vol. 72 of *Tai Ping Jing (Canon of Eternal Peace)* written in the Eastern Han Dynasty, its invention being traceable probably to the end of the Western Han Dynasty.

Relatively efficient production and better quality are the advantages of steel puddling. Cast iron is the new product of steel puddling, an iron that can be continuously produced in large quantities. Steel puddling is carried out in the liquid and semi-liquid state and large amounts of cast iron can be continuously produced. Part of its impurities can also be removed. The process of steel puddling is divided into two stages: cast-iron smelting, then steel smelting. Steel puddling can therefore be said to be the beginning of two-step steel smelting and as such it is of epoch-making significance. England's invention of steel puddling in the middle of the 18th century played an important role in the Industrial Revolution. Marx expressed his great admiration for this process, saying that no compliment would overstate the extreme importance of the innovation.

Steel from Decarburized Cast Iron

Steel production by decarburization of cast iron was done early in the Warring States Period. White cast iron castings were first made, then subjected to decarburized annealing in an oxidizing atmosphere to reduce their carbon content to within the limits of steel and until very little or no graphite was separated. Their crystalline structure was similar to that of modern steel.

Sickles, types I and II axes, and small iron ploughshares, all among the iron implements stored in the Han and Wei dynasty cellar in Mianchi, had all undergone decarburized annealing. Some are well treated, as for example No. 197 small iron ploughshare whose edge is composed of ferrite, core of ferrite and pearlite, with a carbon content of about 0.3 per cent and no separation of graphite but with granular pearlite. No. 528 sickle is not so well treated. It consists of ferrite and pearlite with a small amount of graphite seen on close observation. All demonstrate comparatively advanced technology. Quite a number of the iron implements have been decarburized and then had their blades carburized or forged. The blade surface of No. 471 of type I axe is pearlite containing 0.7-0.8 per cent carbon; somewhat below the surface is fine pearlite containing 0.5-0.6 per cent carbon, while the core contains less pearlite with a carbon content of 0.3-0.4 per cent and no graphite to speak of. Different parts of implements were subjected to different processes of decarburization and carburization according to their different performance requirements, so that each part has its peculiar composition and structure. Certainly the ancient Chinese had some concept of the effect of carbon content on metals.

The iron adzes of the early Warring States Period excavated from a cement plant in Luoyang of Henan Province had undergone incomplete decarburization and so must belong to the early period of fabricating steel by decarburizing cast iron. This technique matured in the Han Dynasty, as shown by many excavations.

The significance of the invention of steel by decarburizing cast iron may be realized from the fact that the ancient world had no cast steel, and the smelting and further processing of forged steel was very inefficient, while the product contained many impurities. China very early (5th-4th century B.C.) utilized the superiority of cast iron in being easy to shape and low in impurities. The ancient Chinese iron workers' decarburization annealing enabled them to fabricate castings similar to those of today both in composition and performance.

"Co-Fusion Steel"

The so-called "co-fusion steel", made by smelting cast iron with wrought iron, is used for the blades of knives and swords. It is a high-quality steel of high carbon content. Cast iron has a high carbon content and low melting point, wrought iron has many oxidized impurities. At the melting temperature of cast iron, these two interact, and after smelting become steel. The proportion of cast and wrought iron could be changed to obtain the required composition of "co-fusion steel".

Many were the unknown heroes in this early work, but one whose name is known from historical documents was Qiwu Huaiwen, who lived around A.D. 550. He melted cast iron and poured it on wrought iron until steel resulted. The knife he made was called *su* iron knife. Quenched with animal fats and urine, this knife could cut 30 suits of armour. The earliest record of co-fusion steel was written by Tao Hongjing of the Liang period of the Southern Dynasties. Tao said that co-fusion steel is "mixing and smelting cast iron and wrought iron to fabricate knives and sickles".

There were several different processes of co-fusion steel. One was to bind cast and wrought iron plates together, sealed with clay, and smelt them in a furnace. A second method described by Shen Kuo in his *Dream Stream Essays* was to place cast iron on top of wrought iron. The cast iron melted first and dripped on the wrought iron. This is recounted in *Tian Gong Kai Wu* (*Exploitation of the Works of Nature*) by Song Yingxing. The third method belongs to a higher stage in the development of co-fusion steel, called *su* steel, and its advantages were fully exploited.

The main steps in processing *su* steel are as follows: Wrought iron is put into the furnace and heated with a blast, then one end of a cast-iron block is placed obliquely in the furnace opening to be heated. The blast is maintained so that the temperature continues to rise. When a furnace temperature of about 1,300°C is reached, the cast iron will melt and start to drip, while the wrought iron is softened. The cast-iron block is then held in tongs from the outside to let the molten cast iron drip evenly onto the wrought iron, which is being steadily stirred to cause strong oxidation. After dripping, the blast is stopped and the steel mass taken out and hammered to remove impurities. This dripping should generally be done twice.

Su steel smelting has two advantages: (1) During the pouring process there is strong oxidation which results in separation of slag and iron, the wrought iron being soft in structure and containing large amounts of oxide impurities, and also silicon, manganese and carbon. (2) After oxides in the wrought iron cause oxidation of the carbon in the cast iron, iron will be deoxidized, giving higher rates of metal reclamation. In Vol. 1 of *Wu Hu Xian Zhi* (*Annals of Wuhu County*) of 1522, there appears a statement about "initially forged wrought iron", which probably refers to this kind of iron. The chief difference between "*su* steel" and the previous two processes of "co-fusion steel" is that the previous two processes employ wrought iron plates. Since plates must have denser structure, they could not have been "initially forged wrought iron".

Before the invention of crucible steel smelting in 1740, solid and semi-liquid state steel smelting were generally used in all steel-producing countries and it was difficult to separate iron from slag. China's co-fusion steel, with higher productivity, better control of composition and separation of slag and iron, was indeed a superior steel smelting process that was rare in ancient times.

Non-Ferrous Metal Smelting

The Six Alloys

The three "six alloys" are six rules for compounding bronze that appear in *Kao Gong Ji* (*Artificers' Record*). The original quotation is as follows:

Bronze has six alloys: 6 parts copper mixed with 1 part tin for making bells and large cooking vessels, 5 parts copper mixed with 1 part tin for axes, 4 parts copper mixed with 1 part tin for making dagger-axes and halberds, 3 parts copper mixed with 1 part tin for large knife blades, 5 parts copper mixed with 2 parts tin for making arrowheads, 2 parts copper with 1 part tin for making mirrors.

Available archaeological data document China's mastery of the technique of cold forging and founding of copper as early as the Xia period (c. 21-16 century B.C.). By the end of Xia and the beginning of the Shang period (16th-11th century B.C.), China was smelting and founding copper, so that in the mid-Shang period the country entered a highly developed Bronze Culture. Bronze artifacts excavated to date include many ceremonial utensils, weapons, daily utensils and some production tools (including handicraft tools and agricultural implements). The bold and stately Si Mu Wu Ding cooking vessel and the elaborate four-goat wine vessel are the finest of bronze artifacts. "Six compositions" appeared on this basis as the world's earliest knowledge of alloys. It formulates the relationship between alloy performance and alloy composition.

The basic idea of "six compositions" agrees with modern scientific principles. An alloy containing about 14 per cent tin is hard and ductile and has good tonal quality. When such an alloy is quenched in cold water at about 600°C, its tonal quality is further improved, and this is a suitable composition for making bells and large cauldrons. Axes, being tools that must withstand greater impact force than bells, have somewhat less tin than bells and arrowheads. The dagger-axe and halberd are weapons subjected to complex stress including tension, compression and torsion, and they must be very ductile, while knives and swords must be strong as well as sharp. Arrowheads are small and do not break easily, so sharpness is the main consideration, and they have the highest tin content of all weapons.

China's "six compositions" were technically and socially significant, and though its bronze-making technique came fairly late, it developed rapidly for two reasons (besides the factor of natural resources): 1. Early mastery of the high-temperature technology of metal smelting. 2. Knowledge of alloys. Without this knowledge about alloys only copper would be procured, by deoxidizing smelting of metals. If by chance some bronze resulted, it would have been a crude product.

Smelting of Brass and Zinc

Historical documents and excavated objects divide the use of zinc in China into approximately three stages.

The first stage comprises the pre-Han Dynasty period when zinc went into copper alloys as an accompanying ore. Copper utensils of this period contain only a few ten-thousandths in weight of zinc, and few objects of any sort have a higher zinc content.

The second stage includes the period from the Western Han to the Song (960-1279) and Yuan (1271-1368) dynasties. Zinc oxides (calamine) purposely added to the copper-smelting furnace were reduced and dissolved into the copper, producing brass.

In the third stage, i.e., after the Ming Dynasty (1368-1644), metallic zinc was produced by calamine. The "Chapter on Metals" in *Tian Gong Kai Wu* written in the Ming Dynasty gives a detailed account of the production process of metallic zinc. Ten *jin*[1] of calamine is placed in a clay jar which is sealed with clay. This is dried, laid on coal and burnt with wood. The calamine melts into a mass. After cooling the jar is broken and the zinc removed. The weight decreases by 20 per cent. This account is fragmentary since it fails to mention the reducing agent, but the basic principle and implements are similar to the modern method of zinc smelting.

Obviously, metallic zinc was produced on a large scale in China during the Ming Dynasty.

The production of brass and zinc has special significance. Brass has good mechanical properties, is easy to process and found and has a beautiful metallic lustre.

Nickel Brass (Copper-Nickel Alloy)

The use of copper-nickel alloy or "China silver" goes very far back in the history of China. At first a kind of copper-nickel mineral

---

[1] A *jin* now equals 1/2 kilogramme or roughly 1.102 pounds.

was employed, then later nickel ore and copper were mixed and smelted.

The term *baitong* (white brass) first appears in *Hua Yang Guo Zhi* (*Record of the Country South of Mount Huashan*) written by Chang Qu of the Jin Dynasty (265-420). This book records that in the Dongchuan district of Yunnan Province is Tanglang Mountain, which produces "silver, lead, white brass and medicine".

In Vol. 10 of *Chun Zhu Ji Wen* (*Record of Things Heard at Spring Island*) written by He Wei of the Song Dynasty it is recorded that "white arsenic powder" could "dissolve two ounces of copper into a silver-like substance"— the earliest record of man-made nickel brass.

After the Yuan and Ming dynasties, the production of copper-nickel alloy developed rapidly. *Ge Wu Cu Tan* (*Simple Discourses on the Investigation of Things*) says that copper and "arsenic stone could be mixed to smelt white brass"; *Ben Cao Gang Mu* (*Compendium of Materia Medica*) says "Copper and arsenic stone can be smelted into white brass"; *Tian Gong Kai Wu* says "to use arsenic powder and other drugs to smelt white brass". Some people believe that the arsenic stone and arsenic powder mentioned above are white nickel ore ($NiAs_2$), others think they are arsenic trioxide (As_2O_3).

There are many people in China and other countries who are at present investigating the production and application of copper-nickel alloy in China. Some scholars believe that as early as the Qin and Han dynasties Chinese nickel-copper alloy had probably been taken to the ancient country of Bactria where it was coined into money. Remaining specimens upon analysis show a composition similar to Chinese copper-nickel alloy (77 per cent copper and 20 per cent nickel). In the 18th century many persons in the West tried to imitate Chinese copper-nickel alloy, which was finally done in 1923 in England and Germany. Many imitations followed this on the market, the most popular among them being "German silver". The westward movement of Chinese copper-nickel alloy had promoted the production of copper-nickel alloy and modern chemical processes of the Occident.

Blasting Technique and Metal-Smelting Fuel

Hydraulic Bellows

These are a type of wind-blowing device used in iron smelting in ancient China. The earliest wind blowers were usually leather bags, which were called *tuo* (bag). One furnace used many of these bags in a row, hence the device was also called *paituo* (row of bags). Driven by hydraulic power, these rows of bags were called *"shuipai"* (hydraulic bellows).

The Chinese hydraulic bellows were the invention of Du Shi who was the governor of Nanyang in the early Eastern Han Dynasty, after observing work practices. The bellows were widely used as a labour-saving device that raised productivity. During the period of the Three Kingdoms (220-265), Han Ji introduced the invention to the state of Wei and used it in the government-owned iron-smelting plant. The hydraulic bellows, replacing the horse-driven and manual bellows, could be used continuously, saved man and animal power and increased productivity threefold, greatly promoting iron smelting. From the post-Han period to the establishment of the People's Republic, hydraulic wind blowers were used widely in China, while improvements on the structural system were constantly made.

Few records are available on the structure of the Han Dynasty water bellows. From the structure of hydraulic trip-hammer and turn-over wheels of the same period, it may be surmised that the hydraulic bellows must have been a wheel-axle transmission device. The earliest account of its structure appears in *Wang Zhen Nong Shu* (*Agricultural Treatise of Wang Zhen*) written in the Yuan Dynasty. The two types were the vertical-wheel and the horizontal-wheel. Through axles, levers and sometimes with transmission belts, they changed the circular motion into a linear back-and-forth motion used to open and close the wind fan to produce the blast. One revolution of the water wheel opened and closed the wind fan several times, increasing the wind-blowing velocity.

Wind blowers developed through different stages, from the earliest leather bag to the wind fan and finally to the bellows. The

Illustration of horizontal-wheel hydraulic bellows in *Wang Zhen Nong Shu* (*Agricultural Treatise of Wang Zhen*), a book of the Yuan Dynasty (1271-1368).

wind fan was probably invented before the 10th century. The movable furnace mentioned in *Wu Jing Zong Yao* (*Collection of the Most Important Military Techniques*) written in 1044 and the hydraulic bellows in the *Agricultural Treatise of Wang Zhen* were all wind fans, with, of course, improvements in structure. Mention of the piston bellows appeared first in *Exploitation of the Works of Nature* written in 1637.

Hydraulic wind blowing increased wind capacity, pressure and penetration in the furnace and thus improved smelting and increased the height and effective volume of the furnace chamber. Production increased. Further, cast-iron smelting requires adequate wind-blowing and furnace capacity, hence the invention of the hydraulic bellows was highly significant in smelting and in man's use of natural power.

The Development of Metal-Smelting Fuels

Charcoal is the earliest metal-smelting fuel used by man, and it has many merits. 1. It is easily available; 2. It is porous, which gives the material column within the furnace better air penetration to

strengthen the wind-blowing capacity and pressure of ancient bellows;
3. Charcoal is relatively low in sulphur and phosphorus and is
even today the ideal fuel for high-grade cast iron smelting. The
drawback of charcoal is limited natural supply, and from very ancient times man sought for new fuels.

Coal, which was used comparatively early for handicrafts and
daily life, was used rather late in China for metallurgy. Coal was
reportedly found at a Han-Dynasty iron-smelting site in Tieshenggou
of Gongxian County, indicating that it was experimented with
for iron smelting. The earliest written record about the use of coal
in iron smelting appears in the Chapter on Rivers in *Shui Jing Zhu*
(*Commentary of the "Waterways Classic"*), written by Li Daoyuan
of the Northern Wei Dynasty. It tells of a mountain 100 kilometres
north of Kuqa in central Xinjiang whose coal was used for iron
smelting and which provided iron to wide surrounding areas. More
accounts of the use of coal in metal smelting appeared after the Song
Dynasty, marking the beginning of its wide use for iron smelting in
China.

But coal is not a perfect fuel for smelting either. 1. It contains
large amounts of impurities such as sulphur and phosphorus which
combine with the iron and make it brittle. 2. These large amounts
of impurities produce large amounts of slag, which damages the
furnace. 3. Coal is not porous and so does not allow for air penetration; it is unstable at high temperatures, and it explodes easily.

Coke, produced by the dry distillation of coal, retains the merits
of coal but does not have its shortcomings. Coke is still the main
fuel today in metallurgy.

The earliest record of the use of coal in China appears in the
book *Wu Li Xiao Shi* (*Small Encyclopaedia of the Principles of Things*)
written by Fang Yizhi in 1664. It says that coal is to be found everywhere, and that when coal that emits bad odours is sealed and burned it becomes coke; further, that coke could be used for smelting
and in daily life. There are also accounts about coke production
in *Jie An Man Bi* (*Essays of Jie An*), *Yan Shan Za Ji* (*Records of
Mount Yanshan*), and *Hui Li Zhou Ji* (*Annals of Huili Prefecture*).

The transitions from the use of charcoal to coal and from coal
to coke both marked great changes in metallurgy.

Founding

Hua Jueming

> *Fire burned in a thousand furnaces,*
> *Ore was melted into utensils.*

These are lines often quoted from *Yong Ye Fu* (*Ode to Founding*) written by the poet Cao Pi of the Jin Dynasty (265-420) that well describe the flourishing scenes of founding in ancient China. Founding played an outstanding role in metal processing of those days and had wide social influence. Chinese characters like *mofan* (pattern, model), *taoye* (mould), *rongzhu* (smelting) and *jiufan* (submit to moulding) are all derived from terminology in founding. The labouring people invented a typically Chinese founding craft by production through the generations, who handed it down. The three most important branches of this traditional craft were clay-mould casting, iron mould casting and molten-pattern casting.

Clay Mould Casting

From the later Neolithic period, China entered an era when copper and stone were both used. Among the early copper utensils excavated in Tangshan of Hebei Province were some that were forged and some that were cast. This shows that mould casting was done very early in China.

There are no few accounts in ancient Chinese literature about the Kun Wu (a tribe living in the north of the present Puyang County of Henan Province) making pottery and casting copper, and about Yu the Great casting the nine *ding* (large sacrificial vessels). Recent

archaeological excavations also indicate that the Chinese people
were moulding bronzes in the Xia Dynasty (c. 21st-16th century
B.C.). The earliest moulds were made of stone. But stone was
a difficult material to form into moulds, nor could it withstand high
temperatures and so, as pottery making developed, clay moulds
came into use. For 3,000 years after that, casting by clay moulds
was the main founding method, to be followed by casting from sand
moulds which flourished in modern machine manufacturing.

In the early Shang Dynasty (c. 16th-11th century B.C.), clay
moulds were already used to cast small implements of copper —
adzes, chisels and such items as small bells and vessels. This period
in founding is marked by remains excavated in Erlitou of Yanshi
County, Henan Province. Bronze founding developed later, as
marked by excavations at Erligang of Zhengzhou city. Two large
square vessels weighing 64.25 and 82.25 kilogrammes respectively
were excavated recently in Zhangzhai, also of Zhengzhou, showing
that the technique of copper founding was highly developed in the
mid-Shang Dynasty. From single and double moulds, the craft
went on to using many moulds and cores combined to form complex
founding moulds that made possible the casting of pieces more than
50 kilogrammes in weight.

After King Pan Geng moved the capital of the Shang Dynasty
to Yin in the 14th century B.C., the technique of bronze founding
reached its zenith as typified by the Yin site excavated in Xiaotun
of Anyang. Tens of thousands of bronze utensils excavated or
passed down through the ages are important historical artifacts
that demonstrate the talent and intelligence of the foundry slaves.
In technique and artistic value, as well as craftsmanship, they are
recognized throughout the world today.

In order to produce bronze casting of complex shape and exquisite relief, the ancient foundrymen undertook such important
technical measures as selecting pure and highly refractory sand from
the locality for the moulds. The sand, which was used on the surface of the moulds, was washed and precipitated in water, leaving a
fine pure green clay of high plasticity and strength. Moulds made
of it yielded articles that were fine and distinct in detail. The sub-

strate clay was of a rougher texture, or else mixed with sand and plant tissue to reduce shrinkage and increase permeability.　All clay materials underwent long periods of drying in shaded areas and repeated beating to make them highly homogeneous so that they did not crack during moulding and heating.

The pattern-making process was based on the principle of casting by stages to achieve complex configurations.　That is, the main body was cast first, then auxiliaries (such as animal heads, columns, etc.) were cast on the body.　Or the auxiliaries (such as the ears and legs of large vessels) were first cast separately and then cast together with the main body.　The famous four-goat wine vessel (excavated in Ningxiang County of Hunan Province) was cast in stages, a method that can be traced back to the Erligang period, while the basic method was fully developed by the Xiaotun period.　During the Spring and Autumn Period (770-476 B.C.) the method of casting by stages consisted mainly of first casting the auxiliaries and then the body of the utensil.　Sacrificial vessels of Xinzheng and the large sacrificial vessels and kettles of the Warring States Period (475-221 B.C.) were cast mainly by this method, exemplifying the unique artisanship of ordinary working people who resolved complex fabrication problems by applying simple processing principles.　The method is also a key to understanding bronze founding technology in the Shang and Zhou (c. 11th century-221 B.C.) dynasties.　It is inaccurate to classify a complex utensil like the four-goat wine vessel as a casting by the lost-wax method.　It is even more fallacious to describe the bronze vessels of the Shang and Zhou dynasties as profound wonders whose craftsmanship could not be surpassed.

Furthermore, the drying, heating and assembling of cores, achieving evenness of wall thickness to attain simultaneity in solidifying and preheating of moulds to facilitate founding — all were mature fabrication methods during the Shang and Zhou dynasties, methods that laid the technical foundation not only of clay-mould casting but also of metal-mould and melting-mould casting for succeeding generations.

But, under the social conditions of the exploiting classes owning the means of production and controlling the state machine, the scientific and technological creations of the labouring masses were always used by the ruling classes to suppress and deceive the populace in order to maintain their luxurious and decadent life. The Shang and Zhou dynasty bronzes are mostly ceremonial implements, musical instruments, weapons or chariots; very few are production implements. Moreover many bronze castings were buried as soon as they were made. Such waste and ravage of labour greatly hindered the development of social productivity, and clay-mould founding remained at a one-off type stage for more than 1,000 years until the Spring and Autumn Period when a durable clay mould was used for casting pickaxes of copper.

The early and wide application of the "stack casting" method is another outstanding achievement of ancient Chinese clay-mould founding. This method means the assembling of many mould blocks or pairs of mould plates in a series of layers, the founding being done via a common pouring passage. Several tens or even a hundred castings could be done in one founding. Modern development of this method occurred only with the appearance of large-scale machine production where large amounts of small castings (e.g., piston rings and chain links) are needed. Due to its relatively high rate of productivity, lower cost and economy in mould-making and founding area, it is still in wide use. The earliest Chinese stack-castings are the knife-shaped coins of the state of Qi during the Warring States Period. Highly symmetrical and exchangeable mould plates were made from copper pattern boxes, every two plates being combined to form one layer. These knife-shaped coins were then cast by overlaying these layers and stack founding. In the Han Dynasty (206 B.C.-A.D. 220) this method was widely used for making coins and components of horse carts.

In recent years, many such moulds and furnaces have been excavated in the provinces of Shaanxi, Henan and Shandong. In a mould-heating kiln at a Han Dynasty founding site at Xizhaoxian Village of Wenxian County, Henan Province, 16 types of stack founding moulds consisting of more than 500 sets in 36 specifications were excavated, providing valuable data for understanding this fabrication process. Of elaborate structure, finely crafted and easy to clean, these moulds have inner pouring basins only 2-3.5 millimetres thick. The products of this moulds have a surface finish of the 5th grade (the total gradation of surface finish is 14) and a metal recovery rate of 90 per cent, which is very near the manufacturing level of Foshan, Guangdong Province, of modern times.

From the Tang and Song dynasties, the fabrication of large and massive castings by clay moulds developed rapidly. The iron lion of the Five Dynasties (907-960) in Cangzhou of Hebei Province, the iron tower of the Northern Song Dynasty (960-1127) in Dang-yang of Hubei Province and the giant bell of the Ming Dynasty in Beijing's Great Bell Temple are all huge castings known the world over.

Song Yingxing (1587-?) records two founding methods of large castings in his book *Tian Gong Kai Wu* (*Exploitation of the*

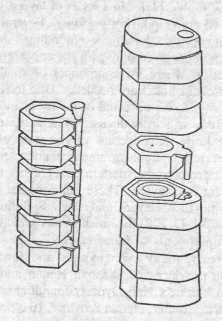

Reconstruction of hexagonal stack founding moulds of the Han Dynasty unearthed in Wenxian County, Henan Province.

Works of Nature). One is by using many moving furnaces to pour one after another, the other uses many furnaces to pour in a trough. Under the conditions of manual production this is a sophisticated fabrication method that demands skilled labour and good co-ordination. Even today the successful founding of a 40-ton casting is no easy matter. Coghlan, the English historian of metallurgy, praised the ancient Chinese people for their exceptional talent in founding massive castings.

Founding by Metal Moulds

The materials used for making moulds have evolved from stone, clay and sand to metal. The moulds became increasingly durable, from being used only once to several times, until there evolved the permanent (i.e., metal) type. This was very significant in the development of founding. In 1953 iron moulds were excavated in Xinglong of Hebei Province, showing that as early as the Warring States Period China was using metal moulds made of white iron for casting. These iron moulds include 87 hoes, sickles, axes, chisels and cart components, most in whole sets. The sickles and the chisels are in one mould; the hoes and axes use metal cores. Their structures are compact and have peculiar features. The shape of moulds fit closely with that of casting, thus the thickness of the wall is even and convenient for heat dissipation. Handles are attached to the moulds, facilitating gripping them and also strengthening the moulds. These are typically Chinese metal moulds that took definitive shape at that time. In recent years many iron moulds of the Han Dynasty have been excavated in Nanyang, Zhengzhou and Zhenping of Henan Province, Mancheng of Hebei Province and Laiwu of Shandong Province. Variety increased markedly over the Warring States Period, but they remained basically unchanged in type. The storage caves of the Han and Wei (220-265) dynasties in Mianchi of Henan Province contain iron moulds for casting iron plates and arrowheads, and also a large iron plough mould which measures half a metre in length.

Beside iron moulds, moulds made of copper were used to cast

coins during the Warring States Period and the Han Dynasty (e.g., copper moulds for the five-*zhu*[1] coins that were passed down and excavated). Though these played major roles in production, very few accounts of them are found in documents. *Han Shu* (*History of the Han Dynasty*) has the words, "like metal during casting, which was done by the smelter", with the added commentary, "that is the mould for casting". This may be considered as the earliest account of founding with metal moulds.

An opinion once prevailed that the iron mould was invented very early in China but this process was lost and only recently reintroduced to China from another country — an opinion that is not true. For though small agricultural tools like the hoe and sickle were forged instead of cast after the Tang and Song dynasties due to the invention and spread of forging, implements like plough boards were, up to recent times, always cast with iron moulds. Beginning from the Warring States Period, in the Qin (221-207 B.C.) and Han dynasties, the Chinese founding process remained largely unchanged: cast-iron moulds were cast from clay moulds, and iron castings were made from these iron moulds. A series of operational methods was thus formed conforming to scientific principles. The use of molten iron to preheat the moulds (the first few castings could be counted as rejects), using a double layer of loam on the surface of the mould, opening the box in the right time after pouring in the molten metal, employing clay ware aside from metal core and using simple mounting machines are all examples. Using metal moulds had the following advantages: higher productivity, longer service life (they could be used hundreds of times when making smaller castings), uniform specifications, and certainty of obtaining white iron structure (when casting iron objects). These, coupled with the annealing of cast iron, played important roles in the casting of agricultural implements in ancient China. "Irrigation Records" of *Han Shu* describes the construction of irrigation ditches thus: "When men lift the spades it looks like clouds; when the water comes

---

[1]   *Zhu* was a weight unit in ancient China approximately equivalent to two grammes.

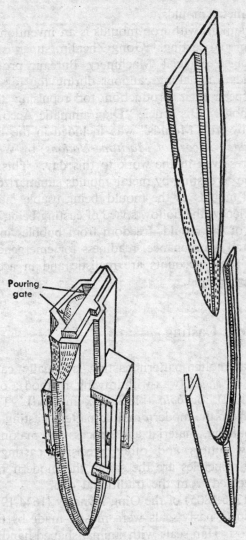

Pouring
gate

**Reconstruction of a recently unearthed
iron plough mould of the Han Dynasty.**

along the ditch it looks like rain." Examination of the many iron
implements excavated leads to the conclusion that large numbers

were cast in metal moulds.

Casting cannons with iron moulds is an invention in traditional Chinese iron-mould casting. Gong Zhenlin, supervisor of weapons in the Zhejiang Provincial Machinery Bureau, proposed the use of iron moulds for making cannons during the First Opium War (1840-42) to speed their production for repulsing the aggressors, and this was done. His article "Diagrammatic Account of Founding Cannons by Iron Moulds" was included in the book *Hai Guo Tu Zhi* (*Illustrated Record of Maritime Nations*) by Wei Yuan (1794-1857) and is preserved in the work to this day. This earliest scientific description of casting by metal moulds summarizes such advantages of metal moulds as the mould being usable many times, low cost, high efficiency, the hollow space of casting being easy to clean, non-humidity of the mould, freedom from bubbles in the castings, easy storage and maintenance, readiness for emergency use in case of war, etc. These accounts are realistic and in accordance with modern casting method.

Molten-Pattern Casting

Traditional molten-pattern casting is usually called the lost-wax method, or *chula* (wax-withdrawing method), or *niela* (wax-kneading method), or *bola* (pluck-wax method). This method is very different from the modern molten-pattern casting in the kind of wax used, moulding, material and process for making turbine fan blades, milling cutters and other precision castings. Yet their manufacturing principles are the same; the modern molten-pattern casting has evolved from the traditional one.

Gui Fu (1736-1805) of the Qing Dynasty (1644-1911) commented: "The Han Dynasty seals were mostly made by the pluck-wax method." Those Han seals with animal-shaped handles, exquisitely shaped, elaborately designed and showing no traces of chiseling were probably made by the lost-wax method. The cover of a Han Dynasty shell container of the Dian ethnic group that was excavated in Shizhaishan of Yunnan Province was also most probably made by this method.

Volume 89 of *Tang Hui Yao* (*Administrative Statutes of the Tang Dynasty*) says that Kaiyuan[1] coins were made from wax patterns, and this is likely the earliest documentary account of the lost-wax method. Among the coins of that period either handed down or excavated is one type bearing fingernail impressions which are said to have been left in the wax mould casting.

Dong Tian Qing Lu Ji (*Clarification of Strange Things*) written by Zhao Xihu of the Song Dynasty (960-1279) records this fabrication process in detail. The wax was sculptured into a pattern which was placed in a bucket-shaped receptacle and immersed in a mix-

Making bell moulds — an illustration from *Tian Gong Kai Wu* (*Exploitation of the Works of Nature*), **1637.**

---

[1] Kaiyuan is a reign title of the Xuan Zong Emperor of the Tang Dynasty which was used between 713 and 741.

ture of fine clay and water. Soft clay containing salt and paper tissue were next applied, and this became the pattern. The wax was later withdrawn and melted bronze poured in. This method was used for small castings. Similar to the lost-wax method, it was used during the Ming and Qing dynasties.

The modern molten-pattern casting is used mainly for small castings, as precision is difficult to maintain if the casting is too large. Ancient molten-pattern casting was mostly used for making such artifacts as bells which did not require the high precision of modern machine components. There is an account in *Tian Gong Kai Wu* of casting the Ten Thousand *Jun*[1] Bell by the lost-wax method. Its mould was made by digging in the earth, the wax consisting of 10 parts vegetable tallow and eight parts yellow wax. Charcoal powder was added to clay to minimize shrinkage, increase permeability and ensure good surface finish. One part of wax was used for 10 parts of copper.

A Bureau of Wax Withdrawing was established in the Yuan Dynasty (1271-1368) to preside over lost-wax casting. The fabrication division of the Qing Dynasty Department of Imperial Household employed professional craftsmen. The bronze lions, elephants, cranes and leopards now preserved in Beijing's Palace Museum and Summer Palace are typical lost-wax castings of high artistic value. Certain structures in the Summer Palace's Bronze Pavilion were also made by the lost-wax method as evidenced by wall inscriptions bearing the names of the moulders Yang Guozhu, Zhang Cheng, Han Zhong and Gao Yonggu.

The above accounts demonstrate that the lost-wax method enjoys a long history in China and has its own peculiar features and art styles.

---

[1] *Jun* was a weight unit in ancient China equivalent to 15 kilogrammes.

Agricultural Machines

Zhao Jizhu

During the 4,000 years of China's cultural history, agriculture has consistently occupied a major part of total production, and as the social economy advanced, the labouring people very early began inventing various types of agricultural implements. Here we shall introduce a few of these.

The Plough from the Han and Tang Dynasties

The ancient Chinese used the spade for field cultivation and sowing seed. In time the spade was developed into a plough. But before the period of the Warring States (475-221 B.C.) the only materials used in field-levelling implements were stone, wood, bone and some bronze. Later, during the period of the Warring States (475-221 B.C.), when oxen replaced human labour for ploughing, and iron was produced, the iron plough came into use. This was a spectacular achievement as it marked a new era in man's history and his struggle to reform nature.

With the wider use during the Han Dynasty (206 B.C.-A.D. 220) of production implements and cultivation methods that were advanced for the time, the plough was further improved and more widely used throughout the country. Among the 100 and more iron ploughs of the Han Dynasty excavated since the establishment of the new China are iron-tipped, sharp double-winged, tongue-shaped trapezoid, and large ploughshares. The structure and shape of

Han Dynasty ploughs can be seen in paintings excavated from Han tombs in Pinglu County of Shanxi Province. These are iron plough-shares which already have mould-boards attached. Iron mould-boards of the Han Dynasty have been excavated in Anqiu County of Shandong Province, Zhongmou County of Henan Province, and Xi'an, Xianyang and Liquan in Shaanxi Province. The invention of mould-boards attached to the ploughshares saved the additional hoe and spade work to crush and loosen the soil and form furrows in the crop fields. Only ploughs fitted with mould-boards can crush and turn the soil to one side, bury weeds for fertilizer and help to kill insect pests. The wooden part of ploughs consists of shaft, handle, bed, adjuster and bar. This standard plough design already existed in the Han Dynasty. There was also a marker to adjust the depth of ploughing. Han Dynasty ploughs had double or single shafts and were generally pulled by two oxen. Though inefficient in turning up the soil and unable to make turns because of its long and straight shaft, the Han Dynasty plough was a great improvement over that of the Warring States.

During the Tang Dynasty (618-907), *Lei Si Jing* (*Canon of Spades*) written by Lu Guimeng described in detail the different parts of a plough together with dimensions and functions. The Tang plough consisted of metal ploughshare and mould-board, with wooden bed, pressing-share, enforcer adjuster, plug, handle, head axle, etc. — 11 parts in all, each with its special function and shape. The mould-board was mounted on the ploughshare forming a curved planar combination device whose function it was to turn and mound up the soil. The plough-bed and pressing-share made the plough-head fast and gave it stability, the enforcer protected the mould-board, while the adjuster and plug fixed the ploughing depth by controlling the angle between the plough shaft and bed. The width of the furrow was controlled by the plough handle. The shaft was short and curved, and its head had a turnable head axle. The draught animal pulled the plough by a plough rope. This was a structurally complete plough, advanced and manoeuvrable. It was easy to steer, the ploughing depth was readily controlled, and it turned up the soil without much effort. In short, it was an efficient plough.

What Lu Guimeng described was a Chinese plough that was structurally fairly complete and more complex than its Qin and Han predecessor. It was basically similar to the modern plough.

After the Song (960-1279) and Yuan (1271-1368) dynasties a number of new types of plough were invented to suit varying conditions in China. In the paddy fields of the south a smaller ploughshare served better than the larger one, which suited the dry fields of the north. Virgin land choked with grass and weeds was opened up by using a plough knife; spongy land was cultivated with a seed plough.

The Han Dynasty Seeder — a Triple Seed Plough

The Chinese were using seeding machines in the Warring States Period. In the time of the Emperor Wu Di (140-85 B.C.) of Han, a triple seed plough which sowed three rows at once was invented by Zhao Guo on the basis of the single and double seed ploughs. This was pulled by an ox, which one person drove while another held the seed plough. With it 6.6 hectares of land could be sown in a single day — a great increase in efficiency. The Emperor Wu Di decreed that this advanced seeder be employed throughout the country.

A reconstructed model of this Han Dynasty triple seed plough is now an exhibit at the Museum of Chinese History in Beijing. In structure, the three lower small ploughshares or plough feet are used for ditching. Their central hind part is hollow; the distance between them is one furrow. The lower parts of the three wooden plough feet with hollows are inserted in the cavity of the ploughshare and their upper parts are connected with the seed trough. The front of the lower part of the seed trough has a square hole connected to the plough funnel on the front. The lower part of the back of the plough funnel has an opening with a board which is fastened by a wedge. To prevent the seed being bunched and blocked at the opening, a bamboo slip is suspended on a peg in the plough handle, its front end inserted in the lower part of the plough funnel and made fast; a piece of iron is attached midway. Shafts are mounted on each of the two sides of the seed plough with space between to accommodate one ox. Behind is the seed-plough handle.

Before sowing, the board at the opening of the plough funnel had to be adjusted according to the kind of funnel used, the size of the seed and the humidity of the soil, to allow a suitable rate of seed flow into the plough funnel. The man holding the seed plough also controlled the plough handle and so the depth to which the plough feet sank into the soil, i.e., the depth of seed sowing. As the seed plough moved forward it also shook the seed out of the funnel into three streams through the seed-plough legs, on through the lower part of the ploughshare and into the soil. The soil was levelled and the seed covered over by a square stick suspended by two ropes attached to a wooden frame behind the seed plough, the stick lying transversely on the ridge and raking the soil as the seed plough moved forward. Thus ditching, sowing and burying were done in one operation. This was followed by a roller to press down the soil and ensure the seed's sprouting and growth.

The functions of the most modern seeders are ditching, sowing, burying and pressing in that order. The triple seed plough developed more than two millennia ago in China accomplished the first three processes, its invention at that early date being no small feat and a milestone in ancient China's agricultural history.

Irrigation Machinery —
the Square-Pallet Chain-Pump

The square-pallet chain-pump, called *long gu shui che* (dragon backbone water-wheel), was one of the most widely known of ancient China's agricultural irrigation devices. In ancient Chinese books it was called *fanche* (turn-over scoops). According to *Hou Han Shu* (*History of the Later Han Dynasty*) this irrigation equipment was invented during the end of the Eastern Han Dynasty (25-220).

At first human labour drove the wheel axle and raised the water for irrigation. Later, with the development of the wheel-shaft there was invented the square-pallet chain-pump which was driven by animal, wind or hydraulic power, and this was widely used throughout the country. According to the type of power supply, there are the following kinds of square-pallet water-wheel.

The man-powered square-pallet chain-pump. This is turned mainly by the foot, though some use the hand. Lin Qing of Qing Dynasty (1644-1911) wrote *He Gong Qi Ju Tu Shuo* (*Illustrations and Explanations of the Techniques of Water Conservancy and Civil Engineering*) which contains a fairly detailed description of the square-pallet chain-pump. Aside from the balustrade and the frame, the body of the machine was made of wooden boards in the shape of a trough 6-7 metres long, 13-23 centimetres wide and about a third of a metre high. A rail board mounted in the middle of the trough was of the same width, but both ends were a third of

The man-powered square-pallet chain-pump.

a metre shorter than the trough to leave room for installing the large and small wheel shafts. Above and under the rail board, square-pallets were connected by wooden chains along the whole circumference, making it look like a giant backbone, and so it was called "dragon-backbone water-wheel". Four wooden pedals were mounted on either side of the upper large wheel shaft which was mounted between the wooden frames on the bank and could be turned by a man leaning on the wooden framework and treading the pedals. This moved the lower square-pallet boards upwards along the wooden trough, the square-pallets scooping water up to the shore and into ditches which led it into the crop fields. The square-pallet boards then revolved over the upper large wheel and moved downwards along the tail board, turning around the lower shaft and scooping up more water. This cycle was repeated so long as the wheel was trod.

The amount of water raised by the man-powered square-pallet chain-pump was limited by human strength; but it could be used near any water source and was operable by one or two men — features that gave it wide application.

The animal-powered square-pallet chain-pump. Early in the Southern Song Dynasty (1127-1279) the square-pallet chain-pump began to be powered by draught animals, a new stage in its development. The structure of its water-raising system was similar to the man-powered one, but the power mechanism was different. A vertical cogwheel was mounted on the upper horizontal shaft of the chain-pump; near it was installed a large vertical shaft, and in the middle of that a large horizontal cogwheel, the cogs of the horizontal and vertical cogwheels being engaged. A large horizontal pole connected to the vertical shaft was pulled around the shaft by an ox. The transmission of the two cogwheels turned the chain-pump and scooped the water up. The greater strength of the draught animal over man lifted more water to a greater height.

The hydraulic square-pallet chain-pump. This type of pump is described in *Wang Zhen Nong Shu* (*Agricultural Treatise of Wang Zhen*), which was written in 1313, placing this mechanical invention at about the beginning of the Yuan Dynasty, or nearly 700 years

ago. Essentially the same otherwise as its predecessors, the driving-device of this one was installed beside turbulent rivers. A large wooden frame was first erected and a rotating shaft mounted at its mid-point. An upper and a lower large horizontal wheel were mounted on the shaft. The lower horizontal wheel was the water-wheel on which were mounted fan boards that turned the water-wheel as they were hit by the flowing water. The upper horizontal wheel was a large cogwheel which engaged with the vertical cogwheel on the upper shaft of the chain-pump, which was installed in a deep ditch dug in the river. The flowing water rotated the water-wheel, the horizontal cogwheel brought the vertical cogwheel into rotation, and this turned the chain-pump, raising water from the deep ditch to the shore from where it flowed into the field for irrigation.

If the water source was sufficiently high, a large vertical water-wheel could be installed on the axle of the chain-pump to turn it, obviating the two large cogwheels.

This application of a pair of large wooden cogwheels to the water-powered irrigation machine to transmit the rotation of the water-wheel to the shaft of the chain-pump for scooping water up into the fields was a great step forward in Yuan Dynasty mechanics and also a major achievement in utilizing natural power for the benefit of mankind.

An Advanced Grain-Processing Machine — the Hydraulic Trip-Hammer and Watermill

After grain is threshed, it must be polished or ground in order to be edible by man. The Chinese had in ancient times used such grain-processing equipment as the rotary mill, roller mill, trip-hammer, winnow and sieve. Later they invented the hydraulic trip-hammer and watermill. These were important inventions as they were more efficient and more widely used than previous agricultural machinery.

The hydraulic trip-hammer for hulling rice appeared at the end of the Western Han Dynasty (206 B.C.-A.D. 24). *Huan Zi Xin Lun* (*New Discourses of Master Huan*) written by Huan Tan in the Han Dynasty describes this huller.

The motor power of the hydraulic trip-hammer was a large vertical water wheel with fan boards attached. The length of the wheel axle depended on the number of hammers, while alternate board-catches were mounted on the wheel axle. If for example a trip-hammer had four board-catches, four trip-hammers would have 16 board-catches, used to trip the pole of the hammer. Every trip-hammer had an upright wooden pole on one end of which a cone-shaped stone was mounted. The threshed grain was placed in the stone mortar beneath the stone hammer, the flowing water turned the water-wheel, and the board-catches on its axle tripped the tips of the hammer pole making the trip-hammer rise and fall to hull the rice. This equipment could be used continuously.

The hydraulic trip-hammer could be installed anywhere on the shore of rivers and streams, the installation differing according to the specific water potential. If it was relatively small, the water was directed by wooden boards to flow with increased velocity along the sides of the water-wheel. The number of trip-hammers to be driven was determined by the hydraulic power: more trip-hammers could be installed in cases where hydraulic power was abundant; fewer where it was weaker. When more than two trip-hammers were used, the installation was called a connected trip-hammer. These were often used, generally with four trip-hammers.

The rotary mill was used for grinding rice, wheat, beans, etc. It consisted of two cylindrical stones of a certain thickness. In the centre of the lower stone was a short vertical iron shaft, and in the centre of the upper millstone was a hole the size and shape of the shaft. When the upper and lower millstones were brought together, the lower stone was stationary while the upper one revolved around the shaft. Cavities on both faces of the millstones, called grinding cavities, were in contact. Teeth were chiselled on both faces. The upper millstone had a mill hole through which the grain flowed into the cavity and was evenly distributed. The milled grain came through the gaps in the teeth and flowed into the milling pan. After sifting

Geared water-power mill.

out the bran, flour was left. There were rotary mills of the manual, animal-powered and hydraulic types. The latter, or water-mill, was invented as early as the Jin Dynasty (265-420). The water-mill, driven by a horizontal water-wheel, has the upper millstone mounted on the vertical shaft so that flowing water turns the wheel. This type of mill suited localities where water power was ample. For places where water impact was slight and the water amount large, a water-mill was equipped with a vertical wheel as its driving unit. In this case a cogwheel was mounted on the shaft of the vertical wheel, the cogwheel engaging with a horizontal cogwheel mounted below the mill shaft. The rotation of the water-wheel was transmitted through the cogwheels to the mill. Both types were simple in construction and widely used.

With progress in technology, a more complicated water mill was invented where one water-wheel drove several mills simultaneously. This was called geared water-power mill. This mill is mentioned in *Agricultural Treatise of Wang Zhen*. Its large water-wheel was mounted vertically and required a large volume of fast-flowing water to drive it. The wheel shaft was very thick and of suitable length. Three cogwheels were mounted on the shaft at specified intervals, each cogwheel meshing with one on the mill. The three cogwheels on the mills each meshed with the wooden cogs of the mills on each side of those in the centre, the water-wheel turned the mills in the centre through cogwheels, and the middle mills turned the side mills. In this way one water-wheel drove nine mills at once.

Various Types of Carriages; the South-Pointing Carriage and Odometer

Zhou Shide

More than 4,600 years ago the Chinese people already had the carriage, and about 4,000 years ago the ancient Chinese Xue clan was renowned for making them. The book *Zuo Zhuan* (*Zuoqiu Ming's Chronicles*) says that Xi Zhong of the Xue clan had held the post of Chezheng (officer in charge of carriages) in the period of Xia (c. 21st-16th century B.C.). *Mo Zi* (*The Book of Master Mo*), *Xun Zi* (*The Book of Master Xun*) and *Lü Shi Chun Qiu* (*Master Lü's Spring and Autumn Annals*) all state that "Xi Zhong made carriages", indicating that he had made a name for himself in this trade.

During the Shang Dynasty (c. 16th-11th century B.C.) Chinese carriage craftsmen were already fashioning rather sophisticated two-wheeled carriages, a fact evidenced also in Chinese ideograms. The word *che* (carriage) on bone and tortoise shell inscriptions was

[ideograms] , [ideograms] and [ideogram] , showing that Shang

period carriages had two wheels with spokes [symbol] , *yu* (carriage box)

[symbol] , *yuan* (shaft) [symbol] , *heng* (transverse bar) [symbol] , and *e*

(yoke) [symbol] . Shang Dynasty excavations substantiate this elaborately designed two-wheeled carriage with a shaft, one transverse bar and two yokes, and a carriage box.

Carriages of the Shang period excavated in Dasikong Village

of Henan Province have wheels with 18 spokes and rectangular carriage boxes. The craftsmanship of Shang Dynasty carriages was still praised in the Spring and Autumn Period (770-476 B.C.). The reconstructed model of a Shang carriage exhibited in the Museum of Chinese History in Beijing shows a rather sophisticated two-wheeled carriage requiring fairly high carriage-making technique. The ancient Chinese two-wheeled carriage was usually drawn by two or four horses, the general practice being to use two horses with two others pulling from either side.

During the Spring and Autumn and Warring States periods (770-221 B.C.) the high-chassis carriage appeared, called *chaoche*. The chronicle *Zuo Zhuan* says: "The prince of Chu climbed on the *chaoche* to watch the troops of Jin state." *Chaoche* was a carriage with a high chassis and small tower constructed on top. Special attention was given to reinforcing weak places of the carriages, *jiafu* (reinforcement) being used on the wheels, and most of the large carriages excavated from tombs of the Warring States Period have this *jiafu*. The carriage-wheel spokes were all obliquely attached to the hubs, rather advanced construction for that time.

The wheelbarrow, or single-wheeled cart, was extensively used during the Eastern Han Dynasty (25-220) and the Three Kingdoms Period (220-280), being a very economical means of transport especially adapted to move through narrow alleyways and field paths and also suitable for wide roads. The wheelbarrow was indeed a very important invention in the history of transportation. According to the studies of many scholars the "wooden oxen and gliding horses" as artistically described in the historical novel *San Guo Yan Yi* (*Romance of the Three Kingdoms*) were none other than a type of wheelbarrow. Historical records state that in the Three Kingdoms Period a certain Pu Yuan invented a wheelbarrow to be used in warfare. Pu Yuan was a subordinate officer under Zhuge Liang, the prime minister of the state of Shu. When Zhuge Liang led his troops in a northward march, Pu Yuan fashioned "wooden oxen" to transport food and forage for the army. During this period the four-wheeled carriage also appeared. The book *Gu Yue Fu* (*Treasury of Ancient Songs*) contains such verses as

I made you a carriage with four wheels;
It will carry you to Luoyang Palace.

During the Southern and Northern Dynasties (420-589) large carriages drawn by 12 oxen appeared. *Hou Wei Shu* (*History of the Later Wei Dynasty*) says: "The large *lounian* [towered carriage] is drawn by 12 oxen, the small *lounian* is also drawn by 12 oxen." At about the same time, the *moche* (mill-cart) appeared. This carriage had a millstone mounted on it which rotated when the carriage moved. *Gu Jin Tu Shu Ji Cheng* (*Collection of Books, Ancient and Modern*) says: "[A certain] Shi Hu has a mill-cart with the millstone mounted on the carriage. When the carriage has travelled 10 *li*[1], then 10 *hu*[2] of grain will be ground." There are also accounts about using petroleum as lubricant. *Shui Jing Zhu* (*Commentary on the "Waterways Classic"*) says: "In Gaonu County there was the Weishui River with oil on the surface which could be used to lubricate the carriages." Likely this petroleum was used as a lubricant for carriage bearings and the hydraulic trip-hammer. According to *Liang Shu* (*History of the Liang Dynasty*), Hou Jing, who came from a commoner family in the Liang Dynasty (502-557) during the Southern Dynasties, invented a 100-*chi*[3]-tall *louche* (tower carriage), besides *dengchengche* (city-mounting carriage), *jiedaoche* (stair carriage) and *huoche* (fire carriage). All these were scores of metres high.

Commenting on the three-wheeled carriage which appeared during the Five Dynasties period (907-979), *Lei Shu Zuan Yao* (*Summary of Books*) Vol. 30 says: "Lin Zhiyuan sometimes rode on a three-wheeled carriage, commanding a servant to pull it." This is obviously an ancient prototype of the rickshaw, or human-powered carriage, but triple-wheeled.

During the Ming Dynasty (1368-1644) Chinese carriage-makers turned out eight-wheeled carriages for transporting architectural

---

[1] A *li* was then about 440 metres.
[2] 1 *hu* = 100 litres.
[3] 1 (Liang) *chi* = 0.244 metre.

materials. *Gu Jiu Tu Shu Ji Cheng* says that after Mao Bowen became Minister of Works and took charge of the Tian Shou Shan engineering project, he designed and had made eight-wheeled carriages to move the stone pillars used in the project. This new-type vehicle was suited both for transporting large stone blocks and being drawn over mountain roads.

The sail-wheelbarrow that appeared in the Qing Dynasty (1644-1911) utilized wind power to help the vehicle along. In the same era an iron-armoured car called *tiejiao* was made. *Nan Yue You Ji* (*Travels in the South*) Vol. 1 says: "There was a *tiejiao* in a big building in Shiziqiao." This kind of iron-armoured car had four wheels, each wheel being one *chi* in diameter. The box of the carriage was covered with iron plates. This was a carriage specially designed for security.

Especially important in mechanical structure were the south-pointing carriage and odometer. From the period of the Three Kingdoms, almost all Chinese histories mention these types. Not until the Yuan Dynasty (1271-1368) however, in *Song Shi* (*History of the Song Dynasty*), was a detailed description given of its internal gear structure.

The south-pointing carriage is two-wheeled with a single shaft. A wooden human figure stands in the carriage with one outstretched arm pointing south, and the arm continues pointing south no matter which direction the carriage moves, due to the gear system. During the Song Dynasty (960-1279), Yan Su in 1027 and then Wu Deren in 1107 constructed a south-pointing carriage. We will describe the basic principle of the device made by Yan Su. There are two foot-wheels each six (Song) *chi*[1] in diameter and 18 *chi* in circumference; two small wheels each three *cun*[2] in diameter; attached to the foot-wheels are two vertical subordinate wheels each 2.4 *chi* in diameter and 7.2 *chi* in circumference. These each have 24 teeth three *cun* apart. There are also two small horizontal wheels on the right and left sides. These are 1.4 *chi* in diameter and have 12 teeth.

---

[1] 1 (Song) *chi* = 0.319 metre.
[2] 1 *cun* = 1/10 *chi*.

In the middle is a large horizontal wheel 4.8 *chi* in diameter and 14.4 *chi* in circumference with 48 teeth three *cun* apart.

The "foot-wheels" rotate and are in contact with the ground; the small wheels are pulleys. The vertical subordinate wheels are a gear attached to the foot-wheels. The large horizontal wheel in the middle is a large gear mounted horizontally midway of the carriage box. The shaft goes through this large horizontal wheel and carries the wooden figure, which turns along with the large wheel. The two small horizontal wheels are gears installed on either side of the large horizontal wheel to transmit motion. If the carriage moves straight forward, the two small horizontal wheels are disengaged from the large horizontal wheel, and so the rotation of the two foot-wheels will be independent of the large horizontal wheel. If the carriage turns left, the front end of the vehicle's pole will move towards the left and its rear end will move towards the right, the cable of the rear end of the pole moving towards the right through the pulley to let the small horizontal wheels on the right side fall so that contact is made with the large horizontal wheel in the centre. The large horizontal wheel in the centre, which is controlled by the carriage wheel on the right side, will rotate towards the right, this motion being just sufficient to counterbalance the effect of turning left. The outstretched arm of the wooden figure will always point in the same direction, namely south. If the carriage turns 60 degrees to the left, for example, then the large middle horizontal wheel will rotate 60 degrees to the right and the wooden figure go on pointing south.

The south-pointing carriage is simple in structure. The crucial part is the large horizontal wheel in the middle and the design to connect or disconnect the subordinate wheel attached to the foot-wheel. Still, it demonstrates skill and creativeness on the part of ancient Chinese labouring people and machine designers.

Ji li gu che (*li*-recording drum carriage) is the original Chinese term for the odometer. There are two records in *Song Shi* (*History of the Song Dynasty*) in the section "Records of Carriages and Dresses" about making the odometer. Lu Daolong constructed the odometer in 1027, and Wu Deren reconstructed it between 1107 and 1110.

Lu Daolong's odometer had:

2 foot-wheels each 6 (Song) *chi* in diameter and 18 *chi* in circumference;

1 vertical wheel (gear) 1.38 *chi* in diameter and 4.14 *chi* in circumference with 18 teeth 2.3 *cun* apart;

1 lower horizontal wheel (gear) 4.14 *chi* in diameter and 12.42 *chi* in circumference with 54 teeth 2.3 *cun* apart;

1 "turning-like-the-wind" wheel (gear) with 3 teeth 1.2 *cun* apart;

1 middle horizontal wheel (gear) 4 *chi* in diameter and 12 *chi* in circumference with 100 teeth 1.2 *cun* apart;

1 small horizontal wheel (gear) with 10 teeth and 1 upper horizontal wheel (gear) with 100 teeth.

Some dimensional inaccuracies occur in the book.

The whole system of gears in the odometer moves or stops with the carriage. As soon as the wheels start to rotate, the whole system of gears is thrown into action, and once the wheels stop the whole system of gears stops.

The foot-wheels are 6 *chi* in diameter, one revolution covering 18 *chi*. If the foot-wheels make 100 revolutions, then the carriage will cover 1,800 *chi*, which is exactly one *li* in the Song Dynasty time. The number of teeth of the vertical, lower horizontal, "turning-like-the-wind" and middle horizontal wheels are 18, 54, 3 and 100 respectively. When the carriage covers one *li*, therefore, the middle horizontal wheel will make $100 \times 18/54 \times 3/100 = 1$ revolution. A catch that acts like a cam wheel is attached to the middle horizontal wheel to pull the arms of the wooden figure, which in turn will strike the drum. The wooden figure will thus strike the drum once for each *li* of ground covered. If a small horizontal wheel of 10 teeth and a upper horizontal wheel of 100 teeth are added, then the upper horizontal wheel will make one revolution for each 10 *li* of distance covered and a catch attached to the upper horizontal wheel will pull another wooden figure's arm to strike a bell once.

Mr. Wang Zhenduo has reconstructed models of the south-pointing carriage and the odometer as shown in the illustrations.

The odometer has itself a reduction train of gearing to slow down the rotation so that the last axle will revolve only once when the carriage travels one *li* or 10 *li*. The train of gearing in the south-pointing carriage is simpler, but its ability to automatically connect or disconnect the gearing system makes it more sophisticated than the odometer. Historical accounts place these two types of carriage as early as the Han and Wei dynasties, when China was still in the era of early feudalism.

They show the high level of mechanical engineering of China 2,000 years ago and embody the outstanding achievements of ancient Chinese technology.

The Great Wall

Zhang Yuhuan

Snaking over the vast expanse of northern China, the Great Wall extends east-west through loess plateaus and desert plains, rising to the rims of mountains and descending into river basins. It is one of the architectural marvels of the world both for its magnificence and for the tremendous amount of labour its construction involved.

The building of the Great Wall began over 2,200 years ago in the period of the Warring States (475-221 B.C.). A number of cities had already risen, such as Xianyang in the state of Qin, Handan in Zhao, Xiadu in Yan and Daliang in the state of Wei. The extensive northern plain became the scene of contention by the rulers of nomadic tribes in their southward expansion, among them the Xianyun,[1] Linhu,[2] Loufan,[3] the Eastern Hus and the Huns. Wars became ever more frequent among the feudal states, which were trying to absorb each other.

The earliest walls were erected along the northern borders of the warring states by their feudal rulers as a defence against surprise attacks from the north. These walls were later extended to boundaries between the states themselves for defence against each other.

Remnants of walls built by the states of Yan, Zhao, Wei and Qi

---

[1] Xianyun, known also as Rong or Di at that time, was referred to as Xiongnu or Huns in the Qin and Han dynasties.

[2] Linhu, a nomadic tribe, also called Zhanlin, who in the period of the Warring States inhabited the areas of present-day Shuoxian in Shanxi and Inner Mongolia.

[3] Loufan, an ancient tribe in what is Shanxi today.

remain standing to the present day.   The walls of Yan, from Dushikou eastward to beyond Liaoning, defended against the Huns and
the Eastern Hus.   Those of the state of Zhao, extending from Linhe
County in Inner Mongolia in the west to Yuxian County in Hebei
in the east, were to ward off the Linhus and the Loufans.   The state
of Wei built a wall from the Great Northern Bend of the Yellow
River in the north all the way down to northwestern Shaanxi and
reaching as far south as Mount Huashan, to guard against incursions
by the Huns and from the state of Qin.   The walls of Qi, starting
west from the Yellow River in Shandong and skirting Mount Taishan down to Zhucheng County, were prevention against possible
assaults from the states of Wu and Chu.

Conquering all six states outside his realm, Qin Shi Huang
(the First Emperor of Qin) built up the first feudalist empire under
centralized authority in Chinese history.   To defend against possible
aggression by the Hun slave-owners, Qin Shi Huang had the walls
of Yan, Zhao and Wei linked up to form one great barrier.   Basic
completion of the work consumed 10 years' labour by 300,000 men
and resulted in what was to become known as the Great Wall of
Ten Thousand *Li*.[1]

The still-existent walls of the state of Wei typify wall structure
in the Warring States Period.   Nearby South Maliangzhuang Village in Hancheng County, Shaanxi Province, they take the form of
two parallel north-south walls 160 metres apart.   The wall on the
south is seven metres thick at the base and four at the top, with
a remaining height of four metres.   The smaller one on the north
is five metres thick at the base and 3.5 metres at the top.   As it remains today, its height is about three metres.   The walls are built
entirely of rammed earth.   The beacon tower, located about 250
metres south of the southern wall, is also constructed of earth.   It
is square in shape, each side measuring seven metres, and tapering
at the top to a height of nine metres.   On each side of the watch-
tower midway from the ground are the ends of three built-in square
timbers, presumably for the purpose of consolidation.

---

[1]   1 *li* = 1/2 kilometre or roughly 1/3 mile.

The linked-up wall of the Qin Dynasty (221-207 B.C.) rose from Lintao (Minxian) County, Gansu Province in the west, followed the Yellow River to Linhe in Inner Mongolia, then curved north to Yinshan Mountains and on south to Yanmenguan, Daixian and Yuxian in Shaanxi, where it joined the walls on the northern border of the state of Wei. It ran from there through Zhangjiakou eastward to the Yanshan Mountains, Jinzhou and on eastern Liaoning.

The wall's western extremity at Lintao was built of rammed earth, in some places mixed with small quantities of rocks. It is 4.2 metres thick at the base and 2.5 metres at the top, with a remaining height of about three metres. The walls are obviously well tamped, probably with tampers of handy size, judging from the small tamping marks still observable.

The walls of the succeeding Han Dynasty (206 B.C.-A.D. 220) were far superior to their predecessors in size and appearance. The walls and their beacon towers have survived to this day. Besides rebuilding the walls of Qin, the Han rulers built others at Shuofang (Erdos Right Rear Banner, Inner Mongolia), began large-scale construction of the Great Wall west of Liangzhou, set up the

A section of the Great Wall built in the
Western Han Dynasty (206 B. C. -A. D. 24).

four prefectures of Liangzhou (now Wuwei), Ganzhou (now Zhang-ye), Suzhou (now Jiuquan) and Shazhou (now Dunhuang).

The Han rulers expended enormous manpower and material on building this "Great Wall West of Liangzhou". The purpose was dual: to guarantee through traffic along the Gansu Corridor so as to maintain rule over minority peoples in the western regions; and to cut the communication links which the Huns had with the Western Regions.

The Great Wall west of Liangzhou was in three sections, all within Gansu Province. The northern section rose in Ejin Banner and continued southwestward through Dafangcheng to Jinta County, from where the mid-section continues by way of Pochengzi and Qiaowancheng to Anxi County. The southern section extends from Anxi County, passes to the north of Dunhuang, goes all the way down to Dafangpancheng and the Yumenguan Pass and into Xinjiang. Describing these three sections of the Great Wall, begun during the reign of the Emperor Wu Di of the Han Dynasty, the Han bamboo strips excavated in Juyan County say they were "punctuated by a beacon tower at every five *li*, a platform at every 10 *li*, a fortress at every 30 *li* and a walled stronghold at every 100 *li*". The beacon towers have been found, however, to be scarcely three *li* apart and the walled strongholds occurring at each few dozen *li*.

Han Dynasty walls were quite different from those of Qin. The section at Yumenguan southwest of Dunhuang, for example, is 3.5 metres thick at the base and 1.1 metres at the top. The ruins are still four metres high. To counter alkaline corrosion, layers of reeds 6 centimetres thick were built in from 50 centimetres upwards. These were arranged at intervals of 15 centimetres, laid crosswise. These reed layers are intact today. The walls were built of earth fetched locally and mixed with rocks, the whole consolidated by ramming.

Several hundred of the beacon towers built alongside the wall, some inside and others outside, still remain. Of these, some 200 are found in Jinta County and Ejin Banner, Gansu Province. Here, each beacon tower is square at the base, which measures 17 metres on each side. The four sides slope inwards at the top, which is 25 metres from the ground. Some of the beacon towers are built of rammed earth, others of clay bricks, while a combination of these materials is used in still others. The clay bricks are $38 \times 20 \times 9$ centimetres, placed in layers with one intermediate layer of reeds to each three layers of bricks. Most of the beacon towers remain, with only their corners corroded, indicating that the building in of reeds fortified the walls against destruction by the elements.

Holes left in the beacon tower walls point to the use of scaffolds during construction. Two rows of such holes, four in each row, occur in the wall of a beacon tower in Tiancang Commune, Jinta County. The first row is 14 metres above the base, the second 0.8 metre above the first row. The holes, arranged at intervals of 1.2 metres, apparently have been left from a scaffold supported on the outside by wooden poles planted in a single row.

Tremendous difficulties were certainly encountered in rigging up sheds, providing food and shelter for the workers and transporting supplies and material for the building of the gigantic wall over the barren Gobi desert, uninhabited for hundreds of miles around. Still greater hardships must have been endured by the builders in the Northwest where the weather is fickle, with blinding sandstorms and freezing temperatures most of the year. Untold amounts of agonizing labour must have gone into the building of the Great Wall.

The lofty, unforgettable Great Wall as seen today at Badaling and Juyongguan near Beijing is the result of grand-scale reconstruction over a period of 100 years during the Ming Dynasty (1368-1644).

The Great Wall as built in the Ming Dynasty begins at Jiayuguan, Gansu Province, in the west and terminates at Shanhaiguan on north China's east coast. It has a total length of 12,700 *li*. Divided into four sections, the Great Wall in Ming times was divided into four garrison zones:

The Liaodong Garrison Zone, with headquarters at Liaoyang in Liaoning Province, guarded the section from Fenghuangcheng to its terminal Shanhaiguan in the west — a length of 1,950 *li*;

Xuanfu Garrison Zone, with headquarters at Xuanhua in Hebei Province, held the section from Juyongguan in the east to Datong in the west, a total length of 1,023 *li*. Flanking Beijing, the capital, on the north, this section was of such vital importance that nine walls were erected one within another.

The Datong Garrison Zone with headquarters at Datong, Shanxi, was responsible for the 647 *li* of the wall from Pianguan eastward to Tianzhen, both in Shanxi.

The fourth, the Yulin Garrison Zone with headquarters at Yulin in Shaanxi commanded the 1,770 *li* of the wall between Qingshuihe, Inner Mongolia, in the east and Yanchi County in Ningxia.

To strengthen the defence, three more garrison zones were later set up:

Ningxia Garrison Zone with headquarters at Yinchuan, Ningxia, controlling the 2,000-*li* section from Yanchi in Ningxia westward to Jingyuan in Gansu;

Gansu Garrison Zone with headquarters at Zhangye. This controlled the section from Lanzhou westward to the Great Wall's beginning at Jiayuguan, a total length of 1,600 *li*;

Jizhou Garrison Zone with headquarters at Jixian County in Hebei Province which controlled the 1,200 *li* from Shanhaiguan westward to Juyongguan.

The walls in the Jizhou Garrison Zone were specially built to defend the capital city of Beijing. The area around Juyongguan was protected by triple walls on which were the three "outer passes" of Pianguan, Ningwuguan and Yanmenguan, and the three "inner passes" of Juyongguan, Zijingguan and Daomaguan.

The Ming walls were divided into the eastern wall and the western wall.

The eastern wall, stretching east from Shanxi, was built in mountains, curving and undulating with their peaks and ravines. The wall structure is six metres thick at the base, sloping inwards to 5.4 metres at the top. The top of the wall is flanked on the outside — the northern side from which the threat came — by battlements two metres high, and on the inside by parapets one metre high. The wall itself rises to a height of 8.7 metres. Guardhouses stand in the wall about 70 metres apart. Some are hollow and served as barracks for the garrison troops, while others are solid blocks. Ditches and openings on top of the wall provided drainage and for the disposal of rain water. Mounting-steps are built at intervals of 200 metres on the inner, south side of the wall. Both sides and the top of the wall are faced with bricks on the rammed earth. Brick kilns were set up to supply bricks as near the site as possible. Yet the difficulty of hauling bricks up mountains could not compare with that involved in the wall construction at Badaling and other eastern sections which are of 18 layers of giant flagstones. Here the toil and hardships can scarcely be imagined.

The western sector of the Great Wall, lying west of Shanxi, as it was rebuilt in Ming times, is entirely of rammed earth unfaced with bricks. It is four metres thick at the base and 1.6 metres wide at the top, with battlements 0.8 metre high. A passageway 1.2 metres wide on the top of the wall provides for the movement of troops. The total height of the wall is 5.3 metres.

Beacon towers built on either side of the wall, some detached and standing on mountain peaks while others form part of the wall, are 12 metres high and square in shape, the eight-metre sides sloping inwards towards the top.

The colossal amount of earthwork involved in building the wall may be realized from some figures. The wooden tamping frames were each four metres long. Eighty cubic metres of earth was tamped in for each tamping-frame length, while 800 cubic metres of earth went into each beacon tower.

The builders of the Ming Dynasty walls were organized into brigades according to a stone tablet titled *The Beginning of the First Brigade* unearthed at the north wall of Jiayuguan in the 19th year of the Jiaqing period (A.D. 1540). Six brigades made up one construction unit.

This reconstructed Great Wall is replete with passes fortified by strongholds, all located in key strategic positions. The two most famous passes are Jiayuguan and Shanhaiguan, the two ends of the Wall. The western terminus at Jiayuguan is located 70 *li* west of Jiuquan on the road leading to Xinjiang. The walled fortress is square, its sides totalling 660 metres in length. On the eastern and western walls of the fortress are city gates, each protected by an extra wall. The walls of the fortress are 12 metres high, with guardhouses at the corners. Watchtowers guard the southern and northern sides of the walled fortress, while the city gates, such as Guanghua Gate, Rouyuan Gate and the Gate of Jiayuguan Pass, are crowned by towers. These stately and impressive gate-towers overlook the plateau below. The wall top, parapets, battlements and steps leading to the top are all brick-laid, while other parts of the wall are ramparts of rammed earth. The Great Wall rises from the foot of the Qilian Mountains, reaching the walled fortress of Jiayuguan after a few hundred winding metres. The combination of topography, wall and gate constitute a powerful shield of defence known as The Martial Barrier of All Under Heaven.

The eastern terminus of the Ming reconstruction of the Great Wall at Shanhaiguan is situated on the Hebei-Liaoning provincial border. The walled fortress of Shanhaiguan, with mountains on its west and the sea of Bohai on its east, is joined by the Great Wall from the hills above. This is a key juncture, a narrow pass on the trunk line of communication between north China and the northeastern provinces.

A section of the Great Wall and beacon tower reconstructed in the Ming Dynasty (1368-1644).

The Shanhaiguan fortress is a square walled compound with a gate in each of the four sides, only the eastern gate being shielded by an outer wall. Each of the four gates has a gate-tower. The fort's walls, 12 metres high, are built of rammed earth faced with bricks and remain solid to this day. From the gate-tower one can view the rolling mountains to the north and the sea to the east. Shanhaiguan has been known from ancient times as The First Gate Under Heaven.

As a gigantic architectural work representing the toil and knowledge of countless labouring people the Great Wall commands visitors' great admiration and respect.

Bridges

Huang Mengping

The construction of bridges was very important in ancient China's splendid civilization. Over the centuries the people laboured to build the myriad bridges linking the vast land masses of China. Impassable rivers and canyons became thoroughfares, the greater number of bridges being found in south China where rivers, streams and creeks abound. "Go out the door of your home, and you'll see two bridges," say the residents of Suzhou in Jiangsu Province, known as "the land of waters". Such popular lyrics as *Anchoring by the Maple Bridge at Night*, *Ode to Willow in Snow by the Bridge over the Ba River* and other poetical works eloquently describe the beauty of landscapes enhanced by the lovely arches of bridges.

Great variety is shown in bridges constructed since ancient times. Besides the three main styles — beam, arch and suspension bridges — there are the cantilevers, stone-beam and movable bridges, as well as many kinds of truss-beam structures. No few show an originality in architectural technique that is worthy of scientific study. Many bridges surviving from ancient times remain in use today. Their construction conforms with modern engineering principles and reflects ancient bridge builders' knowledge of architectural mechanics, pedology and hydrology.

Development of Bridge Construction Technique

Chu Xue Ji (*The Primary Anthology*), a collection of classic literature dating from Zhou through early Tang, compiled by Xu

Jian and others in the Tang Dynasty (618-907), contains the following record: "King Wen of Zhou made a boat bridge over the River Weishui." This was 3,000 years ago, the first bridge recorded in Chinese history. It was a temporary pontoon bridge which King Wen provided for his bride to cross the river, and it was removed after the royal wedding. By 257 B.C. King Zhaoxiang of Qin built a large pontoon bridge across the Yellow River, one that exhibited fairly high technique in bridge building at that time. It showed that in large rivers where the depth of the water and swift current made it difficult to build piers, pontoon bridges were preferred.

Easily built and removed, the pontoon bridge was suitable for temporary or military use. Very often it preceded the construction of a permanent bridge, as the construction and use of the pontoon type helped in understanding the river and the changes it underwent. Examples were numerous of truss or arch bridges of stone being preceded by pontoons. Over the past 1,000 years many a large pontoon bridge had spanned the Yangzi and Yellow rivers. Their planning and design were varied to suit the local conditions, utilized local material and adapted to seasonal changes.

The earliest non-floating bridges were the wooden truss type. As the material was not strong enough for ever larger spans, timbers were piled crosswise on the piers, each layer extending outward to make up the length of the beam and increase the span. This was like the modern cantilever bridge, which has large spans with short beams supported by the cantilever. When the number of layers was more than the piers could bear, diagonal wooden stays were placed between the wooden beam and the pier itself. The slanting stay served the dual purpose of strengthening the beam and keeping the cantilevers in place. This type of bridge with truss beams and diagonal stays was the prototype of the modern cantilever bridge.

Next were bridges built of stone, and Chinese stone bridges figure prominently in the history of world bridge-building. The first record of stone-bridge construction appeared over 2,000 years ago. Says *San Fu Jiu Shi* (*Tales of Three Cities*):

To the north of the city of Changan, two *li* from the city

gate, there is a bridge which was built by the First Emperor of Qin. The bridges built in the Han Dynasty followed the Qin model. The bridge is 60 *chi*[1] wide and 280 *bu*[2] long, with 68 arches, 750 piers and 220 beams. It is named Stone Pillar Bridge.

This is obviously the 400-metre-long, 68-arch stone pillar bridge of Changan, now Xi'an in Shaanxi Province.

Archaeologists have determined from excavations that besides bridges with stone piers and beams, stone-arch bridges made their appearance at least as early as 250 B.C. This is supported by the stone-arch structure of a gate in the tomb of Han Jun, a noble who lived at the end of the Zhou Dynasty (770-221 B.C.), discovered at Luoyang in Henan. Five centuries later a stone-arch bridge named "Travellers' Bridge" is mentioned for the first time in the Chapter on Rivers in *Shui Jing Zhu* (*Commentary of the "Waterways Classic"*), as follows:

> ... then the river curves eastward and feeds into a mountain gorge known as the "Brook of Seven *Li*"[3]. ... It was spanned by a stone bridge, which is the Travellers' Bridge. ... Situated some six or seven *li* from the Luoyang palaces, the bridge is built entirely of great stone blocks. It is rounded underneath so that not only does the water flow, but also large boats can go through. ...

Stone-arch bridges were built in great numbers mainly because of the non-decaying and pressure-resisting properties of the stone. However, with the demand for larger spans and consequent increase in the weight of the bridge itself plus that of the traffic, the strain fell on the arch in a more complicated manner. To cope with this, Chinese bridge builders created a new type of stone structure, an outstanding example of which is the stone bridge at Anji, Zhaozhou

---

[1] A *chi* of the Tang Dynasty was equivalent to 0.294 metre.
[2] A *bu* of the Tang Dynasty was equivalent to 1.47 metres.
[3] A *li* was then equivalent to 540 metres.

(now Zhaoxian County in Hebei Province). This brought stone bridge building technique to a new high level.

Besides wood and stone, the ancient Chinese bridge builders used such materials as bricks, bamboo, rattans and iron. The suspension rope bridge which first appeared in southwestern China was the product of construction materials most easily available in the locality.

In the course of centuries the ancient Chinese technique of bridge building, which evolved and developed through practice, resulted in the great variety of splendid bridges today.

Arch Bridge Models

The Anji Bridge at Zhaoxian County in Hebei, the Lugou Bridge southwest of Beijing and the Jewel Belt Bridge over the Daidai River south of Suzhou are three classical arch bridges famous through the centuries. Highly individual in style and standing elegant and grand, these bridges truly exemplify the ancient Chinese arch bridge.

The Anji Bridge, also known as the Zhaozhou Bridge, spans the Xiaohe River. It is a large one-arch stone bridge built in the period of Daye (605-616) of the Sui Dynasty (581-618). Over 1,300 years old, it has withstood nine fierce wars, eight major earthquakes and innumerable floods. It stands today, firm and in unscathed magnificence.

The Anji Bridge is singular in design and structure and graceful in appearance. The span of the arch is 37.37 metres; the rise from the chord-line to the crown a mere 7.23 metres; the ratio of the span to the height of the arch is not more than 5:1, making the height of the bridge far less than the radius of the arch. The entire bridge proper is within the arch itself, a very practical type of bridge.

The gradualness of the arch is advantageous for both pedestrian and vehicle traffic. The Anji Bridge is admired in China and other countries for its balance and stateliness. Its single span was unequalled in its time.

The brilliance of the bridge engineering involved, however, is best shown by the two small holes at either end of the arch spandrel. This was in contradistinction to the usual practice of filling the spandrel with earth, and created the world's first spandrel arch bridge some 1,200 years earlier than in the West. Dr. Joseph Needham of Britain pointed out that the bridge builder Li Chun's work "has a priority of more than a millennium, for not until the railway age (the seventies of the nineteenth century) did comparable, if larger, Western bridges arise. . . ."[1] Needham highly commends the "spandrel-arch school" initiated with the Anji Bridge and its signal influence on bridge engineering. In Dr. Needham's words: "Li Chun's spandrel-arch construction was thus the ancestor of those many bridges of reinforced concrete which dispense with all filling between the arch ring and the deck, connecting them only by vertical pillars or a reticulate construction of concrete members."[2]

The Anji Bridge, which has stood firm for over 1,000 years, is a successful example in bridge engineering that fully displays the following advantages of spandrel-arch construction: 1. the weight of the bridge is reduced and filling material saved; 2. its subsidiary arches allow greater passage of water in flood season; 3. it is more

Beijing's Lugou (Marco Polo) Bridge.

[1] J. Needham, *Science and Civilisation in China*, Cambridge University Press, 1971, Vol. 4, p. 180.
[2] *Ibid.*, p. 179.

elegant in appearance than filled spandrel arches; 4. its structure conforms with the principles of modern mechanics so as to be always most favourably utilized under its own weight and its traffic load. The spandrel-arch structural technique, first applied in the Anji Bridge, ushered in a new era in the history of stone-arch bridge building.

Where the river is too broad for a single span, multi-span segmental bridges were designed, an outstanding example being the Lugou (Marco Polo) Bridge built in the Kin Dynasty (1038-1227) 800 years ago. The Lugou Bridge is located on the Yongding River southwest of Beijing. It has eleven spans with a total length of 265 metres. Its structure is singular in being what is known as the "continuous bridge". Each pair of neighbouring spans rest on a common pier, all spans co-ordinating to form an integral whole, so that when one span receives a load, it is "assisted" by all the others. The load is thus borne by the entire bridge as it passes over.

The Yongding River, which the Lugou Bridge spans, translates as "Forever Fixed" River. Because of its frequent disastrous flooding, however, it has been given another name — Wuding, or "Unfixed" River. Freezing over in winter, its ice-floes can be catastrophic in spring. The bridge builders took this into consideration and constructed the piers accordingly. Besides being securely "rooted" in the riverbed, the Lugou Bridge piers have pointed "beaks" on the upstream side, enabling them to cope with the fury of the summer floods and the ice-floes in spring. They have done this for

Jewel Belt Bridge at Suzhou.

1,000 years and stand intact today. Though the river it spans is justifiably called "unfixed", the Lugou Bridge has shown itself to be definitely "fixed".

Of the countless number of classical Chinese arch bridges, the Baodai (Jewel Belt) Bridge built at Suzhou in the Tang Dynasty is worth special mention. With its prospering economy and flourishing culture, Suzhou was then the scene of very busy traffic both by water and by land. Consequently the region was particularly noted for its bridges and roads. This is most vividly portrayed in the popular verse by Bai Juyi, the great Tang poet: "Three hundred ninety bridges with crimson railings".

With 53 spans and a total length of 317 metres, the Baodai Bridge was planned with such originality as to meet the needs of both the traffic over it and that beneath its arches. Through the dynasties of Tang, Song, Ming and Qing, when the bridge underwent repairs and rebuilding, special attention was paid likewise to the design of the three middle spans, which rise above all others for ships to pass through. This bridge is singular in multi-span stone bridge design.

Different from the Lugou Bridge, in which each pair of neighbouring spans rest on one common pier, the Baodai has each of its spans supported by two abutments on either end, so that damage to one span does not affect the one adjacent, and repairs can be done separately. At the same time, each pair of abutments rests on the same pier, so that the spans are linked up to make the bridge an integrated whole.

Five Examples of Beam Bridges

The beam bridge has the longest history in bridge engineering whether in China or elsewhere. Of the stone beam bridges in China three may be mentioned as master works, namely, the Luoyang Bridge and Anping Bridge in Fujian and the Ba Bridge at Xi'an, Shaanxi.

Situated at the estuary of the Luoyang River at Quanzhou in Fujian, the Luoyang Bridge lies on the highway between Fuzhou and Xiamen. Built in the Northern Song Dynasty (960-1127), the bridge has a total length of 834 metres and a seven-metre-wide deck for traffic. Many feats of daring were required to lay the foundations of the bridge that was to span the turbulent waters of the river where its stream met the sea's tidal surge. As a final resort, an underwater dam was built as foundation by throwing tens of thousands of square metres of stones in the river for the length of the bridge. This dam, over 20 metres wide and half a kilometre long, raised the river-bed by more than three metres. The piers were constructed on the dam. This method has developed into the grillage foundations of modern bridge engineering.

Luoyang Bridge in Fujian Province.

The Luoyang Bridge marked a new stage in the history of China's beam bridge building. Following its pattern, many similar bridges were built in various parts of China, and bridge building flourished in the Southern Song Dynasty 900 years ago.

Following up a century later was a five-*li* bridge crossing sea water. This was the Anping Bridge at Jinjiang in Fujian Province. It is similar to the Luoyang Bridge in that it crosses an arm of turbulent sea, but it is four times as long, lying on some 300 piers. What was remarkable was that such a giant project was completed in one year, making the bridge unprecedented not only for its length, but for the speed of construction. The piers vary in shape according to their positions in the river. Those situated in the mainstream have a cross-section pointed in one or both directions to reduce resistance to the current. The rest of the piers are rectangular in cross-section. Five pavilions on the deck provide pedestrians with a place to stop and rest. These and many other features of the bridge demonstrate that despite the speed of its construction the bridge was designed and built with great circumspection and attention to detail.

The most famous megalithic beam bridge is the Ba Bridge at Xi'an. Situated 20 *li* east of the city centre, the bridge was originally built in the period of Qin and Han over 2,000 years ago. Repaired and reconstructed over the dynasties, the bridge as it stands now was built in the last dynasty, that of Qing. Its deck was replaced with beams of reinforced concrete in 1955. The bridge now has a total length of 386 metres with 64 spans, each not exceeding four to seven metres in breadth. This great density of piers does not offer excessive resistance to the current, however, for all 64 piers are of flat stones set vertically and parallel with the stream-direction, reducing the wash of the current upon the foundation. This explains why the Ba Bridge foundation remains intact after 2,000 years, though the surface structure has been several times rebuilt or replaced.

Suspension and Movable Bridges

In southwestern China many turbulent rivers and mountainous gorges are innavigable and cannot be spanned by pier-supported bridges. What was used for communication between the two banks were catenary cables, the predecessor of the suspension bridges. This type, after the arch and beam bridges, became the third main bridge type. The Luding Iron-Chain Bridge and the Zhupu Bamboo-Cable Bridge are ancient examples.

The Luding Bridge in Luding County, Sichuan Province, where the Red Army forced the Dadu River in 1935 on the Long March, spans the Dadu River, which rushes between the steep crags and precipices of the Erlang and Snow Mountains. With a total length of 101 metres and a width of three metres, the bridge was built of nine chains of iron links nine centimetres in diameter. These chains are suspended from the cliffs on the banks of the river. Wooden planks were laid on the chains to form the deck, with two supplementary chains stretched on either side as a railing. All 13 chains are attached to iron columns socketed in the foundations of the abutments forming a single-span giant chain bridge.

The Luding Iron-Chain Bridge in Sichuan Province.

The builders of the bridge first tried to use boats for hauling the iron chains across the river, but were forced to give up the attempt due to the great weight of their load. After all such attempts failed, the chains were finally delivered by first throwing stout ropes across. The chains were attached to hollow bamboo tubes which had been slipped on the cables, so that the chains could be moved along the cables by pulling the bamboo tubes across. In this way the heavy chains were delivered to the opposite bank.

The Zhupu Bridge, situated at the upper end of the Dujiangyan canal in Sichuan Province, was rebuilt in the Qing Dynasty (1644-1911) after its destruction towards the end of Ming (1368-1644). It is made up of 10 bamboo cables each about 10 centimetres thick,

with transverse wooden planks as deck and three additional bamboo
cables on either side for railings. The bridge is 320 metres long.
To prevent excessive sagging of the cables at mid-point, intermediate
piers were built with log-cribs planted on foundations driven deep
into the river-bed. The piers mark the division of the bridge into
nine spans, of which the largest is 61 metres in breadth. A bamboo
grove in the vicinity provides material for cable replacement.

In Chaozhou, Guangdong Province, is the Guangji Bridge
which may be considered the forerunner of the movable or swing
bridge seen in many parts of world which opens up to allow tall
ships to pass.

The Guangji Bridge, also known as the Xiangzi Bridge, was
built in the Tang Dynasty. This 580-metre bridge is situated at
a key point of communication where the three provinces of Fujian,
Jiangxi and Guangdong have a common border. As navigation
is heavy on the Hanjiang River which the bridge spans, the builders
divided it into three sections, the middle one being a movable pontoon which could be thrown open for the passing of ships, while the
two end sections were fixed.

The Hanjiang River is known for raging rapids and catastrophic floods in the rainy season, while its flow varies in the dry and
flood seasons by 100 times. The construction of piers obviously
posed an extremely difficult problem. It took 57 years to build the
nine piers of the western section and another 16 for the nine on the
eastern section. The pontoon in the middle was made up of 24
boats. This not only facilitated navigation but saved the builders
the further years of hard labour constructing piers in mid-stream.

Bridges of Special Structure

Ancient Chinese bridges present a great variety of elegant shapes
because they were designed with an eye to both practicality and
aesthetics. Several bridges of special structure particularly demonstrate this.

A wooden bridge of unique style appears in the antique painting *Riverside Scene at Qingming Festival* now preserved in the Palace Museum of Beijing. Depicted is a scene of happy festivity on Qingming (Pure Brightness) in Bianjing (now Kaifeng, Henan), capital of the Northern Song Dynasty. The bridge, named "Rainbow", might be expected to be an arch; it is actually constructed of piled-up straight beams with five short, straight and round timbers laid crosswise to make the ribs of an arch. These in turn are connected by beams between the ribs and secured with ropes to form the body of the bridge. Zhang Zeduan, the painter of the picture, calls the "Rainbow" "an arch of piled beams". This masterpiece of design resulted in a long-span bridge made of short beams, a unique example in the history of bridge engineering and of wood structure in Chinese and world architecture.

In front of Jinci Temple, an ancient building in Taiyuan, Shanxi, shaded by cedars 3,000 years old, is a peculiar cross-shaped beam bridge which bears the fancy name of "Flying Girder over Fish Swamp". The bridge is made up of two stone beams measuring 18 metres by 6 metres and 6 metres by 4 metres, which rest on two

The "Flying Girder over Fish Swamp" Bridge in front of Jinci Temple in Taiyuan.

rows of stone columns arranged east-westward and south-northward respectively. These stone beams serve as piers. Viewed from above the bridge was said to resemble a bird with outstretched wings over the fish pond, hence its name. Flanked by railings of white marble the bridge harmonizes perfectly with the landscape to form a magnificent and novel picture.

Another unique type of bridge features variety in the bridge axis. This is seen in Wanshou Bridge (Longevity Bridge) in Fujian. Built in the Yuan Dynasty (1271-1368), the bridge is 800 metres long and has 36 spans. Its axis curves convexly in the direction of the current, so that the bridge appears slightly bent. This was caused by the stones thrown in to form the foundation deviating from a straight line due to the rush of the current. The piers were built on this foundation. This unplanned peculiarity resulted from a situation that existed once the foundation was laid.

An L-shaped bridge of the Southern Song Dynasty (1127-1279) built in Shaoxing, Zhejiang Province, was also the result of circumstances. The bridge deck was made five metres higher than the roads on either bank to allow ships to pass beneath. Traditional bridge engineering would have required ramps leading straight down from the bridge, but on one end this would have necessitated the destruction of many dwellings. To avoid this, the builders made a turn in the bridge and built the ramp on that end parallel to the river bank. The result was the L-shaped bridge.

More fascinating are zigzag bridges of various designs built over the ponds or lakes of Chinese gardens. On Kunming Lake of Beijing's Summer Palace is the Xing Bridge, which assumes the shape of an inverted swastika, or *wan* sign, signifying good fortune. This is also known as Wan Bridge. At the eastern gate of Zhongshan Park in Xiamen, Fujian Province, is a bridge shaped like the character for man (人). There is a bridge at the Chenghuang Temple in Shanghai with nine zigzags and another on West Lake of Hangzhou with the same number, while Zhuozheng Garden in Suzhou is decorated with a five-zigzag bridge. Many others are found at scenic spots in various parts of China, enhancing the charm of the landscape with their elegance.

Still more spectacular are the zigzag bridges built on the sea. Shuzhuang Garden situated at the southern tip of Gulangyu Isle in Xiamen is also famous for a zigzag bridge stretching out to sea. Its bends carry pavilions where visitors to the garden can find shade while viewing the sea waves or the setting sun. The following poetic lines were written in its praise:

Across three thousand fathoms
Our beauteous bridge spreads;
Over its stone balustrades
The moon its ray sheds.

The Beauty of China's Stone Bridges

Through the millennia as many as three million stone bridges were possibly built in China. These numerous architectural wonders are also remarkable for their beauty.

These bridges, whether large or miniature, were often named as their shape suggested, e.g., Rainbow Bridge at Wuxian County in Jiangsu Province and Jewel Belt Bridge at Suzhou. The former is described by the couplet:

A rainbow lying on the waves,
A turtle's back reaching the clouds.

Jade Belt Bridge in the Summer Palace, credited with being the jewel among arch bridges, has a deck resembling a camel's hump. The view of the palace gardens from the top of the bridge is quite different to that from below. Its deck and railings are of white marble.

Outside the eastern gate of Guilin city in Guangxi, southern China, recognized as a prime scenic spot, is a stone bridge with nine arches and crowned by two pavilions on its deck. Harmonizing with the adjacent landscape, it has been called "the bridge veiled in misty rain".

Baiquan Bridge (Bridge of a Hundred Springs) in Xinxiang County, Henan Province, appears like a slender line floating on the waves. One walking along it may feel as though strolling on the water. It is also called "One-Line Bridge".

Dr. Joseph Needham, speaking highly of the aesthetic quality of Chinese bridges, writes:

> . . . no small part of the genius of their [the Chinese] civilization lay in a subtle combination of the rational with the romantic, and this had its result in structural engineering. No Chinese bridge lacked beauty and many were remarkably beautiful.[1]

Another artistic feature of Chinese stone bridges is the carving that embellishes practically all large and medium-sized ones. In addition to such renowned masterpieces as the Lugou, Luoyang, Anping and Zhaozhou bridges, also exemplary are Jimei Bridge over the Xiaohe River (Zhaoxian County, Hebei), the stone-arch bridge at Huoxian County, Shanxi, the Tianxian (Fairy) Bridge at Tongzi, Sichuan, the Jin Ao Yu Dong (Golden Turtle and Jade Column) Bridge in Beijing and the Seventeen-Arch Bridge at Beijing's Summer Palace.

Especially remarkable are the stone carvings on Lugou Bridge. An old saying of Beijing is that "you can't count the lions on Lugou Bridge". These exquisitely carved lions number several hundred. Guarding the parapets and the monumental columns at the bridge-heads, all gaze differently — most playfully — at passers-by. One is likely to be so fascinated with them as to forget to count them. Lugou Bridge is credited by Marco Polo in his 13th-century travel book as the most beautiful stone bridge in the world. And Portugese travellers in the 16th century expressed the opinion that "in all the world there be no better workmen for buildings than the inhabitants of China".

Another artistic aspect of Chinese stone bridges lies in the supplementary structures. These include pavilions on bridge decks, large bridge towers and monuments on the ramps, which besides being masterpieces of art serve practical purposes. Five-Pavilion Bridge at Slender West Lake at Yangzhou in Jiangsu Province has

---

[1] J. Needham, *Ibid.*, p. 145.

five pavilions on its deck. The bridge rides on fifteen arches, each of which encloses a reflection of the moon on moonlit nights. Across the Linxi River at Sanjiang in Guangxi is the Chengyang Bridge. A wooden cantilever structure, each pier of the 100-metre bridge is crowned by a five-storey pagoda-like tower some ten metres tall and occupying five square metres. These provide excellent views for travellers.

Listed above are but a few examples of China's thousands of ancient bridges. One characteristic common to all is that none fail to conform to scientific principles of engineering in structure or to satisfy modern basic technological requirements. With the equipment and methods available at the time, ancient Chinese builders worked wonders in the construction of innumerable bridges. On the ancient Jiangdong Bridge at Zhangzhou in Fujian Province, a monolithic beam 20 metres long and weighing 200 tons was successfully hauled up and set in position entirely by manual labour, a feat that commands admiration to this day.

Countless ancient bridges have borne the load of heavy traffic from the date of their construction to this day. Nor has nature's destruction put them out of service. Many of these ancient stone bridges now carry truck and bus traffic far heavier than could have been foreseen. The ingenuity and architectural knowledge of the ancient Chinese bridge engineers and working people can scarcely be overestimated.

The Tang Dynasty Capital of Changan

Ma Dezhi

Changan, the capital of the Tang Dynasty (618-907), was world-famous for its huge scale and careful city planning. From the 7th to the 9th century Changan was an important world trade and cultural centre which contributed much to the exchange and development of ancient civilizations.

Situated in the centre of the fertile Guanzhong Plain, Changan included what is now Xi'an and several nearby towns. The great Tang capital lay south of the famous Weishui River, north of the Zhongnan Mountains, east of the Fenghe River, and west of the Bashui and Chanshui rivers. Mild climate, rich resources and attractive scenery made the Guanzhong Plain an economic centre in ancient China.

Eight centuries earlier, in 221 B.C., the first national feudal state was founded by Qin Shi Huang, the First Emperor of Qin, upon his unification of China. Xianyang, the Qin capital, was located on the north bank of the Weishui River. Fourteen years later (207 B.C.) the Qin empire collapsed and that of Western Han was established with its capital, Changan, built south of the Weishui River. This was the first mention of the name Changan in history. In A.D. 25, the capital was moved to Luoyang by Eastern Han. Some 200 years later, state power in China was again decentralized with the downfall of Han and it was not until A.D. 581 that China was reunited under the Sui Dynasty, which located its capital southwest of the ruins of the Western Han capital Changan. Yang Jian, or the Emperor Wen Di of Sui, entrusted its planning and construction to Yuwen Kai, a celebrated architect of the time. Work began in A.D. 587 and was speedily completed, the royal family and officials moving into the new palaces of the capital, named Daxing City, the following year. Daxing covered the unprecedentedly vast urban area of 84 square kilometres, seven times that of present Xi'an, which was built after 1368, during the Ming and Qing dynasties. Daxing, renamed Changan, was made the capital of the Tang Dynasty at its founding in A.D. 618, from which time the only changes in the layout of the city have been minor alterations or expansions. However, the city grew in wealth as the Tang Dynasty prospered. Surpassing the Sui Dynasty in economy, culture, and foreign trade, Changan soon joined the ranks of the world's largest and richest cities. Tang Dynasty Changan has remained an object of international scholarly interest for its planning, construction and administration, and much research has been done in these fields.

Palace and Imperial City

The construction of Changan started with the Palace, extended to the Imperial City and ended with the Outer City. The Palace, or royal residence, was situated midway of the northernmost part of Changan city. Archaeological surveys show it to have been 2,820 metres from east to west, 1,492 metres from north to south, and 8,600 metres in circumference. History records that the Palace walls were 35 *chi*[1] (10.29 metres) high, the Palace itself being separated by inner walls into three parts. The western part, called Yeting Palace, was where the palace maids learned their various arts. The eastern part, or Eastern Palace, was where the crown prince lived and attended to state affairs. The middle part was called Daxing Palace in the Sui Dynasty, and renamed Taiji Palace in Tang. It was also known as the Western Imperial Palace and the Principal Palace. This included several main halls where emperors lived, conducted state affairs and received officials. The principal throne room was called Taiji Hall. To its north were the Liangyi and Ganlu halls and a

---

[1] *Chi* in the Tang Dynasty was 0.294 metre.

score of other palace buildings. The largest southern gate to the Palace was Chengtian Gate. This and Zhuque Gate of the Imperial City and Mingde Gate of the outer wall all lay on the capital's central axis. North of the Palace were the Xuanwu and Anli gates leading to the Western Forbidden Garden.

Daxing Palace (Taiji Palace in Tang times) is the only palace we have from the Sui Dynasty. Two more palaces, Daming Palace (or Eastern Imperial Palace) and Xingqing Palace (or Southern Imperial Palace) were built in the Tang Dynasty, the three royal architectural ensembles being known as the Three Imperial Palaces.

Daming Palace became the place where emperors attended to state affairs with the ascension to the throne of Li Zhi, the Emperor Gao Zong of Tang. The principal throne room was Hanyuan Hall, where important state ceremonies were also held. North of Hanyuan Hall were the Xuanzheng and Zichen halls, where emperors conducted state affairs. And there were 30 other buildings including the Yanying and Linde halls in the Daming Palace. The spacious Linde Hall was where the emperor received foreign envoys and feasted bevies of officials. Excavated remains of Daming Palace mark it as a masterpiece typical of Tang Dynasty palace architecture.

Li Longji, Prince of Jin, who mounted the throne in 713, made a palace of his former residence at Xing Qing Fang in the second year of the Kaiyuan reign (A.D. 714) and called it Xingqing Palace. Expansion of the former residential halls into a number of throne halls began in 726 and was completed two years later.

South of the Palace, the Imperial City, also known as Zicheng, i.e., Smaller City, was occupied by the various central government ministries and other offices. Zicheng is 2,820 metres from east to west, the same width as the Palace, 1,843 metres long and over 9,000 metres in circumference. There were three gates on the southern side of the Smaller City, with Zhuque Gate in the centre, and Anshang Gate and Hanguang Gate to the east and west respectively. In the eastern wall of the Imperial City were the Yanxi and Jingfeng gates, while in the western wall were Anfu Gate and Shunyi Gate. Five north-south and seven east-west thoroughfares formed a rectangular grid system of streets inside the Imperial City. The northernmost east-west street, which lay between the Palace and the Imperial City, was known as Hengjie, i.e., Cross Street. History records that Hengjie was a boulevard 300 *bu*[1] or 441 metres in width, though excavations reveal it to have been narrowed down later to 220 metres. The widest boulevard in the capital, the section of the street in front of Chengtian Gate was used as a public square where important outdoor ceremonies were held.

It was an innovation of the Sui planners to place the Palace and the Imperial City at the northern end of the central axis of the capital instead of putting them among commoners' living quarters as in capitals of former dynasties. The definite detachment of the royal from the common was obviously done in consideration of the rulers' safety. The co-axial series of walled enclosures, with the Palace and the Imperial City on the central axis, symbolized the supreme authority of the feudal ruler.

"Chess Board and Vegetable Plots"

The Outer City was also known as Luocheng, i.e., Greater City, which bordered on the east, south and west sides of the Imperial City. The Outer City was a perfect rectangle. On excavation, the east-west dimension was found to be 9,721 metres and that north-south 8,651 metres. The four sides of the Outer City had a total length of some 36.7 kilometres, the walls being six metres high. There were a dozen gates altogether, three on each side. Since the middle section of the northern wall bordered the Palace, the three gates in the north wall were west of the Palace. Each gate had three doors except for Mingde Gate which had five. The left door was for entrance and the right for exit. This conformed to the regulation that traffic kept to the right, as every street was two-way. Eleven streets ran north-south, and fourteen east-west. The street constituting the central axis of the capital was the famous Zhuque Street, 150 metres wide. All other streets leading to the gates were more than 100

---

[1] *Bu* in the Tang Dynasty was 1.47 metres.

metres in width, while those along and inside the city walls were at least 25 metres.

There were drainage ditches on both sides of the street. Those along Zhuque Street were 3.3 metres wide and 2.1 metres deep. Since they were open ditches, a bridge was necessary at every crossing. The ditches were flanked by neat rows of trees. Straight boulevards shaded by abundant trees added to the grandeur of Changan.

The rectangular grid system of streets divided the capital into wall-enclosed blocks, each of which was called a *li* (ward) in the Sui Dynasty and *fang* in Tang. The *fang* flanking the Imperial City had in their four walls central openings joined by intersecting streets. There were also alleys and narrow streets along and inside the walls. Many princely or high official residences and temples were built amidst common houses in the *fang*. But the privileged lived either in prosperous districts within short distances of the Palace, ministries and markets, or in scenic areas.

There were many temples in Changan, nearly every *fang* having at least one Buddhist or Taoist temple and some having three or four. There were also places of worship built and frequented by merchants or travellers from Persia and other foreign countries. The temples in the capital were large and magnificent. Among the most famous were Xingshan Temple in Jingshan Fang and Xuandu Monastery in Chongye Fang, both along Zhuque Street and both centres for excursions. Qinglong Temple, which propagated the Buddhist esoteric doctrine, was situated in Xinchang Fang inside Yanxing Gate in the southeastern part of the capital. This temple was remarkable for its elevated, beautiful surroundings. Two Japanese priests, Kukai and Ennin, studied there and later returned to Japan as masters of the esoteric doctrine in Japan. Pagodas were important in Buddhist construction, and Dayan Pagoda in Cien Temple and Xiaoyan Pagoda in Jianfu Temple still fascinate sightseers today.

Every *fang* was furnished with a number of small shops for daily necessities such as food and drink. There were also inns and handicraft workshops. For instance, in Jinggong Fang there was a lane named Zhanqu where felt goods were manufactured and advertised.

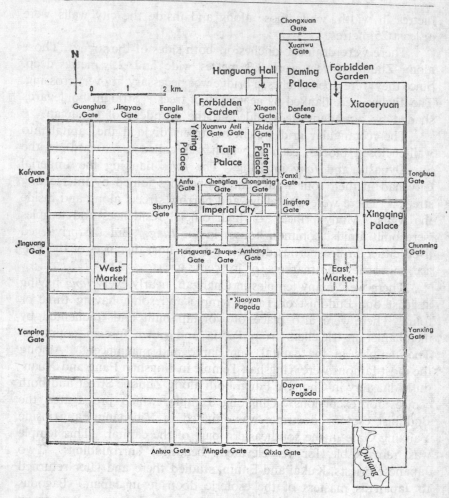

Reconstructed plan of the city of Changan in the Tang Dynasty (618-907).

In other words, each *fang* was an independent small "city", over a hundred such "cities" separated by walls and shaded boulevards making up the Tang capital. Bai Juyi (A.D. 772-846), noted poet in the Tang Dynasty, writes of the city layout thus:

Crossed like lines on a chessboard the boulevards and streets,
Neat are a hundred wards quite like vegetable plots.

The two big shopping centres in Changan were East Market and West Market, called Duhuishi (Metropolitan Market) and Lirenshi (Profit-Makers' Market) in the Sui Dynasty. The two busy markets were both on Zhuque Street. East Market displayed products of 220 trades, mostly valuable items and curios. West Market was even more prosperous with its shops selling clothing, spun silk fabrics, balances for weighing, fodder, and leather horse trappings. Most of the foreign merchants from the western regions traded at West Market. Business at both markets was conducted under the supervision of government offices — Shishu (Bureau of Commerce) and Pingzhunju (Bureau of Standards and Price Stabilization).

The four main streets crisscrossing the two markets divided each into nine rectangular areas accommodating nine different trades, with metre-wide aisles within. Some of the aisles had brick-lined covered sewers beneath that led to the open ditches along the streets. Since most business was done at these two markets, large crowds gathered, and the two sites served also as recreation centres for Changan's inhabitants.

Handicraft industries flourished in Changan. Besides the workshops run by the government, there were many that were operated by individuals in the *fang*.

Waterways in Changan

Several canals were dug to carry water for washing (drinking water came from a network of wells in the *fang*), drainage, boating, afforestation and improvement of local climate. Two streams were led in from the south. The one from the Jiaoshui River was called Yongan Canal. This meandered by West Market on the east and flowed out through the northern city wall into the Forbidden Gardens, to empty finally into the Weishui River north of the capital. Both banks were green with willows. The canal from the Xueshui

River, known as Qingming Canal, lay east of Yongan Canal and flowed west of Anhua Gate (west gate in the south city wall) into the city, then on northward into the Imperial City and Palace. It ended in Three Seas (actually man-made lakes) west of Taiji Palace.

A canal was dug to lead the Chanshui River water westward, and this was called Longshou Canal, or Dragon Head Canal. This canal soon forked, one branch flowing through the east city wall near Chunming Gate, then turning north and flowing into Qingxing Palace and swelling into Dragon Lake. It then flowed further westward into the Imperial City, turned north into Taiji Palace and again expanded into Shanshui Lake. It then proceeded further northward into one of the Three Seas — East Sea. The other branch flowed north outside the city, passed the northeast corner of the city wall, then turned west into Eastern Forbidden Garden in Daming Palace where it broadened into Longshou Lake. This fork flowed westward through Daming Palace into Western Forbidden Garden, continuing along the city wall till it joined Yongan Canal near Guanghua Gate.

The canals were first dug in the Sui Dynasty to lead water to the royal gardens for ornamental rivulets, screens, pools, torrents and cascades. The abundant water supply inspired urban aristocrats, bureaucrats and wealthy merchants to follow suit and indulge in private garden construction with water as the leitmotif. These private gardens soon became favourite pleasure haunts for officials and scholars. In the first year of the Tianbao reign (A.D. 742) in the Tang Dynasty a navigation canal was built to lead a branch of the Jueshui River eastward through the west city wall near Jinguang Gate. The canal flowed by West Market and ended in a large pool alongside the street bordering the Market on the east. Charcoal and timber from the Zhongnan Mountains were carried to the capital in ships on the canal.

Rise and Fall of Changan

Planning and construction of Daxing started in the Sui Dynasty, after China was unified for the second time.

When construction of Daxing as the Sui capital was completed, it was one of the world's biggest cities. The planning was far-sighted, comprehensive and carefully carried out in scheduled steps. A complex of factors had to be considered such as topography, water resources, tree planting, communications, defence, city administration and urban economic life. The original plan, as shown from remnants found of it, foresaw and solved various problems of the capital of a newly unified feudal state, a feat impossible without a high degree of national culture and technology. The plan encompassed over a thousand years' rich experience in capital designing, especially that gleaned in building Luoyang in the Eastern Han Dynasty (A.D. 25-220), Yecheng in the Kingdom of Wei (A.D. 220-265), and Luoyang again under the Northern Wei Dynasty (A.D. 386-534). The Sui capital thus benefited from ancient Chinese city designers' accumulated wisdom over the centuries.

Changan in the Tang Dynasty reached the zenith of feudal grandeur. The city was world-famous for its majestic layout, magnificent buildings, neat blocks of urban districts, straight boulevards, green trees, clear water, dense population and dazzling display of Chinese and foreign goods. It became almost a paragon of prosperous economy and splendid culture which won the admiration of many neighbouring countries. Its highly developed handicraft industries and commercial system supported a complex of brisk, sophisticated urban activities. Changan in the Tang Dynasty was the finest but also the last national capital composed of walled urban districts.

But within the prosperity of Changan lurked the factors of its downfall.

> Behind these vermilion gates meat
> and wine go to waste,
> But along the road are bones of
> men who have frozen to death.

These unforgettable lines of Du Fu (A.D. 712-770) bespeak the cruelty behind the facade of a feudal capital in full flush. The peasant rebellion led by Huang Chao and others was to shake the rule of the Tang emperors to its very foundations.   Changan began its decline, and at the beginning of the 10th century this great capital was laid waste amidst the turbulence of civil strife.

A Wooden Pagoda More Than 900 Years Old

Yang Hongxun

Sakyamuni Pagoda, built of timber more than 900 years ago in Yingxian County of Shanxi Province, is the oldest wooden pagoda in China. Called Yingzhou in the 11th century, Yingxian County is on the outskirts of Pingcheng (now Datong city), then the capital of the state of Liao (916-1125). The pagoda was built by order of the Xing Zong Emperor of Liao and completed in the second year of the reign of Qingning, or 1056. Embodying the superb wisdom and creative ability of medieval builders, the pagoda is the principal structure among the architectural ensemble of Fo Gong Si, or Buddha Palace Temple. The pagoda, octagonal in shape, measures up to 30.27 metres between the two front veranda columns on the two opposite parallel sides. The outward appearance of the pagoda is a five-storey structure with six protruding eave lines (the lowest storey is double eaved), and in each of the first to fourth storeys there is a mezzanine floor, making a total of nine storeys inside. At 67.31 metres in height, the pagoda is the tallest wooden building still in excellent condition today. Weathering nearly a thousand years of Mongolian sandstorms, earthquakes and wars, the pagoda is an architectural wonder representative of Chinese civilization.

Buddha Palace Temple was formerly located in the centre of the Liao city of Yingzhou, symbolizing the importance attached by the Liao rulers to Buddhism in their government. The city wall we see today in Yingxian was rebuilt during the Hongwu reign (1368-98) of the Ming Dynasty (1368-1644). The west and north sides have been moved about half a kilometre towards the centre so that the temple is now in the northwest of the city.

The wooden pagoda retains the distinctive Chinese traditional features of multi-storey structures. In the Han (206 B.C.-A.D. 220) and succeeding dynasties, Buddhist temples mushroomed with the propagation of that faith in China. But Chinese temple buildings, perhaps with the exception of the Indian-inspired pagodas, have never been much different from ordinary civil buildings. According to *Luo Yang Qie Lan Ji* (*Description of the Buddhist Temples of Luoyang*) (547), the prototypes of the wooden pagodas during the period of the Three Kingdoms (220-280) were dwarfed by the Yingxian pagoda, the former being merely multi-storey square buildings, each storey smaller than the one below. But for the finials and discs atop they would be the same as the square multi-storey buildings in vogue in the Han Dynasty. During the Southern and Northern Dynasties (420-589), Buddhism thrived under the aegis of the feudal rulers. As wood construction techniques improved continually, wooden pagodas became ever taller. An example is

This wooden pagoda at Yingxian, Shanxi Province, was built more than 900 years ago.

the formerly well-known wooden pagoda in Yong Ning Si (Temple of Perpetual Tranquillity) in Luoyang, believed to have been more than 100 metres high. It was built in the Northern Wei (386-534) period. The Yingxian pagoda however excels the Luoyang structure, the design of pagodas having become octagonal from the end of the Tang Dynasty (618-907). Its plan provides a better balanced distribution of stress. Moreover, there is no central column to break the space. Tie-beams and posts, branching out parallel and transverse to the vertical central axis, with tenons and mortises, pegs and dowels, are used to fill and bind together the columns and cross-beams so that all the members are mutually linked up like a huge cage, while the unbroken space in the centre is reserved for placing Buddhist statues and ornaments. The framework with central open space also provides greater shearing strength and ushered in a new epoch in wood construction.

The Yingxian pagoda is built on a platform foundation of brick-encased rammed earth in two stages. The lower, larger stage is

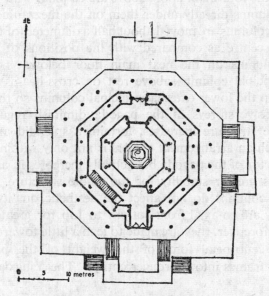

Plan of the Yingxian pagoda.

square and the upper, smaller stage is octagonal in shape. Three rings of columns stand on the upper octagon — the internal principal columns, the external principal columns, and the shorter veranda columns supporting the eaves. The principal columns are linked up by cross-beams and tie-beams to form a giant hollow vertical cylindrical structure as stated. On the ground floor an additional column is attached to each of the internal and external principal columns in the eight corners.[1] The internal and external principal columns are buried in walls a metre thick. The lower parts of the walls are brick aprons while the upper parts are made of adobe blocks. There is a damp-proof layer of planks between the bricks and the adobe blocks. The additional columns are for greater shearing resistance at junctures of cross-beams and corbel-brackets atop the principal columns; the brick and earthen walls are against torsion, to ensure the stability of the whole framework.

As stated, there are mezzanine floors between the ground and the first to fourth floors inside the pagoda. The external principal columns on the four main floors above are co-axial with the external principal columns directly under them on the mezzanine floors, but the co-axial columns are moved about half a diameter (of the columns) towards the centre as compared with the positions of the external principal columns on the next main floor below. The drawn-in external principal columns above rest on cross-beams and corbel-brackets atop the lower external principal columns so that the areas of the successive storeys of the pagoda diminish regularly as the height of the structure increases, forming a splendid and beautiful profile. Such an arrangement of tiers is not only within the permitted safety factor of the pagoda but actually increases it, as less weight above means greater stability for the whole structure. And since the internal columns do not affect the aesthetic quality of the silhouette, they are co-axial from bottom to top for greater propping strength. Moreover, they are made to lean a little towards the centre so that the centripetal force of the weight of the pagoda itself squeezes the frames into a stronger entity. The Yingxian pagoda is

---

[1] The additional columns were probably added in during later repairs.

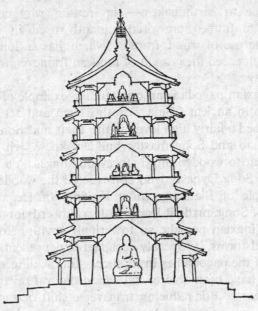

Cross section of the Yingxian pagoda.

a perfect unity of strength and beauty still inspiring architects today.

Wood construction has developed in China for at least 6,900 years as proved by the excavation at Hemudu of Yuyao County, Zhejiang Province. Relics of long huts raised on wood frames composed of poles and beams with tenons and mortises have been unearthed there. Wood frames have the advantage over structural walls in many ways. In a house with the roof supported by a wood frame composed of poles and beams the walls are merely partitions which bear no part of the weight. They may even collapse and not cause the roof to cave in. In fact the house can stand without any walls at all, providing complete freedom for placing any number of doors and windows as desired. Such walls are quite different from load-bearing walls in which openings are restricted for safety, to say nothing of doing away with the walls completely. Flexibility in space treatment leaves the designer a free hand in suiting the building to various practical uses. Moreover, framework buildings are much less vulnerable to earthquakes — the reason why concrete, steel and aluminium frames are predominantly used in modern sky-scrapers. The use of wood frames, which has a long history in China, has influenced the growth of modern frame construction in a number of ways.

The Yingxian pagoda is an outstanding example of wood frame craftsmanship in ancient China. Both design and actual construction had reached a fairly high level by the Liao and Song (960-1279) dynasties. Liao and Song master builders knew well that wind is a serious threat to wooden multi-storey buildings. Yu Hao, master builder in the Song Dynasty who constructed the wooden pagoda in Kaibao Temple of Bianliang (now Kaifeng of Henan Province), then capital of Song, made it lean a little windward for greater resistance. The Yingxian pagoda, so toweringly inviting to Mongolian snow and sandstorms, is superbly designed to defy them. The greatest triumph of the pagoda lies in its successful tackling of the lateral load so as to have been able to withstand natural ravages of nearly a thousand years. For reducing transverse shift or torsion caused by wind and seismic shocks, a large number of slanted struts are used to fix the complicated structure of beams. These combinations of struts and beams fall into two categories according to their functions. One is to stabilize the internal and external column systems respectively. The other is to keep the two systems in their relative positions firmly. The various bracing planes grasp and support each other to form an integral space framework which reacts with a combined rigidity when a load is laid on any of its components.

Under each of the mezzanines is a trussed girth. On the inner ring of the truss is another girth composed of four layers of tie-beams in the form of the walls of a square log-cabin. As a whole, the girths under and above the mezzanine floors form a rigid ring. Four such rigid rings laid evenly among the five main floors greatly strengthen the pagoda against lateral shear. Outside, each ring appears to be a girdle of balcony roof-line — a most ingenious treatment.

In the hollow central space on each of the main floors is one or several images of Buddha. Between the internal and external columns is an octangular passage not thwarted by slanted struts.

The four cardinal sides due east, south, west and north on a main floor, each side being three bays or panels wide, have in each a door and latticed windows which open onto a balcony with railing around the pagoda for sightseeing. On the other four sides, there were formerly diagonal bridgings hidden behind panel walls of mats and plaster. The bridgings were of course parts of the whole framework and the panel walls were for artistic contrast with the doored and windowed sides in appearance. Unfortunately, in later repairs the bridgings and panel walls were removed to make room for more doors and windows, to the detriment of the structure. As an after-effect, the pagoda has since twisted a bit northwestward.

The stairway inside the pagoda meets the demand of easy climbing while minimal harm is done to its structure. To mitigate steepness, two flights instead of one are arranged in a storey. Struts on the mezzanine floors are removed where the quarter-landings go through. As this might cause a structural weakness, the staircases are laid along every other side of the eight sides with quarter-landings between, so that the entire stairway looks like a huge, evenly broken spiral.

Other details in the structure such as the well-proportioned parts and the perfect tenon-and-mortise joints are praiseworthy. More than 60 kinds of corbel-brackets are used in the pagoda — a rich display of ancient wood craftsmanship. They are no sedulous apings of still earlier masters but superb models for practical structural purposes. They are works of art as well as balanced components of the whole structure.

Murals decorate the ground floor. Since the two rings of thick walls open only on the north and south, the golden image of Buddha inside gleams in semi-darkness to create a mystic religious atmosphere. The upper floors are better lit, as there are no walls there but a door and windows on all sides. Railings between the internal principal columns keep sightseers and possible worshippers at a respectful distance from the images. All the images are probably Liao sculptures though they now appear in garish colours from being repainted subsequently.

Sakyamuni Pagoda at Yingxian is an architectural marvel

that arouses the keen interest of architectural historians as well as daily sightseers. China has the oldest wood structure in the Yingxian pagoda, which stores a wealth of experience. The timber, supporting a tremendous load for more than 900 years, must also be a unique specimen for the reference of researchers in timber mechanics.

Shipbuilding

Zhou Shide

Types of ...

China has one ~~~~~ istories of shipbuilding in the world. Wooden junks alone as describe~~~~~ historical records varied greatly in type, being estimated at about 1,000 by the mid-20th century. For coastal fishing alone, 200 to 300 types were noted. The best known sea-going vessels in ancient China were the "sand ship" (*shachuan*), "bird ship" (*niaochuan*), Fujian ship (*Fuchuan*), and Guangdong ship (*Guangchuan*). The best known of these were the "sand ship" and the Fujian ship.

The "sand ship" originated at Chongming, Jiangsu Province, its prototype probably dating as far back as the Warring States Period (475-221 B. C.). The "sand ship" was known in the Song Dynasty (960-1279) as the "flat-bottomed sand-beaters", in Yuan (1271-1368) as the "flat-bottomed ship", while in Ming times (1368-1644) the name "sand ship" became popular.

The loading capacity of the sand ship was generally recorded as 4,000-6,000 piculs (*dan*), roughly 500-800 tons. Other records give its capacity as 2,000-3,000 piculs, or about 250-400 tons. The big sea-going ships of Yuan carried 8,000-9,000 piculs, or more than 1,200 tons. In the reign of the Emperor Dao Guang (1821-1850) of Qing there were 5,000 sand ships in Shanghai alone, while the total number in the country was estimated to be over 100,000. With their great adaptability they were seen cruising both inland and coastal waters. In Yuan and Ming times, from the 13th to 17th century, when coastal navigation reached its zenith of prosperity, the annual volume of shipping exceeded 3.5 million piculs. Sand ships were also active in oceanic navigation. At the beginning of the 10th century Chinese sand ships called at Java, while murals in both India and Indonesia depicted various types of Chinese sand ships. It was thought at the beginning of the 20th century that the sand ships then sailing between North China and Singapore were the same as those Chinese ships that sailed to the Red Sea and Eastern African ports for trade before the Middle Ages.

Early in the 15th century a glowing page was written in the history of world maritime navigation when Zheng He made seven voyages to the South Seas in 20 years, calling at ports in more than 30 countries. On each voyage a fleet of 100 to 200 ships, of which from 40 to 60 were "treasure ships". The fleet carried over 17,000 men. The vessels, built at Nanjing, assembled at the port of Liujiagang in Taicang County, Jiangsu Province, before going to sea. Zheng He's treasure ships were about 150 metres from stem to stern, their rudder posts 11.07 metres long, each carrying 12 sails. These were the largest of the sand ships.

A five-masted "sand ship".

The sand ship had several advantages: 1. Its flat bottom prevented it from running aground, and it was fairly safe in storms. Especially when the wind blew against the tide, the sand ship with its relatively shallow water draught was less affected by the tide and again safer; 2. It was unaffected by wind direction and could sail against a head wind and the current; 3. Broad in the beam and equipped otherwise for steadiness, it was fairly stable; 4. The sand ship was a swift-sailing craft, its multiple masts and many tall sails propelling it efficiently with the wind, while resistance was small since it drew little water.

Square in both bow and stern, the sand ship was popularly called the "square boat". Its spacious deck slanted down to port and starboard for rapid water draining. Abaft the stern was a "false stern" or false transom for carrying the rudder and working the sails. The rudder blade was large, and the rudder could be hoisted and lowered. When the ship set sail, a part of the rudder blade was lowered beneath the hull to increase rudder efficiency and prevent the boat from drifting sidewise; the rudder could be hoisted up in shallow water.

The sand ship had a flat and relatively weak keel 40-50 per cent the thickness of that on other types of ship of similar size. It was the four to six solid wales running lengthwise from stem to stern and the strong lengthwise planking nailed to the bulkheads that accounted for the sand ship's greater strength than that of other seagoing vessels of about the same size. Bulkheads forming many separate water-tight compartments secured the ship against foundering so that it sailed safely in force 7 gales. Plying between China and Africa posed no big problem for the sand ship.

The *Fuchuan*, or Fujian ship, was round-bottomed and known for sailing to the South Seas and on the high seas. It was recognized in the Song Dynasty that "Fujian is unsurpassed in the building of sea-going vessels". Fujian ships formed most of the Chinese naval fleet in Ming.

Tall as a pagoda, the ancient Fujian ship was pointed at the keel and broad at the deck. It curved upwards at both bow and stern, with the sides of the hull strengthened by shielding planks. The ship had four decks:the lowest was for ballast to steady it; the second served as seamen's quarters; the third provided space for the sailors to work the masts and sails; and the uppermost deck was for battle from where archers and gunners could shoot down upon a dwarfed enemy fleet.

With its curved bow and strong armour, the Fujian ship often sped at full sail to ram and sink an enemy vessel, and this was often how a battle was won. Drawing four metres of water, the Fujian ship was an ideal man-of-war where the water was deep.

A sea-going ship of the Song Dynasty was found in the summer of 1974 at Houzhu Harbour in Quanzhou Bay, Fujian. The hull has a cross-section pointed at the bottom and flat and broad at the deck. Its longitudinal section was nearly elliptical, pointed at the bow and square at the stern. Fourteen rows of strakes from keel to deck form the hull. The first 10 rows are double-layer boards, while the rest are triple-layer, with a total thickness of 18 centimetres (eight mm for the innermost layer, five mm for the middle and five mm for the outer layer). The triple-layer strakes strengthened the part of the hull at water level, the principle being similar to the use of wales on the sand ship. The planks, doubled at the ends, meet in some places. The seams are caulked with hemp, flax or bamboo and lime mixed with tung oil.

The reconstructed ancient ship discovered at Quanzhou is 34.5 metres long, 9.9 metres broad and 3.27 metres high, with a displacement of 374.4 tons.

By the 7th century Chinese sea-going ships were well-known for their size, loading capacity, structural strength and seaworthiness. Arabian traders are known to have travelled along the coasts of Southeast Asia aboard Chinese sailing vessels. From the middle of the 9th century still more Asians and Africans sailed on large sea-going vessels built in China, and shipbuilding developed in Song and Yuan to the point that the Chinese were regarded as "the most advanced shipwrights in the world".

The ancient Chinese shipwrights were remarkable for their ability to develop a great variety of models and types to suit different marine conditions. An example was the *fangchuan* (square ship) in the Warring States Period. This was double-bodied, made up of two junks secured together side by side. Steadiness and loading capacity of both cargo and passengers were increased. Similar double-bodied ships were the *fangzhou* (square junk) of the Zhou Dynasty before 781 B.C. The *louchuan* (turreted ship) of Han was tall and stately, while the period of the Three Kingdoms (220-280) saw the appearance of sea-going vessels over 70 metres from stem to stern. Lu Xun, a shipwright of the Jin Dynasty (265-420), built a warship with eight compartments, and Zu Chongzhi, the great scientist of the Southern and Northern Dynasties (420-589), built the "Thousand-*Li*[1] Ship". In Tang there was the *haiguchuan* (sea-hawk) and a new model of grain freighter. The Song produced paddle-wheel boats, the largest measuring 110.5 metres long by 12.6 metres wide. Besides the treasure ships of Zheng He, the Ming Dynasty produced double-head ships, centipede ships, articulated junks, boat carriers and a host of other ships of novel design. The articulated junk was a warship in two sections coupled in tandem. The front section was loaded with explosives and could be hurled at an enemy ship while the hind section disengaged itself and sailed back. The boat carrier had a cavity near the stern to accommodate a small life-boat in which the crew withdrew after the carrier, loaded with explosives and combustibles, charged the enemy fleet. The articulated boat was used in civil transport as well, being capable of negotiating the turns in a twisting stream.

The Chinese shipwrights were good at devising new types of ships by combining the good points of various kinds of vessels. A Song Dynasty ship used in both inland and sea-going navigation combined the bottom of a lake-boat, the deck of a warship and the bow and stern of a sea-going vessel. Again, in the reign of the Emperor Kang Xi in early Qing, a type of freighter built in Fuzhou for timber shipping and known as the "Three Unlikes" was not like the sand ship, bird ship or egg ship but was a new model combining the advantages of all three.

---

[1] A *li* was then equivalent to 540 metres.

The Designing of Vessels

The words "drawings of ships" had appeared in the official archives of Song:

> Wenzhou Prefecture has directed that it was in receipt of two volumes of drawings of ships from the Military Governor's Office. It is ordered that officials be dispatched to buy timbers for the construction of 25 sea-going vessels in each magistrate under the jurisdiction of this prefecture.

The "drawings" mentioned here probably referred to sketches and the quotas for manpower and material.

In *Nan Chuan Ji* (*Notes on Ships of the South*), *Long Jiang Chuan Chang Zhi* (*Records of the Dragon River Shipyard*) and *Cao Chuan Zhi* (*Record of Grain Carriers*) detailed accounts are found concerning manpower requirements as well as the quantity and measurement of timbers needed for building ships, but no mention is found of their designing. Not until early Qing did historical literature now extant deal with ship designing. This, however, followed traditional Chinese ship designs of the late feudal age, which showed ingenuity and originality.

The design of the "arrow pursuit vessel" (*ganzengchuan*) in early Qing represented the zenith of ship designing of the ancient Fujian school. The keel of the ship is curved to a degree determined by the length of the keel. The length of the ship, which was highly vari-

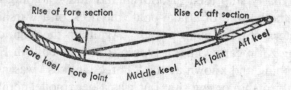

Longitudinal section of the keel on the "arrow pursuit vessel". For every *zhang* (3.33 metres) of the keel there was a fore rise of 5-5.2 *cun* (16-17 centimetres) and an aft rise of 2.6 *cun* (8 centimetres). Excessive rise of the curve was not desirable.

able, also determined the length of the keel, which was divided into three sections in proportion to the total length. Next to be decided was the rise of the curve, the curve of the entire keel being determined by the rise of its fore and aft sections.

The longitudinal section of the keel was drawn thus: Draw one triangle taking the rise of the fore section for height and the fore and aft keel for the two sides, then draw another triangle with the rise of the aft section for height and the middle and aft keels for the two sides. Overlap the common side (the middle keel) of both triangles, then link up the angles with a smooth curve as shown in the figure above. The result is the longitudinal section of the keel. The cross section of the arrow pursuit vessel is designed and drawn as follows: Take the length of the bulkhead (the beam athwartships) for the length of the upper horizontal line. Draw the lower horizontal line parallel to it and proportionately shorter. Join the ends of the upper and lower horizontals to make the sidelines. The distance between the upper and lower horizontals is equal to the depth of the hull minus the "chicken breast". Then there is the "drawing space" which is the distance from the crossing point of the sideline and the lower horizontal line to the outer curve of the hull (measured perpendicular to the sideline). As shown in the figure below, draw the upper and lower horizontals and the sidelines, measure off the "chicken breast" and the "drawing space". Then connect the five points — the two ends of the upper horizontal, the lower end of the

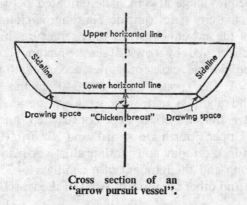

Cross section of an
"arrow pursuit vessel".

"chicken breast" and the two outer ends of the drawing space — with a smooth curve. This curve plus the upper horizontal line forms the cross section of the vessel.

The four cross sections at the four bulkheads, which with the prow bulkhead, wind-using bulkhead, official stateroom bulkhead and stern bulkhead, plus the longitudinal section of the keel make up the major contours of the arrow pursuit vessel.

In addition, a short line may be drawn from the middle of the sideline and perpendicular to it. This line is called "breast line", which serves to inflate the curve near the water level.

This traditional method of ship designing created by ancient Chinese shipwrights is characteristic of the Chinese national style. It is marked by simplicity, efficiency and clever integration of individual parts with the whole. The cross section for example would be entirely altered by a change in the trapezium encircled by the upper and lower horizontals and the sideline, while a change in the drawing space or the breast line would mean partial adjustments. The alteration of the whole plus partial adjustments constitutes a designing method that is not only simple but highly adaptable to the requirements that may be made of it.

Shipbuilding

Zhongbaocun Village at Sanchahe, situated between the Hanzhong and Yijiang city gates outside Nanjing, is claimed by legend to be the site of the ancient shipyard where treasure ships were built. The place is chequered by rows of large rectangular ponds with First Pond, Second Pond, etc., for names, according to their arrangement. Some of the ponds are 200-240 metres long, 27-35 metres wide and about a metre deep. Local residents know this as a place where treasure boats were constructed, and such names as Shangsiwu (Upper Fourth Dockyard) and Xiasiwu (Lower Fourth Dockyard) still in use today substantiate this. A giant rudder pole 11.07 metres long and the remnant of a windlass (4.75 metres long when restored) and other component parts of the vessel were excavated from these large dockyards in 1953 and 1965 respectively.

No few records occur in historical archives about model boat-making and the building of ships in dockyards. Song Dynasty history mentions a shipwright, Huang Huaixin, who undertook the repair of large junks in large dockyards, while a Ming record states: "... issued forth from the dockyard on the 25th day of the month. The dockyard is a place where ships are built." An account by Zhang Zhongyan of the Kin Dynasty (1115-1234) who "made a small boat just a few cun^1 long by hand" tells of the manufacture of models preparatory to the construction of ships. Zhang's account describes more specifically the designing and building of ships than earlier records concerning wooden models for warships. Earlier, in the Qin and Han dynasties, boat-shaped wooden and pottery sacrificial utensils discovered in Han tombs remind one of ship models, or at least verify that ship models could have been produced for

The ruins of a dockyard of the Qin
and Han dynasties at Guangzhou.

---

1 A *cun* was then equivalent to 3.19 centimetres.

ANCIENT CHINA'S TECHNOLOGY AND SCIENCE

the construction of actual-size ships. The shipyard of Qin and Han excavated at Guangzhou city further confirm the use of docks by ancient Chinese shipwrights for building ships and of slipways for launching them.

Excavated at the end of 1974 at Guangzhou city, the ruins of the Qin-Han dockyard present an immense workshop for shipbuilding. Three dockyards parallel one another, while a worksite for processing timber lies adjacent. With the slipway, the dock simulates railway tracks. It consists of sleepers, slipboard and wooden bilge blocks for steadying the ship under construction. The sleepers are of two dimensions, the slipboards adjustable for the breadth of the slipway. On Dock No. 1 the slipway is 1.8 metres wide, sufficient to accommodate ships 3.6-5.4 metres in breadth. Dock No. 2 has a slipway 2.8 metres wide, for ships 5.6-8.4 metres wide. Two parallel rows of wooden bilge blocks on the slipboards support the ships under construction. The 13 pairs of these blocks, arranged in opposite positions, are about one metre high, appropriate to allow the necessary procedures of drilling, nailing, caulking, etc. to be undertaken by workers from beneath the hull.

To the south of Dock No. 1 was found a worksite for processing timbers, with a "timber-bending ox" for baking and bending wooden members of the ship. Excavated at the same time were plumbs used for cutting the planks straight. Also unearthed at the dock were chisels, short axes and caulking knives all of iron, as well as square wooden weights for determining verticality.

China's shipbuilding industry, which grew to considerable proportions during the Qin and Han dynasties, gave rise to fleets of ocean-sailing vessels by Tang and Song, crediting China for over a millennium with wooden sailing vessels seaworthy of the Indian Ocean and the Pacific.

Features of Chinese Vessels

Swiftness characterized both sea-going and river-transport Chinese ships. The "sand ship" of Jiangsu achieved speed by virtue of its many masts and sails, the height of the masts, which enabled it to receive full wind, and the small draught of water, which avoided resistance. The "bird ship" with its pointed stem and long stream-lined hull was in the same speed class as the "sand" and "brave" ships. Such inland-water vessels of the Ming and Qing period as the "fast-delivery boat" of Huaiyang and the "red boat" of Jiangxi were remarkably swift-sailing.

A second feature of ancient Chinese ships was security against foundering. Tang vessels well caulked with a lime and tung-oil mixture did not leak. The eight-bulkhead warship of the Jin Dynasty (265-420) is thought to have consisted of eight watertight compart-ments, and though this lacks substantial proof, it is certain that the possibility existed technically. From the 10th to 14th century, the watertight compartments of Chinese ships were an outstanding feature both in China and abroad. Many personages friendly to China commended Chinese ships for their safety due to their water-tight-compartment construction, which kept the ship afloat even if one or even two compartments were damaged. Similar ship con-struction did not exist in the West until the 18th century.

Adaptability to various marine conditions was a third feature of ancient Chinese ships, as evidenced by many ship models. The flat-bottomed "sand ship" is an example of a vessel suited to the north China coasts where sandbanks abound. Even after being stranded for some time the "sand ship" remained seaworthy. There were other types of inland-water ships that were likewise highly adaptable.

Steadiness was a fourth feature. The Great Dragon Boat of the Song Dynasty was ballasted with 400 tons of iron in her hull to keep the ship steady. The Fujian ship had four decks, of which the lowest was filled with earth and stones for ballasting, giving steadiness.

Before the 9th century, during the Tang Dynasty, the "sea hawk boat" had one or several floating boards attached on each side to steady the ship. These were the prototype of broadboards or lee-boards. The number of these floating boards was four to six on either side later, in Song. It was in Ming that the number was reduced to one pair, which became the leeboard known in China as *qiaotou*. In Ming and Qing, two blocking boards similar to the bilge keels on modern ships were added to the bottom of the hull for stabilizing the ship. The appearance of the blocking boards marked major progress in shipbuilding. The sand ship carried bamboo "safety baskets", normally hung at the stern. In a storm these were filled with stones and lowered below water level to reduce the rocking.

The Propulsion of Vessels

Another outstanding feature of ancient Chinese nautical technology was a full and effective use of wind power.

In manpower [propulsion, the development from the oars and paddles to the scull was a major innovation, since the latter is twice or even three times as efficient. The ruins of the shipyards of Qin and Han, the turret ships of Han, the invention of the scull as an improved means of propulsion, the introduction of the stern rudder and the use of sails, all point to a mature shipbuilding technique by the Han Dynasty.

Chinese sailors had from Han and Wei times been setting their sails at various angles and trimming them to make the best use of the wind.

A Song Dynasty writer said in the 13th century, "Of all the eight quarters whence the wind may blow, there is only one, the dead ahead quarter, which cannot be used to make the ship sail."

In fact Chinese sailors have been able to sail against the wind for upwards of 400 years, the first records of this concerns the sand ship: "The sand ship is capable of tacking into the wind." Sailing against the wind, a ship must take a zigzag course in order to move ahead. Sailing into the wind required close co-ordination of the sails with the leeboard and stern rudder.

The double-mast sand ship was equipped with a leeboard on either side, the leeward board being put to use when tacking the wind. Lowered in the water below the bottom of the hull, the leeboard

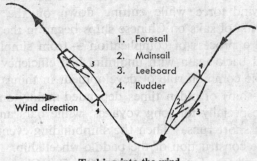

1. Foresail
2. Mainsail
3. Leeboard
4. Rudder

Wind direction

Tacking into the wind.

served to counteract sidewise movement and reduce leeward loss.

Chinese shipbuilders believed in variety in sail shape. The sand ship for example was equipped with different sails for different purposes. Sand ships sailing in inland waters had long narrow sails, while their sea-going counterpart had sails twice as wide but one-third shorter. This was obviously to lower the point of wind impact, probably because of the stronger winds at sea. This sail shape is used on Chinese sailing vessels on the South Seas today to prevent the ship from capsizing in a gale.

The use of sails began with single-mast, single-sail ships, which had small sails and moved slowly. Later the number of sails increased to two and up to 12. It was also found that the sail was most efficient when hoisted to the top of the mast, as expressed in the saying, "The turban on top can lift the ship and make it light and fast." In an unfavourable wind the "turban on top" was used to increase the speed, while "pinned flowers" were attached to the flanks of the main sail to prevent the ship from slanting. This was done mostly in a side wind. A "sail skirt" was added at the bottom of the main sail to further lower the point of wind contact.

Increasing the number of sails, while utilizing the full wind force, made the operation of the masts more complicated and intensified the work of the crew. Any failure to lower the sails in a sudden storm would result in the breaking of the masts and the capsizing of the ship. The shipbuilders began then to simplify the structure to one sail for each mast while enlarging the sail itself. This made possible full use of the wind force while cutting down on the crew's work. This simplifying of sails on Chinese ships began in the 15th century, and followed a process of complication — from simplicity to complication and back again with undiminished efficiency. But is this course not the common rule of development in things?

Chinese ships of olden times depended mainly on the wind for propulsion, especially for long voyages, while oars and sculls were employed on short runs. Then the shipbuilding event of a millennium was the construction of the paddle-wheel ship. In the Southern and Northern Dynasties (420-589), Zu Chongzhi built the "thousand-*li* ship" which actually travelled some hundred *li* in a day. This was apparently a type of paddle-wheel boat. Later, in the Tang Dynasty, Li Gao invented a paddle-wheel boat which is specifically mentioned in historical records. "A warship was built, which is propelled by two tread-mill paddle-wheels. It flies through storms as fast as though under sails." By the Southern Song Dynasty paddle-wheelers had developed considerably. A fleet of such ships were built by Gao Xuan, a general under Yang Yao who commanded a rebel peasant force on Dongting Lake. The large paddle-wheel ship of Song was 70-100 metres long and carried 700-800 men. The paddle-wheel ship of Yang Yao had a two- to three-storey deck-castle and was manned by over 1,000. She drew 3-4 metres of water. The number of paddle-wheels increased from four for early boats of this type to eight, then to 24 and 32. The "flying-tiger warship" propelled by the same device built at the time was a four-wheel boat with two axles, each paddle-wheel having eight oar-blades. Paddle-wheel ships were occasionally seen in southern China till the early 20th century. Driven by manpower, the paddle-wheel ship was costly compared with sailing boats, and for this reason it never gained very wide popularity despite the great numbers once built.

The paddle-wheel ship, also known as the exposed-wheel ship, was brought about by the evolution of oars into paddle-wheels, changing the intermittent action of propulsion to that of revolving (uninterrupted) action. It therefore marked a major step forward in the technique of vessel propulsion. Dating from the time of Li Gao in the Tang Dynasty, China's first paddle-wheel boat preceded its counterpart in the West by seven to eight centuries.

Besides the problem of propulsion, an equally serious question facing the navigators was orientation, a question with two aspects: guiding the ship, and telling its bearings. Clay model ships excavated from a Han Dynasty tomb in Guangzhou had rudders at the stern, indicating that stern rudders were used in China around the beginning of the Christian era. Further developments were the auxiliary rudder and the second auxiliary rudder. Chinese shipwrights meanwhile varied the structure of the rudder to produce one that could be hoisted and lowered, a balanced rudder, a fenestrated rudder and other types. The balanced and fenestrated rudders are marked by a change in the length of the power arm, reducing the effort needed to turn the rudder. Balanced rudders came into general use in China from the Song Dynasty.

The Technique of Maritime Navigation

Yan Dunjie

With a very long coastline bounding much of China's vast territory, the Chinese people developed the skill of sailing on the sea in very early times.

Maritime Navigation by the Heavenly Bodies

Chinese sailors in ancient times learned to orient themselves on the sea by observing celestial bodies. It is mentioned in *Huai Nan Zi* (*The Book of the Prince of Huai Nan*) that travelling aboard ship at sea, one could tell east from west by locating the polar star. A similar remark is found in *Bao Pu Zi* (*Book of Master Baopu*) by Ge Hong (284-364) of the Jin Dynasty (265-420). Ge Hong states that travellers on land who lost their way were guided by the south-pointing chariot, and if they lost their way on the sea they looked at the polar star. Fa Xian, a monk of the Eastern Jin Dynasty (317-420) who returned from a visit to India by sea, said that on board ship "we found ourselves in the midst of boundless waters, at a loss in telling east from west. We advanced by observing the sun, the moon and the stars." This "dependence on stars at night and the sun in the daytime" continued till the Northern Song Dynasty (960-1127), when the Chinese mariner learned to "look at the compass on a cloudy day". It was in the period of Yuan (1280-1368) and Ming (1368-1644) that nautical astronomy developed to a degree where it was possible to determine the geographical parallel of latitude from the altitude of stars, and this was the parent of Chinese ancient nautical astronomy. The method was known as the "star-aiming technique" using a device called "star-aiming plates".

Fashioned of ebony, these consisted of 12 square plates, the largest of which measured about 24 centimetres square, the rest being each smaller than the other by two centimetres, till the smallest was about two centimetres square. There was also a small ivory square with a cleft on each side equal to 1/4, 1/2, 3/4 and 1/8 respectively of the side length of the smallest ebony plate.

To range the polar star with the star-aiming plates, the sailor held with outstretched arm one of the plates by the middle of one side while looking up at the sky. He held the upper edge of the plate level with the polar star and its lower edge at the horizon. A string kept the plate at a constant distance from his eyes. The size of the plate showed the altitude of the polar star at that particular location. For different altitudes of the star different plates were chosen, to be supplemented with the sides of the ivory piece. The degree of latitude was calculated from the altitude of the polar star.

During the Yuan Dynasty, Marco Polo travelled overland to China, remained for more than 20 years, then returned to Italy via Persia, to where he sailed in a Chinese ship, going across the Indian Ocean then turning west. In *Travels of Marco Polo*, sailing on a Chinese ocean-going vessel is described, and mention is made of taking the altitude of the polar star when the ship sailed into the Indian Ocean from the Strait of Malacca. Certainly the star-aiming technique was used.

In all seven voyages to the South Seas made by Zheng He and his fleet in the early 15th century, "star-aiming records were kept on both the outgoing and return trips" as mentioned in a notation, manifesting Chinese navigators' mastery of the technique at that time. Ming sailors usually took the polar star as guide, but when it was not visible they took the Huagai (the twin star β and γ of Ursa Minor) in low latitude degrees below 6°N.

Mariners in Ming chose guiding stars for observation, determining the bearing of the vessel at night by the azimuth of the guiding star and its altitude from the horizon. This was then known as star sighting and is in the scope of the star-aiming technique.

The voyage from Guli (Calicut) on the west coast of India to Dhufar in Oman, on the east coast of the Arabian Peninsula, may be cited as an example of recorded navigation in Ming by the star-aiming technique. When the ship sailed from Calicut, the altitude of the polar star was taken at 6°24' (equivalent in modern figures, the same hereinafter). She turned northwestward and made 900 kilometres to Mangalore on the west coast of India, when the polar star was taken at 8°. Then she sailed northwest by north, went 1,500 kilometres and took the altitude of the polar star at 10°. Thereafter she sailed westward and slightly north and went 2,100 kilometres to reach Dhufar, when the altitude of the polar star was 12°48'.

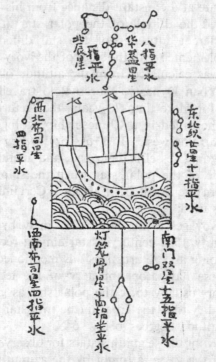

Diagram showing altitudes of guiding stars between Sri Lanka (Ceylon) and Sumatra on the return voyage to China.

The degrees of latitude corresponding to these altitudes, calculated by the method then known to the Chinese pilots, fall in basically with the latitude of the above positions on a modern map. Likewise, the route of the voyage is analogous to that followed by modern ocean liners, all of which points to a fair level of development of nautical astronomy in the Ming Dynasty.

Ming mariners formulated rules governing the rising and setting of celestial bodies. Extant are the timetables for the sun and the moon as found in a late Ming manuscript on navigation routes. The mariners composed songs called *Timing the Rise and Setting of the Sun* and *Timing the Rise and Setting of the Moon*. The first runs as follows:

> *Up at seven and down at five*
> *The sun goes in the First and Ninth Moons alive.*
> *For the Second and the Eighth Moon —*
> *Up at six and down at six in the afternoon.*
> *In the third and the seventh the sun glows*
> *From five to seven it goes.*
> *From four to eight it takes cover*
> *In the Fourth and Sixth Moons of summer.*
> *From nine to three it shines merely*
> *In mid-winter, so cherish it dearly.*
> *For Moons Ten and Twelve it shines more —*
> *Up at eight and sinking by four.*
> *From three through nine the day is light,*
> *So the Fifth Moon is the most bright.*

Such was the way of telling time by the sun for the 12 lunar months as recorded in doggerel form. Though slightly inaccurate, it was a practical rule-of-thumb guide for navigators.

Ming books on navigation record the presence of personnel charged with various duties on board Chinese ocean-going vessels, including those specially for astronomical observation. According to *San Bao Tai Jian Xia Xi Yang Ji Tong Su Yan Yi* (*Popular Story of the Voyages of the Three-Jewel Eunuch in the Western Oceans*), a very popular Ming novel, "ten astrologers to observe the stars" were aboard the ship. "On every ship," it says, "there are three heavenly platforms, arranged one above the other. On each are stationed 24 soldiers whose duty is to observe meteorological changes in the day and stars and constellations at night." This observation contained in a novel no doubt reflects the actual practice in maritime navigation during the Ming Dynasty.

Navigation Technique by Terrestrial Observation

The technical achievements of ancient Chinese navigation by terrestrial observation included the invention and employment of such navigational instruments as the magnetic compass, the log line and lead line, as well as the compass course and navigation charts. The south-pointing needle was put to use in navigation soon after its discovery in China. The floating south-pointing needle of Northern Song developed later into what was known as the water compass. According to Zhu Yu in the Song Dynasty (960-1279), Chinese sea-going vessels started using the compass between 1099 and 1102. Xu Jing, who returned from a voyage to Korea in 1123, told of sailing throughout the night guided by the stars, or when it was overcast they used the floating compass. These records are apparently the earliest in the world concerning the use of the compass in maritime navigation.

A Han Dynasty (206 B.C.-A.D. 220) record mentions the nautical compass being divided into 24 equal parts. The same 24 divisions appear in a map by Shen Kuo of Northern Song. This was the division of the 360° compass into 24 points of direction, at an interval of 15° between each two points. When the compass was used, each of the 24 parts was again split, making 48 directions in all, with each part enclosing 7°30'. The latter were called "intermediate points". This compass was much more precise than a 32-direction type imported from the West in late Ming, and Chinese pilots preferred their own to the imported compass.

The ship's compass in ancient times was mounted in a compartment called the "needle room", accessible only to the officer in charge of the compass. According to *Xi Yang Fan Guo Zhi* (*Records of Countries of the Western Oceans*) of Ming times, the officer in charge of the compass was an experienced sailor. He was the chief navigator responsible for the compass, logbook, sea charts, etc. "This is a matter of vital importance, and should not be treated lightly," warns the *Records*, stressing the importance of the compass to ocean-going vessels.

The log line is also recorded early in history. In the period of the Three Kingdoms (220-280), a sailor of the state of Wu wrote a book called *Nan Zhou Yi Wu Zhi* (*Strange Things of the South*) upon returning from a voyage to the South Seas, in which he tells of measuring the speed of the vessel by throwing a piece of wood in the sea from the bow, then running quickly to the stern to see if the wood reached the stern at the same time. This was the primitive log that remained in use till the Ming Dynasty. Then it became more sophisticated by dividing the 24-hour day into 10 "watches", the time being measured by burning sticks of incense. A piece of wood was thrown in the water and a person walked from bow to stern with it. If they arrived at the same time, the measurement was considered accurate. It was called "under-watch" if the wood travelled behind, and "super-watch" if it was ahead of the person, a "watch" being the gauge for 30 kilometres of sailing. The speed of sailing and distance covered were thus measured.

This ancient Chinese device for measuring distance was comparable to the log ship employed in later maritime navigation. The latter was also a slip of wood, but in the shape of a fan, which was fastened to the end of a line as long as the ship and thrown into the water. The time measuring was done by means of the sandglass, which marked 14 seconds in a run. The line was marked into equal divisions, and the speed and distance of sailing were measured by the length of the line. The Chinese incense stick and the sandglass of the West served the same purpose of timing.

By the Tang Dynasty (618-907) at the latest Chinese pilots were using plumb-line devices, one being called the "sinking hook" and another the "line with an iron weight". These could plumb depths up to 20 metres, which was still shallow-water sounding. Following closely upon this, as records show, depth-sounding was done with cords which had a length of 170 metres. This could be classified as deep-water sounding.

Wu Zimu in Southern Song (1127-1279) wrote in his *Meng Liang Lu* (*Records of a Dream of Grandeur*) that Chinese ocean-going vessels carrying goods to foreign lands put out to sea from Quanzhou in Fujian and while on the high seas "they made depth-soundings for upwards of 700 *chi*"[1] — no small feat in depth-sounding technique in China's Song Dynasty.

Charts indicating compass courses to be taken to various places appeared also in Song. Called "needle guides", they were meant to guide navigation by the compass needle. These charts, accumulated over the years, were later referred to as "Needle Scriptures", "Needle Notations" or "Needle Books".

These "needle guides" usually included: the port of departure, sailing direction, distance covered before changing direction, and the places reached. The direction of sailing was marked as "single needle" or "direct needle" when the needle pointed directly to azimuth points. When the needle pointed between two neighbouring azimuth points the direction was marked out with both points and was called the "intermediate needle" as mentioned above. The four different instances of two neighbouring directions being marked together to denote the course of sailing were 1) direct-intermediate, 2) intermediate-direct, 3) direct-direct, and 4) intermediate-intermediate. In some cases the direction was defined as more than two orientations.

The distance covered was measured in "watches". The arrival of a ship at a destination was recorded appropriately in the four following ways: "to be abreast of", "to pass", "to come in sight of" and "to arrive at".

The compass course included also star-aiming records for night sailing and the depths of the sea as they were sounded out.

---

[1]　1 (Song) *chi* = 0.319 metre.

The following are examples from the "needle guide" records of the voyage from Taicang in Jiangsu Province to Japan as excerpted from *Chou Hai Tu Bian* (*Illustrated Seaboard Strategy*), a Ming Dynasty compilation:

— Set sail from the port of Taicang.   Used the needle pointing direct to Yi.[1]   Sailed one watch.   Ship abreast of Wusong River mouth.

— Used the needle pointing direct to Yi and then pointing between Yi and Mao.   Sailed one watch until abreast of Bao-shan.   Reached Nanhuizui.

— Used the needle pointing between Yi and Chen.   Sailed out of harbour.   Sounded depth at 60-70 *chi*.   Took the route via Shanidi.   Sailed three watches.   In sight of Chashan.

— Henceforth used the needle pointing between Kun and Shen, then between Ding and Wei.   Sailed three watches.   Reach-ed Big Seven Mountains and Small Seven Mountains.   Found Tanshan on the northeast.   Below Tanshan the depth was seven to eight *tuo*.[2]

— Used the needle pointing direct to Ding and then between Ding and Wu.   Sailed three watches.   Reached Huoshan....

Several hand-copied manuscripts of needle-guide records have now been discovered, including voyages both to the East and West Oceans, and these are being given further study.   Needle-guide records compiled by Chinese mariners are known to have been used by Portuguese sailors on voyages to Southeast Asia in the early 16th century.

Navigation maps are mentioned in *Gao Li Tu Jing* (*Illustrated Record of an Embassy to Korea*) by Xu Jing in Northern Song.   These

---

[1]   The ancient Chinese navigational compass is marked out with four of the eight trigrams of the Bagua (the Eight Diagrams), eight of the ten Gan char-acters of the denary cycle and twelve Zhi characters of the duodenary cycle. The compass has 24 points, dividing the circle into segments of 15° each. Thus Yi is about 105°; Mao, 90°; Chen, 120°; Kun, 225°; Shen, 240°; Ding, 195°; Wei, 210°; Wu, 180°.

[2]   1 *tuo* = 5 *chi* or 1.6 metres.

charts have unfortunately been lost through the generations. The earliest Chinese sea-chart now extant is the illustration attached to *Hai Dao Jing* (*Manual of Sailing Directions*) of early Ming.

Wu Bei Zhi (*Treatise on Armament Technology*) by Mao Yuanyi of Ming, which runs to 240 volumes, has a map appended titled "Map Showing Voyage of Ships Which Set Sail from the Yards of Treasure Ships, Put to Sea at Longjiangguan and Reached Various Foreign Countries". This is none other than the navigation map of Zheng He. The sea routes and geographical positions depicted on the map correspond with the story of Zheng He's last trip to the Western Seas in the fifth year of Xuande (1430) as recorded in *Qian Wen Ji* (*Traditions of Past Affairs*) by Zhu Yunming of Ming. The map is believed to have been done in the mid-15th century. The navigation map of Zheng He, a well-known relic both in China and worldwide, has been important in studying the history of communications between China and the outside world and of maritime nautical technique in the 15th century.

Mention is made in some late Ming classics of cartographical illustrations such as a "Map Showing Topographical and Hydrographical Features on the Route from Lingshan to Java", "Map Showing Topographical and Hydrographical Features on the Route from Xincun and Java to Manlajia", "Map Showing Topographical and Hydrographical Features of Pengkeng", etc. All these maps are lost, only the literal explanations and legends being preserved. Such captions never fail to warn about danger spots on the sea (e.g., "Here is an isle overgrown with grass." "Here is a grove of reeds."), shoals (e.g., "Beware of shoals in the bay." "Here is a clay beach."), reefs (e.g., "Reef at the harbour mouth. Keep off it." "There are reefs where the waves roll."), and beaches (e.g., "Here is a ridge of sand in the water."), rocks (e.g., "Here are ancient rocks." "Here is an ancient rocky bank.") etc. These remarks are similar to what might appear in modern navigation maps.

Two sea charts of the early Qing Dynasty that have come down to us are a needle-guide map of navigation to countries in the Western and Southern Oceans, a paper map painted in colour and produced in 1711-1715, and a map of the Eastern and Southern Oceans, also a paper map in colour produced in 1712-1722. Both maps are now preserved at the Palace Museum in Beijing.

The Spinning Wheel and the Loom

Gao Hanyu and Shi Bokui

China had a highly developed textile technology, and delicate Chinese silk won a world-wide reputation in ancient times. The credit is not only due to the high skill of the Chinese craftsmen but also to the creation and continued innovation of textile tools and equipment. Here we shall dwell only upon the development of the spinning wheel and the loom.

The Spinning Wheel

It was known in antiquity that fibres such as hemp, silk, wool and cotton had to be spun into yarn before they could be manufactured into textile fabrics. From time immemorial the Chinese used what was known as the distaff, a primitive spinning tool large numbers of which have been discovered in Neolithic ruins excavated in many parts of China. This tool was a small disk of clay or stone, called the distaff disk, with a hole in the centre into which was thrust a distaff rod. For spinning, a segment of hemp or other fibre was first twisted and wrapped on the rod. The spinner then held the rod in one hand, turning the disk towards the left or right with the other, so that fibre was spun into yarn through twisting and drawing. It should be noted that this is an early application of the mechanical principle of the revolving couple and of the power of gravity in textile-making. When the spun yarn reached a certain length it was wound onto the distaff rod. This process was repeated until the rod carried a full package.

Later, after centuries of production development, there appeared the single-spindle hand spinning wheel, which very soon replaced the distaff and became an important tool for early textile production.

The exact date of the first spinning wheel has not yet been determined. Its earliest description in Chinese recorded history is found in *Fang Yan* (*Dictionary of Local Expressions*) by Yang Xiong (53 B.C. -A.D. 18) of the Western Han Dynasty (206 B.C. -A.D. 24), where such terms as "spinning wheel" and "road tracks" are mentioned. The earliest portrayal of working with the single-spindle spinning wheel was seen in Western Han silk paintings found at Yinqueshan in Linyi County, Shandong Province. Similar descriptions are found in Han stone engravings, of which no less than eight plates have been discovered so far. All these plates depict textile-

Spinning with a single-spindle wheel as seen in a Western Han Dynasty silk painting found at Yinqueshan.

making, while four depict spinning wheels. An engraved stone unearthed in 1956 at Honglou Village, Tongshan County, Jiangsu Province, shows lively images of workers weaving, spinning and arranging silk, illustrating the textile craft in the Han period. It is thus apparent that the spinning wheel had become a very popular tool for the production of yarns by the Han Dynasty, and that its first appearance must have been much earlier.

The Han spinning wheel as seen on the stone engravings bears a similarity to the one depicted in *Tian Gong Kai Wu* (*Exploitation of the Works of Nature*) written in the Ming Dynasty (1368-1644). Though simple in structure, the spinning wheel was 20 times more efficient than the distaff. Cord-wheel drive is noted on the apparatus, which confirms the employment of this power transmission method on production machinery as early as 2,000 years ago.

A spinning wheel of this type produced silk yarn or musical instrument strings of various thicknesses with even fineness by twisting or doubling. A Chinese zither of the Han period unearthed in 1972 from Han Tomb No. 1 at Mawangdui, Changsha, Hunan Province had strings believed to have been produced on this kind of spinning wheel. The thickest string, with a diameter of 1.9 millimetres, was probably produced from 37 silk filaments twisted into a single yarn which was then doubled twice into a four-fold yarn. Four of these were finally twisted together to make the thickest string. This means that the bass string of the Han zither was made up of 592 silk filaments. The 25 strings on the instrument are each of different thickness, spun finely and evenly and produce a melodious sound when plucked. The development of such spinning wheels is highly significant in the history of yarn spinning.

Besides spinning yarn, the spinning wheel raised productivity considerably by being of use also in preparing the weft spool in a shuttle, a job hitherto done by a separate tool.

From the dynasties of Han (206 B.C.-A.D. 220) and Tang (618-907), the splendid Chinese silk shipped to the West via the Silk Road earned world-wide admiration. In many places along this ancient international silk trade route there have been excavated satin, damask, gauze and other silk fabrics of very fine quality. The spinning wheel obviously contributed much to the production of these fabrics.

In spinning hemp, silk and cotton yarns, the Chinese were successful in increasing productivity and improving quality. The hand-operated spinning wheel was developed into treadle and hydraulic types which boosted China's yarn-spinning technique to a new stage.

Based on the eccentricity principle, the treadle spinning wheel was an innovation in yarn-spinning tools. The exact date of the first treadle spinning wheel has not yet been ascertained. However, we do have an Eastern Jin Dynasty (317-420) painting by Gu Kaizhi (c. 345-406) in which there is a three-spindle foot-propelled spinning wheel. Later, *Nong Shu* (*Agricultural Treatise*) written in 1313 by Wang Zhen described spinning wheels for cotton with three spindles and others for hemp with three or five spindles, all foot-powered, giving evidence that the treadle spinning wheel was used at that time.

The spinning wheel has been long and widely used in spinning cotton. China's minority nationalities contributed much to the growing and processing of cotton. Particularly in Yunnan Province and Hainan Island they developed a series of techniques for the processing of cotton. In spinning technology, they were skilled in using a spinning device with bamboo wheels 61 centimetres in diameter as well as a 30-40 centimetre size, the smaller ones for the staple fibres of cotton, which are much shorter than silk or hemp fibres.

During the transition period between the dynasties of Song (960-1279) and Yuan (1271-1368) a woman gained fame as an innovator in textile technique. Named Huang Daopo, she was born in Wunijing Village, Songjiang County, Jiangsu Province but later travelled to Yazhou (now Yacheng) on Hainan Island where she learned cotton textile skills from women of the Li nationality there. Returning to her hometown around 1295, she engaged in the textile craft and in collaboration with local women weavers converted the treadle hemp spinning wheel into a new three-spindle cotton spinning frame. She also improved the tools for cotton ginning and shedding. Her innovations brought about a marked rise in the production of cotton yarn, which soon solved the problem of low cotton-textile production in the district. She also summed up and perfected skills in weaving, including yarn interlocking, colour matching, yarn heddling and jacquard weaving. In time the Songjiang district became a cotton-textile area, and the elegant

"Wunijing quilt" sold well across the country.

Besides giving a full account of the hand-operated and treadle spinning wheels, Wang Zhen in his *Nong Shu* described two other new types of spinning wheels, the jumbo and the hydraulic. The former differed from previous types in its number of spindles, which was increased to 32. Efficiency was considerably raised. A common single-spindle spinning wheel could turn out only 3-5 *liang*[1] of cotton yarn a day, while even a three-spindle wheel turned out no more than 7-8 *liang*. For hemp yarns the production of five-spindle equipment hardly exceeded one kilogramme (20 *liang*) per day. The large spinning wheel, used specially for hemp, produced 50 kilogrammes of yarn in 24 hours, so many peasant households were required to pool their hemp harvest to keep one jumbo spinning wheel in operation. Meanwhile, the large wheel was equipped with a power transmission system resembling the belt drive of modern machinery.

Before the invention of the new spinning, modern mechanized spinning never dispensed with spindles and their revolving motion. The only difference was in the greater number of spindles and increased speed made possible by a large power drive. Comparison of the ancient spinning wheel with the modern spinning machine reveals that the bamboo wheel of the former is replaced by a tin cylinder that moves the spindles, while the cord-wheel drive on the jumbo spinning wheel is very similar to what is known as the up-to-date tangential belt drive. The two types of drive mentioned above work basically on the same principle.

The Loom

Reliable information is lacking as to the actual configuation of a primitive weaving tool. Much recent data collected by Chinese archaeologists do, however, give a general picture of the ancient weaving craft. Among Neolithic ruins discovered in 1975 at Hemudu, Yuyao County in Zhejiang Province were distaffs, tubular bone needles,

---

[1] 1 *liang* = 50 grammes.

The lid of a pot for storing shells unearthed at
Shizhaishan, Jinning County, Yunnan Province.

beating-up knives of wood and bone, reeling rods, etc.

How did man learn to weave cloth? Let's first study some terminology concerning the craft. According to *Shi Ming* (*Explanation
of Names*) by Liu Xi, "Cloth is made up of two sheets of yarn — the
warp and the weft — the latter interlaced with the former." This
means that plain cloth is woven by interlacing one set of threads,
the warp for length, with another set, the weft for width. A weaving
scene is depicted by the moulded figures on the lid of a pot for
storing shells which was unearthed at Shizhaishan in Jinning County,
Yunnan Province. The figures represent slaves of the Dian nationality working for the slave-owner. A woman in a coarse cloth
tunic buttoned at the breast with a belt tied at the waist sits on the
floor weaving. Treading the warp beam under her feet, she tightens
the weft with the beating-up knife in her right hand while feeding it
in with her left. She bends low over her work, the tool she is using
being called a "squatting loom" or "waist loom". The figures show
also that the primitive loom already consisted of movements in three
directions — up and down for shedding, left and right for feeding in

I   II   III

IV   V

Figures of women slaves spinning or weaving, as seen on the lid.   Figures II, III, IV and V are weaving.

the weft, and forward and backward for beating up the weft. It may be considered the earliest prototype of the modern loom.

Later, in the process of development and innovation, a new type of treadle loom appeared which featured a slanting member for raising and lowering heddles.   This is shown in a Han stone engraving unearthed at Caozhuang Village, Sihong County in Jiangsu Province, which bears in relief a picture entitled *A Mother Working at Her Loom*.   This depicts a loom with a frame supporting the warp sheet at an inclination of 50°-60° with respect to the horizontal pedestal, a device that enables the weaver to sit and gives her a good view of the warp surface so she can check whether the tension is even or if any yarns have been broken in shedding. The loom is further equipped with a treadle mechanism for actuating shedding motion. The picture shows clearly that the weaver treads on two pedal levers, one longer than the other, to drive the heddle shafts up and down. The heddle is connected to the pedal by a string, which, on being worked by the pedal, pulls the "horse head" (i.e., oscillating lever which has a front part smaller than the back and resembles the head of a horse) and produces a to-and-fro motion to separate the warp sheet into two layers resulting in a heddle-shedding in triangular shape. The weaver can operate the heddle with the feet instead of hands, which makes the work easier and leaves both hands free to be used simultaneously in feeding and beating up the weft. The slanting loom is 10 times more efficient than the early weaving tools, and it

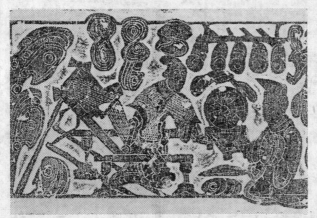

A Mother Working at Her Loom — a Han stone engraving unearthed at Caozhuang Village, Sihong County in Jiangsu Province.

contributed much to increasing cloth production. Historical records reveal that by the time of the Warring States Period (475-221 B.C.) the monarchs exchanged silk presents in quantity far larger than in the preceding Spring and Autumn Period (770-476 B.C.). In the Qin (221-207 B.C.) and Han (206 B.C.-A.D. 220) dynasties, the slanting loom was already popular throughout the vast rural areas along the Huanghe and Changjiang river basins. Moreover, a little instrument is shown in the picture almost dropping to the floor. It is pointed at both ends. This is a shuttle, the tiny, handy instrument used to pass the weft through the warp shed, which further increased weaving speed. A major innovation in weaving tools, the shuttle has been in use through the generations ever since.

In the 13th century, the loom was further improved. In his *Zi Ren Yi Zhi* (*Traditions of the Joiner's Craft*) Xue Jingshi, a one-time carpenter of Wanquan (now Wanrong) County in Shansi, gives the configurations and dimensions of the vertical loom (3), Hua loom (4), gauze loom (5) and the horizontal cloth loom (6), with accompanying detailed description of their structures and principles of operation. The three-dimensional diagrams in the book are clear at a glance and, to quote the author, "furnish 90 per

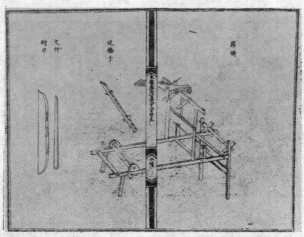

The gauze loom as depicted in *Zi Ren Yi Zhi* (*Traditions of the Joiner's Craft*). The illustration is reproduced from *Yong Le Da Dian* (*Great Encyclopaedia of the Yongle Reign*).

cent of the knowledge a carpenter needs to build a loom". This book on the history of China's textile technology provides valuable data for studying the development of the weaving loom.

The looms found in peasant households were in those days mostly built by individual carpenters and varied greatly in shape and size.

"Each has its own features," comments Xue Jingshi. This caused great inconvenience for the installation and repair of the looms. Xue, summing up his experience in the repair and manufacture of looms over long periods and drawing on the merits of other loomwrights, completed his book only after great effort. It is a book on the construction of looms with detailed explanation and abundant illustrations. "Every part of a loom," says the author, "is singled out and individually described." Just like the designing procedure of a machine in a modern factory, design drawings were prepared for individual parts and for their assembly with spelled-out explanations of measurements of each part and where it was to be fitted. In the preface the book says: "Individual parts have their own names, and they make up one complete machine when put together." Detailed design drawings are provided, for example, for the gauze loom, which was used for the weaving of thin, transparent fabrics with various patterns. The loom was seven to eight *chi*[1] long with its crossbeam 2.4-2.8 *chi* between its two outermost ends. The lease rod on the loom, called the "crow's wing" for its similarity in shape to a bird, guides the warp ends to produce a regular up-and-down motion and a left-and-right crossing so to doup the ground warp threads with themselves into net patterns as designed.

Explanations are given in *Zi Ren Yi Zhi* of the structural principles of the Hua loom and the horizontal cloth loom, and these are dealt with more fully than those of the waist loom and the gauze loom by Xu Guangqi (1562-1633) of the Ming Dynasty (1368-1644) in his *Nong Zheng Quan Shu* (*Complete Treatise on Agricultural Administration*) or Song Yingxing (1587-?) in his *Tian Gong Kai Wu* (*Exploitation of the Works of Nature*). Standard measurements, for instance, are specified for the shuttle, which is 1.3-1.4 *chi* long and 1.5 *cun*[2] wide and 1.2 *cun* thick at the centre, with a shuttle eye to let the weft thread pass through. Thanks to further perfection in the three motions of the loom — shedding, shuttle projection and beating-up — a considerable increase in productivity and improvement of

---

[1] 1 *chi* =0.321 metre.

[2] 1 *cun* = 1 /10 *chi*.

quality was achieved in cloth production. A variety of looms and other weaving tools designed and built by Xue Jingshi were very well-known in the prefecture of Lu'an, Shanxi Province, where the popular utilization of these looms brought about further development in textile production in this already fairly prosperous place, bringing the Lu'an area on a par with Jiangsu and Zhejiang provinces in textile production. In those days a prevailing saying was: "All under Heaven are clothed with the products of either Songjiang in the south or Lu'an in the north."

The Jacquard Loom

The Chinese version of the jacquard loom was invented in antiquity and it contributed much to the elegant and colourful fabrics produced long ago in China. Twills with simple geometric patterns were woven as early as 4,000 years ago. A remnant of jacquard silk fabric with a spiral pattern was found as wrapping for a bronze battle-axe unearthed in a tomb of the royal family of the Shang Dynasty (c. 16th-11th century B.C.) at Dasikong Village, Anyang, Henan Province. It was a single-colour plain back fabric with

Remnant of jacquard-type fabric with geometric pattern unearthed from a Shang tomb in Anyang.

twill warp patterns known as *qi* or damask. By the Zhou Dynasty (c. 11th century-221 B.C.), Chinese weavers were already making *jin* (brocade), a multi-colour jacquard silk material. Historical records indicate this existed even before the Qin Dynasty, such as the remark in *Yi Jing* (*Book of Changes*): "Change at every three and five, vary the number, keep changing and make up the pattern under Heaven." This implies that weavers at that time already had rather sophisticated jacquard-type looms which could generally be used to weave complicated patterns by following the principle of "changing at every three and five".

The silk-textile craft flourished in the Qin and Han dynasties. The government set up the Eastern and Western Textile Offices and chose an officer to be in charge of royal clothing, the textile handicraft workshops employing thousands of workers and the technique of silk jacquard weaving reaching a high level. Among the findings in the excavated Han tomb at Mawangdui of Changsha in Hunan Province were silk *qi* with bird designs and brocade with designs of flowers, waves, dragons and leopards. Also discovered for the first time was pile-loop brocade, which marked a breakthrough in the jacquard technique of brocade. This pile-loop fabric is believed to be the forerunner of the modern *zhangrong*[1] and velvet.

Further study is necessary for determining the exact structure and style of the jacquard-type loom in early Han, yet its structural features can be roughly visualized by analysing the pile-loop brocade. Obviously the product of highly complicated technique, the brocade is an elegant two-sided silk fabric with a double-warp consisting of the four ends in each complete warp. It is 50 centimetres in width, with 8,800-12,000 warp ends. This could not have been done without a jacquard harness device as well as a double-warp beam system to lift the ground and pile warps separately. Moreover, a pile-on-pile loop was produced on the fabric, giving a three-dimensional effect which may justifiably be described as "adding flowers to the brocade" (meaning making a beautiful thing more beautiful), which points to high skill in jacquard-type weaving by early Western Han. Below

---

[1] A fabric with pile on one side produced in Zhangzhou, Fujian Province.

is a record found in *Xi Jing Za Ji* (*Miscellaneous Records of the Western Capital*) concerning the manufacture of brocade and damask in this period.

Huo Guang's wife presented Chunyu Yan with the gift of 24 pieces of Cat-Tail Brocade and 25 pieces of Scattered-Flower Damask. These were then famous products of the Chen Baoguang family. Chen's wife had learned textile technique from him and was employed in Huo's home where she was ordered to weave. There were 120 hooks on her loom. She finished one length of cloth in 60 days, each length being worth 10,000 copper coins.

This simple account fails to provide any information on the structure of the loom used by Chen Baoguang's wife. More detailed in this respect is *Ji Fu Fu* (*Ode to the Women Weavers*) by Wang Yi, an Eastern Han poet. The line about "tall towers facing each other" apparently refers to the elevated jacquard mechanism facing the harness frame of the heddle. Sitting on the three-*chi*-high frame of the jacquard loom, the patterning weaver manipulated the mechanism according to the design of "insects, birds and beasts". Watching from above the uniform and shiny warp ends, she had the feeling of "bending over a crystal pond" and saw the patterns of her creation in full view. "Fish swallowing bait" describes her action of lifting the thousand-some heddle cords, each attached to a bamboo rod at the lower end. Her hand moved with the swiftness of a fish darting for food at the water surface. Some of the warp threads being lifted were tight while others were loose, looking like the astronomers' diagrams popular in Han times, and this inspired the line, "Bending, stretching and moving like constellations." The to-and-fro motion of the beating-up mechanism is described as a kind of alternate pushing and pulling action. The principle and action of the jacquard loom is thus described in verse.

The jacquard loom of such complicated structure could hardly be popularized, nor could it meet the needs of the fast developing feudal economy. Then, in the time of the Three Kingdoms (220-280), a certain young man named Ma Jun who lived in Fufeng (now

Xingping County, Shanxi) in early Wei (220-265) viewed the jacquard loom of his day as too complicated, inefficient, and requiring too high a labour intensity. "It occurred to him," says a memoir, "that the silk loom could be innovated, and this was so simple that anyone could understand it on sight without explanation. The conventional silk loom had 50 heddles with 50 clips or 60 heddles with 60 clips. Master Ma considered this too cumbersome and wasteful of time and labour, so he simplified it to 12 heddles with the same number of clips." The jacquard silk fabric produced on the improved loom was marked for its splendid patterns and great variety of designs, while the loom was much more efficient. Just how Ma Jun's new jacquard loom was built, however, is not known. Judging from the number of its heddle shafts it was similar to the loom depicted in the painting *Geng Zhi Tu* (*Picture of Tilling and Weaving*) by Lou Shu of Southern Song (1127-1279). The jacquard silk fabrics then produced in the kingdom of Wei were already a match for beauty with the nationally famous Sichuan brocade. In 238, Himiko, Empress of Japan, sent an envoy to China, who returned with large quantities of silk brocade of violet ground with dragon patterns. It was then that the Chinese skill

Detail of the painting *Geng Zhi Tu* (*Picture of Tilling and Weaving*) by Lou Shu of Southern Song (1127-1279) showing a jacquard loom.

of jacquard silk weaving and that of printing and dyeing was transmitted to Japan.

The most comprehensive and detailed material now available
about the ancient Chinese jacquard-type loom is found in *Tian Gong
Kai Wu*. It says:

> The jacquard loom measures 16 *chi* from front to back.
> It has an elevated frame for the jacquard mechanism, with the
> comber plate mounted in the centre and the sley feet suspended
> below. . . . The boy weaver sits or stands on the jacquard frame.
> At the back of the loom there is an end beam for winding the
> warp threads. In the middle of it two wooden double-rib rods
> about four feet long have their end points sticking into the two
> ends of the reed.

Here the "comber plate" is the modern comber board; the "sley
feet" are now called sley arms, the "end beam" is now the warp beam,
and the "double-rib rods" are the rods for beating the reeds. The
exquisite Sichuan brocade and the "cloud brocade" of Nanjing are
produced on such jacquard looms.

The most difficult part of the jacquard technique, however, is
"tying the basis of the pattern", or arranging the yarns by the lifting
plan. *Tian Gong Kai Wu* says:

> The most ingenious craftsmen are assigned to tie the pattern
> basis. First a designer draws the intended patterns on paper.
> Then following the design, a planner meticulously measures out
> all the silk threads, ties them and then hangs them on the jac-
> quard tower.

This passage describes that in order to reproduce the designed
patterns on the fabric it is necessary to make such preliminary arrangements that thousands of warp threads can be harnessed up
and down regularly, and wefts of dozens of varieties can be arranged
in proper transverse order; thus a complete memory system for
patterning is formed. This done, the weavers begin work. The
warper and the pattern weaver, working in close co-ordination and

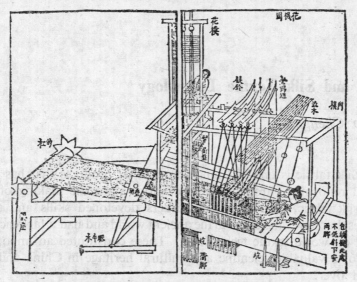

Jacquard loom as illustrated in *Tian Gong Kai Wu (Exploitation of the Works of Nature).*

following the prearranged patterning plan, undertake to feed in the weft yarns thread by thread, and so the elegant, exquisitely patterned fabric is turned out from the loom.

"Tying the pattern basis", therefore, stands as one of the major achievements of ancient Chinese textile workers.

Silk and Silk Textile Technology

Zhao Chengze

Silk fabric is one of China's traditional products that has won world renown. Through the generations since the making of the first silk fabric, Chinese textile workers developed skills and techniques that were unexcelled in the ancient world and had a far-reaching effect on world textile technology. These constituted an important part of a valuable scientific and cultural heritage in China and the world over.

The Role of Silk in Ancient China and the World

China's first silk textiles were produced from wild silkworms, which were later domesticated. According to recorded history, sericulture was practised at least 4,000 to 5,000 years ago in China.

Inscriptions on bones and tortoise shells of the Shang Dynasty (c. 16th-11th century B.C.) bear such characters as "silk", "mulberry", "silk fabric", etc., indicating the significance of silk at that early date. The craft developed to such a degree from Zhou (c. 11th century-221 B.C.) and Qin (221-207 B.C.) that in some subsequent dynasties the peasants were ordered to plant a certain amount of land to mulberry trees and pay taxes in silk. Production increased from dynasty to dynasty, and though figures are lacking on output in the various periods, certain archives give us some idea. One early history book says that the expenses of a single inspection tour to Shanxi and Shandong by the Emperor Wu Di of Han amounted to "over one million bolts of silk". The silk levied for tax and requisitioned by purchase in the Zhejiang area alone during the reign of the Emperor Gao Zong of Song amounted to 1.17 million bolts. These entries give some idea of the importance of silk in the social life and production of ancient China.

China's silk fabrics had a high reputation for fineness and beauty. Besides providing many people with an important clothing material, silk was a major commodity in ancient China's foreign trade. From very early, silk was shipped abroad through what is known in the West as the "Silk Road", and by sea from southeastern China ports. In Western Asia, Europe and Africa, where it was bought, Chinese silk was admired and cherished. In ancient Rome and Egypt Chinese silk was regarded as a treasure "so dazzlingly exquisite that the making of it must have exhausted the cunning of man". They considered it an honour to be clad in such splendid material. Western history has it that when Caesar the Great of Rome once appeared in the theatre dressed in a toga of Chinese silk, his garment became the focus of attention as an unprecedented luxury.

Chinese silk yarn, transported over long distances as the main item of trade for the merchants of many countries, could fetch 12 ounces of gold for one pound. It goes without saying that silk fabric was still more expensive. Chinese silk was in great demand on the Western market till after the 13th century.

Characteristics of Chinese Silk

Most of the silk fabrics of ancient China were made from the filament of the mulberry silkworm. A valuable raw material in the textile industry, this silk has fine qualities that are lacking in any other natural fibre; the silks of the tussah silkworm and those fed on castor or tree-of-heaven leaves do not have these special qualities. A notable textile achievement of ancient China was to separate out the fibre of the mulberry silkworm from all other natural silks. Fine and soft, the silk of the Chinese silkworm is highly lustrous, elastic, takes dye well, and is readily drawn into a filament.

The silk fabrics produced in various periods have all had these qualities of silks from the Chinese silkworm.

Some fabrics are as thin and transparent as gossamer, like cicada wings that are scarcely visible. A silk wrapover-style gown found in 1972 when a Han tomb was opened at Mawangdui outside Changsha in Hunan Province, though 128 centimetres long with sleeves measuring 190 centimetres, weighs a mere 49 grammes. A 9th-century European traveller named Sulleiman noted down that while in China he met a eunuch who wore six layers of silk through which the moles on his body were discernable.

Fabrics other than these gauzes excel in lustre and colour. With a keen eye for colour combination, master craftsmen designed dazzlingly brilliant materials. To create an effect of solidity some fabrics were shot through with gold or silver foil. A few bolts of satin and brocade interwoven with gold thread that were produced in the Qing (1644-1911) and even the Ming Dynasty (1368-1644) are preserved in the Beijing Palace Museum. The gold is still lustrous and to this day the fabric looks rich and splendid, as if fresh from the loom.

Other fabrics are distinguished for weaving skill. Marvels of kaleidoscopic variety, one has a deep, fine pile, another has crepe designs in relief. Open-work in lozenges is a feature of one, while another is patterned with warp cords of different width and thickness. There is an elegance, design and colour in Chinese silk to suit every fabric use and season. It is light, soft and comfortable, warm in winter and cool in summer.

The outstanding traditional designs of Chinese silks which utilize the peculiarities of the textile craft make them a clothing and furnishing material that is at the same time a work of art. Further, splendidly patterned fabrics have had their effect on architectural design and other fields of art. Traditional frescoes on Ming and Qing dynasty buildings known as *biandijin* (brocade covering the ground) and ground designs generally seen on stone sculptures were modelled after patterns on ancient Chinese silk.

The Technique of Producing Chinese Silk

High skill is required in the complicated processes of making Chinese silk. The main steps are reeling, boiling off, reeding, heddling, setting up the jacquard mechanism and tying the pattern basis.

Reeling

This first process after the cocoons are formed refers to beating the cocoons and unwinding the silk. Certain impurities, mainly sericin, or silk gum, must be carefully removed by cleansing before the silk is ready for the loom in order to bring out the soft, long and lustrous qualities of the silk filament.

From the time silk came into use, craftsman in China understood cocoon beating and silk reeling. Records found from the period of the Warring States (475-221 B.C.) and Han Dynasty (206 B.C.-A.D. 220) describe heating cocoons in water to remove the gum, attention being paid to maintaining the temperature of the water and keeping the density of the sericin low. The fire must be slow, they said, and cold water should be added at times to cool the bath down so that the sericin is removed evenly and the silk is perfect. Constant high water temperature disintegrated the cocoons, broke the filament, made reeling difficult and resulted in low-quality silk. It was later specified that the water used for reeling cocoons should look like "the eyes of crabs", and directions were given for changing the water. This should be of a frequency to produce lustrous white silk. Not changing the water often enough produced a shiny floss but of dark colour.

Reeling is done by lifting the loose end of the filament from each cocoon in the pot with a small stick. Then several of these filaments are twisted together to form a thread now known as "singles silk". These threads fall into three categories: the slenderest, which consists of the least number of long, fine fibres; one with a greater number of fibres that are broken yet fairly long; one made up of still more broken fibres and is therefore the thickest. In the time of the Six

Dynasties (A.D. 222-589) the slenderest singles silk consisted of five fibres. In the Song Dynasty (960-1279) the number of fibres was reduced to three.

Boiling Off

Boiling off involves further processes including bleaching, and turns raw silk into boiled silk. Somewhat similar to the preceding procedure, the reeled silk is steeped in a solution of chinaberry-wood or clam-shell ashes or a solution of the dark plum. Before the Han Dynasty, lukewarm water was used, to be replaced by boiling water in Eastern Han (A.D. 25-220). The boiled silk is then dried in the sun and steeped again in the solution. This process does two things: the alkali of the ash and exposure to the sun bleach and whiten the silk; the hot acid or alkaline bath removes the remaining sericin from the silk, leaving it softer and better able to take dye.

Reeding and Heddling

This is done with bamboo reeds on the heddling frame of the loom. It involves dividing the warp threads to make shuttle openings as required for the desired fabric pattern.

A reed is a narrow rectangular frame fashioned of bamboo with small bamboo strips suspended from it at equal intervals. It has been known in ancient times variously as *zhu*, *cheng* and *kun*. The heddling frame is rectangular and made of wooden planks. It has a wooden beam in the middle. Two strings are suspended on either side and parallel to it. The beam, strings and upper and lower sides of the frame are connected by silk cords, forming two interlocked loops. This is the part of the loom which a writer before the Southern and Northern Dynasties (386-581) describes as "bent cords used to control the warp, making it open and close". It was known in ancient times as *fanzi* or "somersaulter". There is only one sley, while the number of heddling frames ranges from two to eight, to be doubled on a jacquard loom.

Reeding means threading the weft by groups through all the reed-dents as required by the design, while heddling is putting the weft through the heddling frame, likewise according to the design requirements. For single-colour weaving, each weft thread traverses only one heddling frame, going through its upper loop, while in multi-colour fabrics it traverses two frames via the upper loop of one and the lower loop of the other.

Setting Up the Jacquard Mechanism, Tying the Pattern Basis

Han Dynasty records describe the jacquard mechanism of the loom roughly as follows: There is an elevated frame called the jacquard tower with the jacquard mechanism mounted on it vertically. This mechanism consists of the harness cord, comber board, warp lines, mails and sley arms. The harness cord, called *daxian* or "main thread", is the equivalent of a heddle on an ordinary loom. The mails are for lifting the warp threads, one going through each. The number of harness cords is determined by the pattern cycle. Each harness cord can lift from two to seven warp threads and is the equivalent of the "tie-up" in modern jacquard technique.

Tying the pattern is what directly produces the design on the jacquard silk. It is therefore also known as "the jacquard basis". This is of two types: the "jacquard basis by the design" and "the basis for the jacquard tower".

Below we outline how "the jacquard basis by the design" is worked out. The design is drawn on a square piece of cloth with equal numbers of warp and weft threads, or it may first be drawn on paper and then reproduced on cloth. Another set of warp threads are overlapped and connected thread by thread with those of the cloth. Then following the positions and measurements in the design, a complete set of weft threads is used to replace the weft of the cloth, reproducing the patterns. Song Yingxing describes this in his *Tian Gong Kai Wu* (*Exploitation of the Works of Nature*) in these words:

First a designer draws the intended patterns on paper. Then following the design, the planner meticulously measures out all the silk threads and ties them up.

The "basis for the jacquard tower" differs from the above in that the warp threads of "the jacquard basis by the design" are connected with an equal number of descending harness cords from the jacquard tower. Then another set of weft threads, somewhat thicker, are run through the harness cords in the same manner as the weft is handled in the "jacquard basis by the design". In this way the patterns are transferred onto the jacquard tower system.

The jacquard mechanism and the procedure of tying the jacquard basis work in co-ordination. When the basis for the jacquard tower is completed, the floating warp threads (the harness cords that bring out the patterns on the jacquard tower) are lifted, setting in motion the entire jacquard mechanism.

Major Silk Fabrics, Their Structure and Development

Chinese silk fabrics are of great variety, each being unique in style and structure. The main representative types are: *jin* (brocade), *sha* (gauze) and *luo* (leno), *ling* (damask) and *duan* (satin), *chou* (silk), *rong* (velvet) and *kesi* (silk fabric with large, stand-out jacquard patterns).

Jin (brocade)

What is *jin*? The term is often encountered in Chinese classics and refers variously to silk fabric woven in coloured patterns, or more likely to multi-colour fabrics with extra warp or extra weft, worked out by compound or complex weaving. Since its manufacture involved relatively high technique, *jin* was the most expensive of silk fabrics. An ancient record says: "*Jin* 锦 (brocade) is none other than *jin* 金 (gold). The making of it takes immense labour, so it costs the same as gold."

Jin is of two kinds: the warp-patterned and the weft-patterned, called respectively warp-*jin* and weft-*jin*. The former is made with two or more sets of warp threads woven with one set of weft. A set of warp consists of two or three or more threads, each of a different colour. If still more colours are desired, an alternate method of colour-stripe weaving would be employed. The weft threads too may be either plain or inlaid weft. The latter serves to separate all the face warp from the ground warp across the width, the pattern being worked out in the face warp. A typical Han Dynasty warp-*jin*, Wan Nian Ru Yi Jin (Brocade of Eternal Good Luck) of Eastern Han (A.D. 25-220) which was unearthed at Minfeng (Niya), Xinjiang in 1959, has 12 lengthwise colour stripes in crimson, white, dark violet, light blue and emerald.

The weft-*jin* has two or more sets of weft woven into one set of warp, which may be of the interweaving or inlaid type, the pattern being worked out with the plain weft on the surface. Specimens of this are the Tang Dynasty brocade stockings unearthed in 1969 at Astana in Xinjiang which bear bird, flower and cloud designs against a crimson background.

Different textile effects are produced in the warp-*jin* from those in the weft-*jin*. With a low weft density, the warp-*jin* is worked with only one shuttle, and this is done with considerable efficiency. The weft-*jin* is more time- and labour-consuming to weave, but with two or more shuttles used, a greater variety of colour and pattern is possible. Both types of brocade made their appearance in China very early, that produced before the third century being for the most part warp-patterned, while that made after the Sui (581-618) and Tang (618-907) dynasties mainly had the pattern brought out by the weft. Some foreign scholars have held that weft brocade was introduced into China from abroad. Actually, the Chinese developed this brocade independently as evidenced by a recent discovery at Bazarek in the Soviet Union of Chinese silks dating from the period of the Warring States among which were brocades with a weft pattern of red and green threads woven in.

Sha (Gauze) and *Luo* (Leno)

The *sha* of ancient China was also of two types according to weave. One, plain with a low warp density, was like the cotton gauze of today. Before Tang times this was called square-holed gauze. The other, like *luo*, belonged to the *sha-luo* weave formation as it was worked by intertwining the face warp with the ground warp, both of relatively low density. Two warp threads were intertwined in some weaves, three in others. Plain gauze was made before the Southern and Northern Dynasties, while from Tang some patterned gauze was made with the jacquard mechanism.

Different from the modern *luo*, most of the ancient Chinese type was made with four warp threads in a set. Two pairs each of face warp and ground warp arranged side by side were intertwined at every other passage of the shuttle. The two pairs of warp form a weaving cycle that produces a line of holes between cycles. The jacquard pattern of such *luo* fabric was not made with the pattern mechanism as such, but was produced by using a varying number of extra harnesses on it.

The modern Chinese *luo*, which was probably first manufactured in Ming (1368-1644), is generally made by twisting the warp once in every three to seven passages of the shuttle in plain weaving, the holes appearing at greater intervals. While the ancient *luo* fabric was delicate, its modern counterpart is sturdier and so each has its particular merits.

Ling (Silk Damask) and *Duan* (Satin)

Ling is a twill weave, whose characteristic is the formation of continuous slanting lines. *Ling*, however, differs from plain twill. It is what is known in modern textile technology as an "alternating twill weave", which is distinguished by a checkerboard design or reversible twill pattern. According to *Shi Ming* (*Explanation of Names*), "*ling* (silk damask) means (icicle) the term denoting the similarity of its patterns to the seams of ice." Its sheen reinforces this similarity.

Duan, or satin, is named for its satin weave. Though developed on the basis of twill, satin does not have apparent twill lines. It is characterized by wider intervals between the surface dots, which in turn are shaded by the long face threads of the warp and weft on their sides. This makes the fabric smooth and lustrous and gives it a pronounced stereoscopic effect, while the ground colour is not likely to turn dull or turbid. For this reason *duan* takes multi-colour complicated designs especially well. This type of fabric first appeared about the time of Song (960-1279) when it was called *zhusi* (ramie silk). *Duan* (satin) is a later name for it.

Rong (Velvet)

Rong is a pile fabric, generally warp-piled in ancient China. The warp was classified as ground or pile warp, the shuttle working up the pile warp once for each three or four trips for the ground warp. Metal wires or slender strips of bamboo, prepared beforehand, are placed in the entry of the shuttle, so that the pile warp forms small loops which are cut with a knife, producing the velvet. According to *Exploitation of the Works of Nature* by Song Yingxing of the Ming Dynasty, there was a type of fabric woven by "concealing bits of thread within the face warp, and after every few inches cutting with a knife to form brilliant silk". This describes the weaving of *rong*. What Song calls "thread" is bamboo in strips or metal wires.

China began producing pile weave very early and did not take the skill from Japan, as has been suggested, because some people called velvet *woduan* (Japanese satin). Among the early Han fabrics unearthed in 1972 at Mawangdui of Changsha in Hunan Province were pile-weave fabrics which were, however, uncut. From the Ming Dynasty on, the highest quality *rong* was made in Zhangzhou, Fujian, the best known of that being Zhang *rong* (Zhang velvet), Zhang *duan* (Zhang satin) and *tianerong* (swan velvet). The Zhang velvet was plain, while the Zhang satin and swan velvet were both jacquard velours. Zhang satin had patterns worked out on a satin ground by the jacquard mechanism, while swan velvet was not made on a jacquard loom, nor had it a satin ground. The pile-cutting for both Zhang velvet and Zhang satin was done at the time of weaving, while that of swan velvet had to be done after it had left the loom, when the design was drawn on the patternless ground and then cut with knives accordingly.

Chou (Silk)

The earliest type of silk produced in China, chou, is a plain weave with two each of warp and weft threads crisscrossed. It is believed that it was at first made entirely from silk waste. Since the Song Dynasty, however, fine silk was used to make chou with single-colour effect, and this was known as anhuachou (hidden-pattern silk). From that time all plain, thin, single-colour silk fabrics were called chou, while similar fabric made from spun yarns was known as fangchou (spun-yarn silk).

Kesi (Silk Fabric with Large, Stand-Out Jacquard Patterns)

The oldest of this type of silk fabrics, known as kesi before the Tang Dynasty, was later called zhicheng (woven fabric) and a number of other similar names.

Itself a plain-weave fabric, kesi is similar to other plain fabrics only in the warp, while its weft is somewhat different. Instead of being woven with a single shuttle through its width, kesi has its weft broken up into a number of segments according to the designed colour and worked out with a corresponding number of smaller shuttles. Says Ji Lei Pian (Chicken-Rib Essays) by Zhuang Zhuo in Song:

> Kesi is produced at Dingzhou. . . . The weaver hangs the refined colour silk on the warp beam and draws flowers, birds, animals or other patterns as desired. The weft is woven by small shuttles. Some areas are left blank and then filled in with silk threads of various colours to make the patterns. . . . Even

if 100 flowers are made, they can all be different for the reason
that the weft is not shuttled across the width of the fabric.

This last remark refers to the segmented weft.

Technically the manufacture of *kesi* is fairly simple, yet the
segmented weft makes possible any desired variation and very delicate
patterns. Especially since the Song Dynasty, which saw the maturity
of the art of painting, *kesi* pictures have closely reproduced original
paintings. Many a masterpiece was woven in silk in its entirety and
minutest detail. This Chinese art form won world recognition.

China's Silk Textile Technique Goes Abroad

China's silk textile craft was taken to other countries in very early
times and has had significant effect on the development of world
textile skill. Chinese silk textile skill had connections with the outside
world even before the Christian era.

Japan was first. Japanese scholars assert that the skills of silk
reeling and the weaving of *juan* (a rough silk fabric) and of *luo* (leno)
made their way to Japan at about the time of Qin (221-206 B.C.)
and Western Han (206 B.C.-A.D. 24). It was then, it is said, that
the four legendary Japanese figures, Anihime, Otohime, Gofuku and
Anaori, who were believed to have founded Japan's silk-textile craft,
reached Japan from China by sea. Similar records exist in China's
historical archives. *Wei Zhi* (*History of the Kingdom of Wei*) in
the chapter "Wo Ren Zhuan" ("History of Japan") says that Japan
was engaging in sericulture and the manufacture of *jianmian* (silk
wadding) by that time — obviously developed from the silk craft
of earlier days.

The influence of China's silk-textile craft on the West probably
also began at about the same time. Following the exploration of
the Western Regions by Zhang Qian in Western Han, the Chinese
began transporting silk west along the Silk Road, introducing this
product of China to some of the Central Asian and European countries. Ancient Western fabrics, woven of flax or wool, were generally

rough and heavy. Impressed by the diaphanous Chinese silk, Western weavers sought to improve their product by taking the Chinese silk fabric apart and weaving the colour threads together with their flax and wool fibres. This is recorded in Chapter 13 of *History of the Kingdom of Wei* in the explanatory notes by Pei Songzhi quoting *Wei Lue* (*Memorable Things of the Wei Kingdom*): the country of Da Qin (the Roman Empire) "often procured Chinese silk yarns for the making of their own fabrics".

Chinese influence was later seen in the structure of the weaving loom. The ancient Western loom had been of the vertical type and, unlike the Chinese loom, could not handle many heddle staves, nor was it equipped with a treadle mechanism for the control of the heddle motion. The weaving of any fabric of complicated structure was impossible on it. In an effort to improve his product, taking Chinese silk as model, the Western weaver in time replaced his conventional loom with the Chinese horizontal type.

In the 5th and 6th centuries, China experienced a long period of political chaos and war, and yet, with the development of communication, China increased her cultural exchange with other countries. Among the no few foreigners who came to China for travel and study, Western records state, were two Persian emissaries who were sent to study the craft of silk textiles. These men returned with silkworm eggs to hatch in Persia. A Japanese historian stated that officials were sent by the Japanese Government to enlist silk craftsmen in the coastal region of Zhejiang to teach the skill in Japan. In the meantime some Chinese had sailed from north China to settle in Japan, where they engaged in the silk-textile industry. These were later known as the Qinshi and Hanshi (the people of Qin and Han), and are often mentioned in Japanese history as the mainstay of the country's textile craft.

The impact of Chinese textile skill on other countries was especially marked from the 7th century. Throughout the period from late Sui (581-618) to early Song (960-1279) Japan imported large quantities of Chinese silk, specimens of which are now preserved in the Shosoin and other Japanese museums. Hakataori, the best-known Japanese silk fabric, was produced using skills learned from

China.

In the 7th to 8th centuries Persian textiles were renowned in the Western world. These, however, were produced borrowing from Chinese textile technique and employing Chinese craftsmen. Du Huan, a Chinese traveller abroad in the Tang Dynasty, writes in his *Jing Xing Ji* (*Record of My Travels*) that when he was in Da Shi (now Iran) in the 10th year of the Tianbao reign (A.D. 751), he witnessed Chinese craftsmen by the names of Yue Huan and Lü Li, both natives of the locality east of the Yellow River, engaged in the weaving of *luo*, now called *chou* (silk fabric). Qiu Chuji, a Taoist monk in early Yuan (1279-1368) who toured part of Central Asia on the order of Genghis Khan, met thousands of Chinese workers employed there in the manufacture of *ling* (damask silk), *luo* (leno), *jin* (brocade) and *qi* (a silk fabric with single colour effect).

The last and most important Chinese textile skill transmitted to Europe was the use of the pattern loom and the pattern basis. It was not known in the West before the 6th century how to make silk fabrics with large woven patterns. And it was not until the 6th to 7th century, when the Chinese pattern loom and the pattern basis technique found their way to the West, that Western weavers began to make pattern fabrics of more complicated type. The Chinese-modelled loom and technique have been used continuously since, remaining basically unchanged through the centuries, with only minor modifications. Even the latest French jacquard loom now widely used the world over is a close relative of the Chinese pattern loom. Except for the pattern basis being replaced by the design plate, the structure is basically the same, while the working principle remains unchanged.

Chronology of Chinese Dynasties

Xia	c. 21st - 16th century B.C.
Shang	c. 16th - 11th century B.C.
Western Zhou	c. 11th century - 770 B.C.
Eastern Zhou	770 - 221 B.C.
Spring and Autumn Period	770 - 476 B.C.
Warring States Period	475 - 221 B.C.
Qin	221 - 207 B.C.
Han	206 B.C. - A.D. 220
Western Han	206 B.C.-A.D. 24
Eastern Han	25 - 220
Three Kingdoms	220 - 280
Wei	220 - 265
Shu	221 - 263
Wu	222 - 280
Jin	265 - 420
Western Jin	265 - 316
Eastern Jin	317 - 420
Southern and Northern Dynasties	420 - 589
Southern Dynasties	420 - 589
Song	420 - 479
Qi	479 - 502
Liang	502 - 557
Chen	557 - 589

Northern Dynasties	386 - 581
Northern Wei	386 - 534
Eastern Wei	534 - 550
Western Wei	535 - 557
Northern Qi	550 - 577
Northern Zhou	557 - 581
Sui	581 - 618
Tang	618 - 907
Five Dynasties and Ten Kingdoms	907 - 979
Song	960 - 1279
Northern Song	960 - 1127
Southern Song	1127 - 1279
Liao	916 - 1125
Kin	1115 - 1234
Yuan	1271 - 1368
Ming	1368 - 1644
Qing	1644 - 1911

Bibliography

Bao Pu Zi　抱朴子
Book of　Master Baopu

Bao Zang Chang Wei Lun　宝藏畅微论
Discourse on the Contents of the Precious Treasury of the Earth

Bei Shi　北史
History of the Northern Dynasties

Ben Cao Gang Mu　本草纲目
Compendium of Materia Medica

Ben Cao Jing Ji Zhu　本草经集注
Commentaries on Materia Medica

Ben Cao Yan Yi　本草衍义
Dilations upon Materia Medica

Bian Min Tu Zuan　便民图纂
Illustrated Compilation for the People's Welfare

Bian Que Mai Shu　扁鹊脉书
Bian Que's Book on Pulse

Bian Que Xin Shu　扁鹊心书
My Understanding of Bian Que

Bin Feng Guang Yi　豳风广义
Comments on the "Book of Odes"

Bin Hu Mai Xue　濒湖脉学
Bin Hu's Study on Pulse

Bo Wu Zhi　博物志
Notes on the Investigation of Things

Bu Xi Yuan Lu　补洗冤录
Supplement to "Manual of Forensic Medicine"

536

Cai Lun Zhuan 蔡伦传
Biography of Cai Lun

Can Fa 蚕法
How to Raise Silkworms

Can Sang Ji Yao 蚕桑辑要
Main Points of Sericulture

Can Shu 蚕书
Book on Silkworm Raising

Cao Chuan Zhi 漕船志
Record of Grain Carriers

Ce Yuan Hai Jing 测圆海镜
Sea Mirror of Circle Measurement

Cha Bing Zhi Nan 察病指南
Guide to Diagnosis

Cha Jing 茶经
Canon of Tea

Cha Lu 茶录
Treatise on Tea

Cha Pu 茶谱
Manual on Tea

Chao Ye Jian Zai 朝野金载
Brief Record of Contemporary News

Chen Fu Nong Shu 陈旉农书
Agricultural Treatise of Chen Fu

Cheng Chu Tong Bian Ben Mo 乘除通变本末
Origins and Details of Various Methods in Multiplication and Division

Cheng Hua Shi Lu 成化实录
Real Account of Chenghua

Chou Hai Tu Bian 筹海图编
Illustrated Seaboard Strategy

Chu Ci　楚辞
Elegies of Chu

Chu Jian Pu　樗茧谱
Information on Philosamia and Cynthia

Chu Xue Ji　初学记
The Primaty Anthology

Chun Qiu　春秋
Spring and Autumn Annals

Chun Zhu Ji Wen　春渚纪闻
Record of Things Heard at Spring Island

Da Yuan Hai Yun Ji　大元海运记
Records of Maritime Transportation in the Yuan Dynasty

Da Zang Jing　大藏经
Tripitaka

Dan Fang Xu Zhi　丹房须知
Indispensable Knowledge for the Alchemical Laboratory

Dan Jing　丹经
Elixir Manual

Dao De Jing　道德经
Canon of the Dao and Its Virtue

Dao Guang Zun Yi Fu Zhi　道光遵义府志
Chronicle of Zunyi Prefecture in the Daoguang Period

Dao Zang　道藏
Daoist Patrology

Di Jing Jing Wu Lu　帝京景物録
Descriptions of Things and Customs at the Imperial Capital

Di Jing Tu　地镜图
Illustrated Mirror of the Earth

Di Zhen Ji　地震记
Record of Earthquakes

Dian Nan Xin Yu　滇南新语
New Talks About South Yunnan

Ding Ju Suan Fa 丁巨算法
Ding Ju's Arithmetical Methods

Dong Guan Han Ji 东观汉记
Han History Compiled at Dongguan

Dong Jing Meng Hua Lu 东京梦华录
Memories of the Eastern Capital

Dong Po Zhi Lin 东坡志林
Journal and Miscellany of Dongpo

Duo Neng Bi Shi 多能鄙事
Various Arts in Ordinary Life

Er Shen Ye Lu 二申野录
Accounts of Ershen

Er Ya 尔雅
Literary Expositor

Fan Sheng Zhi Shu 氾胜之书
The Book of Fan Shengzhi

Fan Yan 方言
Dictionary of Local Expressions

Gan Qi Shi Liu Zhuan Jin Dan 感气十六转金丹
The "Responding to the Qi Method" in Preparing the Sixteenfold
  Cyclically Transformed Gold Elixir

Gao Li Tu Jing 高丽图经
Illustrated Record of an Embassy to Korea

Ge Wu Cu Tan 格物粗谈
Simple Discourses on the Investigation of Things

Geng Zi Xin Chou Ri Ji 庚子、辛丑日记
Diary Written in the Years 1180 - 1181

Gou Gu Yuan Fang Tu Shuo 勾股圆方图说
Theory with Diagrams of the Right Triangle Making Use of
  Circles or Squares

Gu Jin Tu Shu Ji Cheng 古今图书集成
Collection of Books Ancient and Modern

Gu Yue Fu   古乐府
Treasury of Ancient Songs

Guan Yin Zi   关尹子
The Book of Master Guan Yin

Guan Zi   管子
The Book of Master Guan

Guang Can Sang Shuo   广蚕桑说
Treatise on Sericulture

Guang Ya   广雅
Enlarged Literary Expositor

Guang Yang Za Ji   广阳杂记
Miscellaneous Records of Guangyang

Gui Gu Zi   鬼谷子
Book of the Devil Valley Master

Guo Yu   国语
Discourse on the States

Hai Dao Jing   海道经
Manual of Sailing Directions

Hai Dao Suan Jing   海岛算经
Sea Island Mathematical Manual

Hai Guo Tu Zhi   海国图志
Illustrated Record of Maritime Nations

Han Fei Zi   韩非子
The Book of Master Han Fei

Han Shu   汉书
History of the Han Dynasty

He Gong Qi Ju Tu Shuo   河工器具图说
Illustrations and Explanations of the Techniques of
    Water Conservancy and Civil Engineering

He Yi Bian Huo   河议辩惑
On Rivers: Some Erroneous Representations

Huan Zi Xin Lun 桓子新论
New Discourses of Master Huan

Hou Han Shu 后汉书
History of the Later Han Dynasty

Hou Wei Shu 后魏书
History of the Later Wei Dynasty

Hua Jing 花镜
Mirror of Flowers

Hua Yang Guo Zhi 华阳国志
Record of the Country South of Mount Huashan

Huai Nan Wan Bi Shu 淮南万毕术
The Ten Thousand Infallible Arts of the Prince of Huai Nan

Huai Nan Zi 淮南子
The Book of the Prince of Huai Nan

Huang Di Nei Jing (or *Nei Jing*) 黄帝内经
The Yellow Emperor's Canon of Medicine

Huang Di Jiu Ding Shen Dan Jing Jue 黄帝九鼎神丹经决
The Yellow Emperor's Canon of the Nine-Vessel Spiritual Elixir

Huang Di Jiu Zhang Suan Fa Xi Cao 黄帝九章算法细草
Detailed Solutions of the Problems in the Nine Mathematics
  Chapters of Huang Di

Huang Di Ming Tang Jing 黄帝明堂经
The Yellow Emperor's Classic on Acupuncture and Moxibustion

Hui Li Zhou Ji 会理州记
Annals of Huili Prefecture

Ji Fu Fu 机妇赋
Ode to the Women Weavers

Ji Gu Suan Jing 缉古算经
Continuation of Ancient Mathematics

Ji Lei Pian 鸡肋篇
Chicken-Rib Essays

Jia Qing She Hong Xian Zhi   嘉庆射洪县志
Chronicle of Shehong County in the Jiaqing Period

Jian Pu   笺谱
Ornamental Letter-Paper

Jian Yan Ge Mu   检验格目
Forensic Investigations

Jie An Man Bi   戒庵漫笔
Essays of Jie An

Jin Dan Da Yao   金丹大要
Essentials of the Metallous Enchymoma

Jin Gui Yao Lue   金匮要略
Jingui Collection of Prescriptions

Jin Hua Chong Bi Dan Jing Yao Zhi   金华冲碧丹经要旨
Confidential Instructions on the Manual of the Heaven-Piercing
   Golden Flower Elixir

Jin Shu   晋书
History of the Jin Dynasty

Jin Yu Tu Pu   金鱼图谱
Illustrated Book on Gold Fish

Jing Shi Min Shi Lu   经世民事录
Practical Administration of People's Livelihood

Jing Xing Ji   经行记
Record of My Travels

Jing Xiu Xian Sheng Wen Ji   静修先生文集
Collected Literary Works of Master Jingxiu

Jiu Huan Jin Dan Miao Jue   九还金丹妙决
Wonderful Instructions on the Ninefold Cyclically Transformed
   Gold Elixir

Jiu Huang Ben Cao   救荒本草
Famine-Relief Herbs

Jiu Tang Shu   旧唐书
Old History of the Tang Dynasty

Jiu Zhang Suan Shu 九章算术
Nine Chapters on the Mathematical Art

Jiu Zhang Suan Shu Zhu 九章算术注
Annotation on the Nine Chapters on the Mathematical Art

Ju Jia Bi Yong Shi Lei Quan Shu 居家必用事类全书
Guide to Domestic Occupations

Ju Pu 菊谱
Book on the Chrysanthemum

Ju Song 桔颂
In Praise of the Tangerine

Kai Yuan Zhan Jing 开元占经
Kai Yuan Classic on Astrology

Kang Xi Hai Zhou Zhi 康熙海州志
Chronicle of Haizhou Prefecture in the Kangxi Period

Kang Xi Ji Xia Ge Wu Bian 康熙几眼格物编
Kang Xi's Study of the Principles of Things

Kang Xi Zhen Jiang Fu Zhi 康熙镇江府志
Chronicle of Zhenjiang Prefecture in the Kangxi Period

Kao Gong Ji 考工记
Artificers' Record

Lei Shu Zuan Yao 类书纂要
Summary of Books

Lei Si Jing 未耜经
Canon of Spades

Li Ji 礼记
Record of Rites

Liang Shu 梁书
History of the Liang Dynasty

Lie Zi 列子
The Book of Master Lie

Ling Shu (or *Zhen Jing*) 灵枢
Canon of Acupuncture

Ling Xian 灵宪
The Spiritual Constitution of the Universe

Liu Bin Ke Jia Hua Lu 刘宾客嘉话录
Discourses of Liu the Prince's Companion

Long Jiang Chuan Chang Zhi 龙江船厂志
Records of the Dragon River Shipyard

Lu Ban Mu Jing 鲁班木经
Lu Ban's Manual of Carpentry

Lü Shi Chun Qiu 吕氏春秋
Master Lü's Spring and Autumn Annals

Lun Heng 论衡
Discourses Weighed in the Balance

Lun Qi 论气
Discourse on Air

Luo Yang Hua Mu Ji 洛阳花木记
Flowers and Trees of Luoyang

Luo Yang Mu Dan Ji 洛阳牡丹记
Peonies of Luoyang

Luo Yang Qie Lan Ji 洛阳伽蓝记
Description of Buddhist Temples of Luoyang

Mai Fa 脉法
Methods of Pulse Feeling

Mai Jing 脉经
Classic on Pulse

Meng Liang Lu 梦梁录
Records of a Dream of Grandeur

Meng Xi Bi Tan 梦溪笔谈
Dream Stream Essays

Min Guo Shou Guang Xian Zhi 民国寿光县志
Chronicle of Shouguang County in the Republic Period

Ming Shi 明史
History of the Ming Dynasty

Ming Shi Lu　　明实录
Veritable Records of the Ming Dynasty

Ming Yi Bie Lu　　名医别录
Informal Records of Famous Physicians

Mo Jing (or *Mo Zi*)　　墨经
Mohist Canon

Mo Zi (or *Mo Jing*)　　墨子
Mohist Canon

Nan Chuan Ji　　南船纪
Notes on Ships of the South

Nan Jing　　难经
Classic of Difficulty

Nan Qi Shu　　南齐书
History of the Southern Qi Dynasty

Nan Yue You Ji　　南越游记
Travels in the South

Nan Zhou Yi Wu Zhi　　南州异物志
Strange Things of the South

Nei Jing (or *Huang Di Nei Jing*)　　内经
Canon of Medicine

Nei Shu Lu　　内恕录
Clear Conscience

Ni Ban Shi Yin Chu Bian　　泥版试印初编
Initial Notes on Printing with Clay Types

Nong Pu Bian Lan　　农圃便览
Farming and Gardening Manual

Nong Sang Ji Yao　　农桑辑要
Agriculture and Sericulture

Nong Sang Yao Zhi　　农桑要旨
Essentials of Agriculture and Sericulture

Nong Sang Yi Shi Zuo Yao　　农桑衣食撮要
Essentials of Agriculture, Sericulture, Clothing and Food

Nong Shu (or *Wang Zhen Nong Shu*)   农书
Agricultural Treatise

Nong Zheng Quan Shu   农政全书
Complete Treatise on Agriculture

Pan Zhu Suan Fa   盘珠算法
Arithmetical Methods of the Beads

Ping Yuan Lu   平冤录
Wrongs Righted

Ping Zhou Ke Tan   萍洲可谈
Pingzhou Table Talk

Qi Jing   漆经
Book of Lacquer

Qi Min Yao Shu   齐民要术
Important Arts for the People's Welfare

Qian Gong Jia Geng Zhi Bao Ji Cheng   铅汞甲庚至宝集成
Compendium on the Perfected Treasure of Lead, Mercury, Wood
   and Metal

Qian Jin Yao Fang   千金要方
The Thousand Golden Formulae

Qian Jin Yi Fang   千金翼方
Supplement to the Thousand Golden Formulae

Qian Tang Ji   钱塘记
Notes on the Qiantang

Qian Wen Ji   前闻记
Traditions of Past Affairs

Qing Mi Cang   清秘藏
Secret Collections

Qing She Yi Wen   青社遗闻
Recollections of the Qingshe

Qing Wu Xu Yan   青乌绪言
Introduction to the Blue Raven Manual

Qiu Yi Shu Tong Jie 求一术通解
A Thorough Explanation of the Method of Seeking Unity

Qun Fang Pu 群芳谱
Beautiful Flowers

Ri Gao Tu Shuo 日高图说
Theory with Diagrams of the Sun's Altitude

Ri Yong Suan Fa 日用算法
Method of Computation for Daily Use

San Bao Tai Jian Xia Xi Yang Ji Tong Su Yan Yi 三宝太监下西洋记通俗演义
Popular Story of the Voyages of the Three-Jewel Eunuch in the Western Oceans

San Fu Jiu Shi 三辅旧事
Tales of Three Cities

San Guo Yan Yi 三国演义
Romance of the Three Kingdoms

Shan Hai Jing 山海经
Classic of Mountains and Rivers

Shan Jia Qing Gong 山家清供
Fresh Food of a Mountain Dweller

Shang Han Lun 伤寒论
Treatise on Febrile Diseases

Shang Han Za Bing Lun 伤寒杂病论
Treatise on Febrile and Other Diseases

Shang Ke Hui Zuan 伤科彙纂
Collected Treatises on Surgery

Shang Shu 尚书
Book of History

Shen Nong Ben Cao Jing 神农本草经
Shen Nong's Materia Medica

Shi Cha Lu 试茶录
Tea Sampling

Shi Ji   史记
Records of the Historian

Shi Jing   诗经
Book of Odes

Shi Lin Guang Ji   事林广记
Through the Forest of Affairs

Shi Lun   食论
On Food

Shi Ming   释名
Explanation of Names

Shi Nong Bi Yong   士农必用
An Indispensable Book of Agriculture

Shi Shi Xing Jing   石氏星经
Shi's Classic on Stars

Shi Yao Er Ya   石药尔雅
Synonymic Dictionary of Mineral Drugs

Shi Yi De Xiao Fang   世医得效方
Tested Prescriptions of Veteran Physicians

Shi Zhu Zhai Hua Pu   十竹斋画谱
Ten-Bamboo Studio Painter's Manual

Shu Jian Pu   蜀笺谱
Manual on Sichuan Writing Paper

Shu Jing   书经
Book of History

Shu Qu Zi   叔苴子
The Book of the Hemp-Seed Master

Shu Shu Ji Yi   数术记遗
Memoir on Some Traditions of the Mathematical Art

Shu Shu Jiu Zhang   数书九章
Mathematical Treatise in Nine Sections

Shu Xue Tong Gui   数学通轨
Rules of Mathematics

Shui Jing Zhu 水经注
Commentary on the "Waterways Classic"

Shun Tian Fu Zhi 顺天府志
Records of Shuntian Prefecture

Shun Zhi Deng Zhou Zhi 顺治邓州志
Chronicle of Dengzhou Prefecture in the Shunzhi Period

Si Min Yue Ling 四民月令
Monthly Ordinances for the Four Classes of People

Si Shi Lei Yao 四时类要
Classified Outline of the Four Seasons

Si Shi Zuan Yao 四时纂要
Outline of the Four Seasons

Si Yuan Yu Jian 四元玉鉴
Precious Mirror of the Four Elements

Song Hui Yao 宋会要
Administrative Statutes of the Song Dynasty

Song Shi 宋史
History of the Song Dynasty

Song Shu 宋书
History of the Liu Song Dynasty

Su Wen 素问
Questions and Answers

Suan Fa Quan Neng 算法全能
All-Capable Mathematical Methods

Suan Fa Tong Zong 算法统宗
Systematic Treatise on Arithmetic

Suan Jing Shi Shu 算经十书
Ten Mathematical Manuals

Suan Shu 算术
Mathematical Art

Suan Xue Qi Meng 算学启蒙
Introduction to Mathematics

Suan Xue Xin Shuo 算学新说
A New Treatise on the Science of Calculation

Sui Shu 隋书
History of the Sui Dynasty

Sun Bin Bing Fa 孙膑兵法
Sun Bin's Art of War

Sun Zi Bing Fa (or *Sun Zi*) 孙子兵法
Master Sun's Art of War

Sun Zi Suan Jing 孙子算经
Master Sun's Mathematical Manual

Tai Ping Jing 太平经
Canon of Eternal Peace

Tai Ping Yu Lan 太平御览
Taiping Imperial Encyclopaedia

Tai Qing Shi Bi Ji 太清石壁记
Records in the Rock Chamber

Tai Xi Shui Fa 泰西水法
Western Hydraulic Engineering

Tan Yuan 谈苑
Garden Corner of Talks

Tang Hui Yao 唐会要
Administrative Statutes of the Tang Dynasty

Tang Yin Bi Shi 棠阴比事
In the Shade of a Crab-Apple Tree

Tao Shuo 匋说
On Pottery

Tian Gong Kai Wu 天工开物
Exploitation of the Works of Nature

Tian Wen 天文
Astronomy

Tong Ren Yu Xue Zhen Jiu Tu Jing 铜人俞穴针灸图经
Illustrated Manual on the Points for Acupuncture and Moxibustion
  on the Bronze Figure

Tong Yue 僮约
Contract with a Servant

Tu Jing Ben Cao 图经本草
Illustrated Herbals

Wai Ke Zheng Zong 外科正宗
Orthodox Manual of Surgery

Wai Tai Mi Yao 外台秘要
Medical Secrets Held by an Official

Wan Li Shi Lu 万历实录
True Account of the Wanli Period

Wang Zhen Nong Shu 王桢农书
Agricultural Treatise of Wang Zhen

Wei Lue 魏略
Memorable Things of the Wei Kingdom

Wei Shu 魏书
History of the Wei Dynasty

Wei Zhi 魏志
History of the Kingdom of Wei

Wu Ben Xin Shu 务本新书
New Book on the Essentials

Wu Bei Zhi 武备志
Treatise on Armament Technology

Wu Cao Suan Jing 五曹算经
Mathematical Manual of the Five Government Departments

Wu Hu Xian Zhi 芜湖县志
Annals of Wuhu County

Wu Jiang Kao 吴江考
Notes on the Wujiang River

Wu Jing Suan Shu 五经算术
Arithmetic in the Five Classics

Wu Jing Zong Yao   武经总要
Collection of the Most Important Military Techniques

Wu Li Xiao Shi   物理小识
Small Encyclopaedia of the Principles of Things

Wu Shi Er Bing Fang   五十二病方
Prescriptions for Fifty-Two Diseases

Wu Xian Zhan   巫咸占
Astrology of Wu Xian

Wu Xing Zhi   五行志
Records of the Five Elements

Wu Ying Dian Ju Zhen Ban Cong Shu   武英殿聚珍版丛书
Editio Princeps Books of Wuying Hall

Wu Yuan Lu   无冤录
Judicial Infallibility

Xi Jing Za Ji   西京杂记
Miscellaneous Records of the Western Capital

Xi Shang Fu Tan   席上腐谈
Hackneyed Talk on a Mat

Xi Yang Fan Guo Zhi   西洋番国志
Records of Countries of the Western Oceans

Xi Yuan Hui Bian   洗冤汇编
Examples of Redressed Cases

Xi Yuan Ji Lu   洗冤集录
Manual of Forensic Medicine

Xi Yuan Lu Bian Zheng   洗冤录辨正
Amendment to "Manual of Forensic Medicine"

Xi Yuan Lu Ji Zheng   洗冤录集证
In Support of "Manual of Forensic Medicine"

Xi Yuan Lu Jian Shi   洗冤录笺释
Comments on "Manual of Forensic Medicine"

Xia Hou Yang Suan Jing   夏候阳算经
Xiahou Yang's Mathematical Manual

Xia Xiao Zheng 夏小正
Lesser Annuary of the Xia Dynasty

Xian Shou Li Shang Xu Duan Mi Fang 仙授理伤续断秘方
Secret Healing of Wounds, Fractures and Dislocations

Xiang Jie Jiu Zhang Suan Fa 详解九章算法
Detailed Analysis of the Mathematical Rules in the "Nine Chapters"

Xiang Ming Suan Fa 详明算法
Explanations of Arithmetic

Xin Si Qi Qi Lu 辛巳泣蕲录
Tearful Records of the Battle of Qizhou

Xin Tang Shu 新唐书
New History of the Tang Dynasty

Xin Xiu Ben Cao 新修本草
Revised Materia Medica

Xin Yi Xiang Fa Yao 新仪象法要
New Design for an Armillary Clock

Xiu Lian Da Dan Yao Zhi 修炼大丹要旨
Essential Instructions for the Preparation of the Great Elixir

Xu Xia Ke You Ji 徐霞客游记
Travels of Xu Xiake

Xuan He Feng Shi Gao Li Tu Jing 宣和奉使高丽图经
Illustrated Record of an Embassy to Korea in the Xuanhe Period

Xun Zi 荀子
The Book of Master Xun

Yan Pu 砚谱
Manual on Inkstones

Yan Shan Za Ji 颜山杂记
Records of Mount Yanshan

Yan Tie Lun 盐铁论
On Salt and Iron

Yang Can Mi Jue 养蚕秘决
Secrets of Silkworm Rearing

Yang Hui Suan Fa 杨辉算法
Yang Hui's Method of Computation

Yang Yu Yue Ling 养余月令
Monthly Guide to Sericulture

Yang Zhou Shao Yao Pu 扬州芍药谱
The Peonies of Yangzhou

Ye Can Lu 野蚕录
Information on Wild Silkworms

Ye Cao Pu 野草谱
Manual on Wild Edible Herbs

Yi Gu Yan Duan 益古演段
New Steps in Computation

Yi Jing 易经
Book of Changes

Yi Li 仪礼
Rites

Yi Yu Ji 疑狱集
Difficult Cases

Yi Zhou Shu 逸周书
Lost Books of the Zhou Dynasty

Yi Zong Jin Jian 医宗金鑑
Golden Mirror of Medicine

Yin Chuan Xiao Zhi 银川小志
Brief Chronicle of Yinchuan

Yin Hua Lu 因话录
Discourse on the Cause of Things

Yin Yang Mai Zheng Hou 阴阳脉症候
Symptoms of the Yin and Yang Pulse Patterns

Yin Yang Shi Yi Mai Jiu Jing 阴阳十一脉灸经
Eleven Channels for Moxibustion in the Yin and Yang System

Ying Zao Fa Shi 营造法式
Architectural Methods

Yong Jia Ju Lu 永嘉桔录
Citrus Horticulture

Yong Le Da Dian 永乐大典
Great Encyclopaedia of the Yongle Reign

Yong Zheng Dong Chuan Fu Zhi 雍正东川府志
Chronicle of Dongchuan Prefecture in the Yongzheng Period

You Huan Ji Wen 游宦纪闻
Things Seen and Heard on My Official Travels

You Yang Za Zu 酉阳杂俎
Miscellany of the Youyang Mountains

Yu Gong 禹贡
Tribute of Yu

Yu Xiang Xian Zhi 虞乡县志
Chronicle of Yuxiang County

Yuan Qu Xuan 元曲选
Selected Yuan Dramas

Yue Ling 月令
Monthly Ordinances

Yue Ling Zhang Ju 月令章句
Notes to the Monthly Ordinances

Yun Lin Shi Pu 云林石谱
Yun Lin's Manual on Stones

Zhan Guo Ce 战国策
Records of the Warring States

Zhang Qiu Jian Suan Jing 张丘建算经
Zhang Qiujian's Mathematical Manual

Zhe Yu Gui Jian 折狱龟监
Examples of Misjudged Cases

Zhen Jing (or *Ling Shu*) 针经
Canon of Acupuncture

Zhen Jiu Da Cheng 针灸大成
Compendium of Acupuncture and Moxibustion

Zhen Jiu Jia Yi Jing   针灸甲乙经
A Classic of Acupuncture of Moxibustion

Zhen La Feng Tu Ji   真腊风土记
Description of Kampuchea

Zhen Yuan Miao Dao Yao Lue   真元妙道要略
Essentials of the Truly Original Methods

Zheng De Shi Lu   正德实录
True Account of the Zhengde Period

Zheng He Hang Hai Tu   郑和航海图
Charts of Zheng He's Voyages

Zhi Pu   纸谱
Manual on Paper

Zhi Zhi Suan Fa Tong Zong   直指算法统宗
Systematic Treatise on Arithmetic

Zhong Shu Cang Guo Xiang Can   种树藏果相蚕
How to Plant Trees, Store Fruits and Judge Silkworms

Zhong Shu Shu   种树书
Book of Afforestation

Zhou Bi Suan Jing   周髀算经
The Arithmetical Classic of the Gnomon and the
   Circular Paths

Zhou Hou Bei Ji Fang   肘后备急方
Handbook of Medicines for Emergencies

Zhou Li   周礼
Rites of the Zhou Dynasty

Zhou Yi Can Tong Qi   周易参同契
Kinship of the Three and the Book of Changes

Zhu Bing Yuan Hou Lun   诸病源候论
Causes and Symptoms of Diseases

Zhu Jia Shen Pin Dan Fa   诸家神品丹法
Methods of Elixir Preparations

Zhu Sha Yu Pu 朱砂鱼谱
Book on Cinnabar Fish

Zhu Shu Ji Nian 竹书纪年
The Bamboo Annals

Zhuang Zi 庄子
The Book of Master Zhuang

Zhui Geng Lu 辍耕录
Talks in the Intervals of Ploughing

Zhui Shu 缀术
Art of Mending

Zi Ren Yi Zhi 梓人遗制
Traditions of the Joiner's Craft

Zu Bi Shi Yi Mai Jiu Jing 足臂十一脉灸经
Eleven Channels for Moxibustion of the Arms and Feet

Zuo Zhuan 左传
Zuoqiu Ming's Chronicles

Glossary

A

aiqing 艾青
Anfeng Pond 安丰塘
Anfu Gate 安福门
Anhua Gate 安化门
anhuachou (hidden-pattern silk) 暗花绸
Anhui (province) 安徽
Anji 安济
Anji Bridge 安济桥
Anli Gate 安礼门
Anping Bridge 安平桥
Anqiu (county) 安丘
Anshang Gate 安上门
Anxi (county) 安西
Anyang 安阳
Anzhou 安州
ao (decocting) 熬
Astana 阿斯塔那

B

Ba Bridge 灞桥
Badaling 八达岭
badao 八道
baicai (white vegetable) 白菜
bai hu qi ben jian 百虎齐奔箭
Bai Juyi 白居易
Bai Qu 白渠
bai shi hu jian 百矢弧箭
Bai Ying 白英
bai'e 白垩

baijin 白金
baijisui 百圾碎
baila 白蜡
Bailian 百炼
Bailu 白露
Baima 白马
Baiquan Bridge 百泉桥
baitong (copper and nickel, white brass) 白铜
baixiyin 白锡银
baiyü 白玉
Ban Gu 班固
Banpo Village 半坡村
Baodai Bridge 宝带桥
Baoji 宝鸡
Baoshan 宝山
Baqiao 灞桥
Bashui River 灞水
Beijing 北京
ben 本
Bi Sheng 毕升
bian (stone sliver) 砭
Bian Que 扁鹊
"Bian Tu" 辨土
Bian Xin 编沂
biandijin 遍地锦
Bianjing 汴京
Bianliang 汴梁
biao 标
bienao 鳖臑
Big Seven Mountains 大七山

558

Biluochun 碧螺春
bingren 丙壬
bingshi 冰石
Bohai 渤海
bola (pluck-wax method) 拨蜡
"brave ship" 唬船
bu 步
buhuimu (asbestos) 不灰木

C

Cai Lun 蔡伦
Cai Xiang 蔡襄
Cai Yong 蔡邕
Cai Yuanding 蔡元定
Cang Gong 仓公
Canglong 苍龙
Cangshan 苍山
Cangzhou 沧州
Cao Cao 曹操
Cao Chong 曹冲
Cao Pi 曹丕
Cao Shaokui 曹绍夔
caoshayin 草砂银
caowusan 草乌散
Caozhuang Village 曹庄
Censhuichang 岑水场
Chai Zhou 柴周
Chang Qu 常璩
Changan 长安
Changjiang River (Yangzi River) 长江
Changsha 长沙
Changsha Guo 长沙国
changshi (felspar) 长石
Changzhou 常州
Chanshui River 浐水
Chao Yuanfang 巢元方

chaoche (high-chassis carriage) 巢车
chaoqing (roasting out of the green) 炒青
Chaozhou 潮州
Chashan 茶山
che (carriage) 车
Chen (120°) 辰
Chen (star) 辰
Chen (state) 陈
Chen Baoguang 陈宝光
Chen Fuyao 陈扶摇
Chen Haozi 陈淏子
Chen Huang 陈潢
Chen Shaowei 陈少微
Chen Shigong 陈实功
Chen Xianwei 陈显微
Chen Yuanjing 陈元靓
Chen Zhuo 陈卓
Chen Zilong 陈子龙
Chen Ziming 陈自明
Chen Zun 陈遵
cheng (reed) 筬
Cheng Dawei 程大位
Chengdu 成都
Chenghuang Temple 城隍庙
chenglü 乘率
Chengtian Gate 承天门
Chengyang Bridge 程阳桥
Chengqiao 程桥
chensha 辰砂
Chengxintang paper 澄心堂纸
Chenzhou 辰州
Chezheng (officer in charge of carriages) 车正
chi 尺
chong ben yi mo 崇本抑末
chishizhi 赤石脂
Chongling 舂陵

Daye 大冶
Daye (period) 大业
Daye Calendar 大衍历
Dayi 大邑
Deng Ping 邓平
dengchengche (city-mounting carriage) 登城车
Dengfeng 登封
Denglai 登莱
Dengxian (county) 邓县
di (earth) 地
Di (tribe) 狄
Di kiln 弟窑
dian (projection) 点
Dian (nationality) 滇
diaohong 雕红
diluo 地螺
ding (large sacrificial vessels) 鼎
Ding (195°) 丁
Ding Huan 丁缓
Ding kiln 定窑
dingfa 定法
dingshu 定数
dingwei 丁未
Dingzhou 定州
diqi 地气
disang 地桑
diuzhen 丢针
diyuan 地元
Dongchuan 东川
Dongjing 东井
Dongping 东平
Dongting Lake 洞庭湖
Dongzhi 冬至
dou 斗
Dou Cai 窦材
Doubing 斗柄
Douyun 都匀

Du Fu 杜甫
Du Huan 杜环
Du Shi 杜诗
Du Wan 杜绾
du yao yan qiu 毒药烟球
Du Zhong 杜忠
duan (prolonged calcination) 煅
duan (satin) 缎
Duan Chengshi 段成式
Duhuishi 都会市
duhuo 独活
duhuoguan 毒火罐
Dujiangyan 都江堰
Dule Temple 独乐寺
dun huang fan shi 敦煌矾石
Dunhuang 敦煌
Dupangling Ridge 都庞岭
Dushikou 独石口
duzhu 都柱

E

e (yoke) 轭
Eastern Han (dynasty) 东汉
Eastern Hus 东胡
"eight keys" 八纲
Ejin Banner 额济纳旗
Epang Palace 阿房宫
Erdos Right Rear Banner 鄂尔多斯右翼后旗
Erlang Mountain 二郎山
Erligang 二里岗
Erlitou 二里头
Erui 峨蕊
eryuanshu 二元术

F

Fa Xian 法显

Fan Ye 范晔

fanche (turn-over scoops) 翻车

Fang 房

fang 方

fang (ward). 坊

Fang Gan 方干

fang Yizhi 方以智

fangcheng 方程

fangchengshu 方程术

fangchou (spun-yarn silk) 纺绸

fangchuan 舫船

fangfeng 防风

fanggai 方盖

fangji 防己

Fanglin Gate 芳林门

Fangshan 方山

fangzhou 方舟

fanzi ("somersaulter") 泛子，翻子

fei (fertile) 肥

fei (sublimation) 飞

fei kong ji zei zhen tian lei pao 飞空击贼震天雷炮

Fei Zhu 费著

feidaojian 飞刀箭

feiqiangjian 飞枪箭

fen 分

feng (tool) 锋

feng (buried underground) 封

Feng Su 冯宿

Fenghe River 沣河

Fenghuangcheng 凤凰城

Fenghuangshan 凤凰山

Fengjie 奉节

Fenhe River 汾河

fenmu (denominator) 分母

fenqing 粉青

fenzi (numerator) 分子

Five Dynasties 五代

"Flying Girder over Fish Swamp" 鱼沼飞梁

Fo Gong Si 佛宫寺

Foshan 佛山

fu (the negative) 负

fu (subduing the toxicity) 伏

Fu Renjun 傅仁钧

Fuchuan 福船

fufang 负方

Fufeng (county) 扶风

Fugou 扶沟

Fujian Province 福建省

Fuzhou 福州

G

gaitian theory 盖天说

Gan 甘氏

Gan (characters of the denary cycle) 干

ganguozi (crucible) 甘埚子

Ganlu Hall 甘露殿

Gansu (province) 甘肃

Ganyu 赣榆

ganzengchuan 赶缯船

Ganzhou 甘州

Gao Wenhong 高文洪

Ga Xuan 高宣

Gao Yonggu 高永固

Gao You 高诱

Gao Zong, emperor 高宗

Gaonu (county) 高奴

Ge Hong 葛洪

Ge kiln 哥窑

geng 更

Geng Shouchang 耿寿昌

gong string 宫

Gong Zhenlin 龚振麟

Gongcheng Yangqing 公乘阳庆
Gongsun Long 公孙龙
Gongxian (county) 巩县
gou 勾
gougu theorem 勾股定理
gu 股
Gu Kaizhi 顾恺之
Gu Shicheng 顾世澄
Gu Yunqing 顾云庆
Gua (the Bight Diagrams) 卦
Guan kiln 官窑
Guan Zhong 管仲
Guang Wu, emperor 光武帝
Guangchuan 广船
Guangdong (province) 广东
Guangdu 广都
Guanghan 广汉
Guanghua Gate 光华门
Guanghua Gate 光化门
Guangji Bridge 广济桥
Guangxi 广西
Guangzhou 广州
Guangxin 广信
Guanxian (county) 灌县
Guanyin Pavilion 观音阁
Guanzhong (district) 关中
Guanyuan (Ren 4) 关元
Guapian 瓜片
guchuan (sea-hawk) 鹘船
gui (cabinet) 匮
gui (spirit) 鬼
Gui Fu 桂馥
Gui Wanrong 桂万荣
Guijiang 桂江
Guilin 桂林
Guiyu Qu 鬼臾区
Guizhou (province) 贵州
Gulangyu Isle 鼓浪屿

Guo Shoujing 郭守敬
Guo (state) 虢
Guo Zi Jian 国子监
guoshanlong 过山龙
Guoyuanchang 果园厂
Guyu 谷雨
Guzhu 顾渚

H

Haihe River 海河
Hainan Island 海南岛
Haiyang River 海阳河
Hami 哈密
Han (dynasty) 汉
Han Fei 韩非
Han Gonglian 韩公廉
Han Ji 韩曁
Han Jun 韩君
Han Xin 韩信
Han Yanzhi 韩彦直
Han Zhong 韩忠
Hancheng (county) 韩城
Handan 邯郸
Hangou 邗沟
Hanguang Gate 含光门
Hanguang Hall 含光殿
Hangzhou 杭州
Hanjiang River 韩江
Hanlu 寒露
hanluopan 旱罗盘
Hanshi 汉氏
hanshuishi (gypsum) 寒水石
Hanyuan Hall 含元殿
Hanzhong Gate 汉中门
he (standing grain) 禾
He Chengtian 何承天
He Meng 和幪

ji (millet) 稷

ji (tight) 急

ji (whole volumes) 积

ji li gu che (*li*-recording drum carriage) 记里鼓车

ji li huo qiu (barbed fire package) 蒺藜火球

jia (tea) 檟

Jia Dan 贾耽

Jia Heng 贾亨

Jia Kui 贾逵

Jia Sixie 贾思勰

Jia Xian 贾宪

Jia Zi 甲子

jiafu (reinforcement) 夹辅

Jiajing (period) 嘉靖

Jian kiln 建窑

jiancong 减从

jiancongzhu 减从术

Jiangdu 江都

Jianfu Temple 荐福寺

jiang (border) 疆

Jiang Hui 蒋辉

Jiang Ji 姜岌

Jiang Kaoqing 江考卿

Jiang Tao (T. Kiang) 江涛

Jiangdong Bridge 江东桥

jiangfan 绛矾

Jiangjin 江津

Jiangling 江陵

Jiangsu (province) 江苏

Jiangxi (province) 江西

Jianjiang River 浙江

jianmian (silk wadding) 缣绵

Jian Zhen 鉴真

jiao (pouring out) 浇

jiao (string) 角

Jiao Xun 焦循

jiaozi (exchange medium) 交子

Jiaqing (reign) 嘉庆

Jiaxing 嘉兴

Jiayuguan 嘉峪关

Jiazhou 嘉州

Jie, king 桀

jiedaoche (stair carriage) 阶道车

jieqi (solar term) 节气

jijilu (furnace) 既济炉

Jimei Bridge 济美桥

jin (brocade) 锦

Jin (dynasty) 晋

jin (gold) 金

Jin (state) 晋

Jin Ao Yu Dong Bridge 金鳌玉栋桥

Jinan 济南

Jinci Temple 晋祠

jing (well) 井

Jing Fang 京房

Jingchu Calendar 景初历

Jingdezhen 景德镇

Jingfeng Gate 景风门

Jinggong Fang 靖恭坊

Jingshan Fang 靖善坊

Jingshui River 泾水

Jinguan 金关

Jinguang Gate 金光门

Jingxian (county) 泾县

Jingyang (county) 泾阳

Jingyao Gate 景耀门

Jingyuan 靖远

Jingzhe 惊蛰

Jinhua 金华

Jinianming 纪年铭

Jining 济宁

Jinjiang 晋江

Jinning (county) 晋宁

Jinshanwei 金山卫
Jinta (county) 金塔
Jinxian (county) 金县
jinye (potable gold) 金液
Jinzhou 锦州
jishifan 鸡屎矾
jiu (alcohol) 酒
Jiudai Ji 僦贷季
jiufan (submit to moulding) 就范
Jiuquan 酒泉
Jiuyishan Mountain 九嶷山
Jixian (county) 蓟县
Jiyuan Calendar 纪元历
Jizhou 蓟州
Ju (star) 鞠
juan (rough silk fabric) 绢
judu 距度
Jueshui River 潏水
jun (measure of weight) 钧
Jun kiln 钧窑
Juyan (county) 居延
Juyongguan 居庸关

K

kai dai cong ping fang fa 开带从
  平方法
kai fang chu zhi 开方除之
kai fang zuo fa ben yuan tu 开方作法
  本源图
kai lian zhi mou cheng fang 开连枝
  某乘方
kai ling long mou cheng fang 开玲
  珑某乘方
kai ping fang chu zhi 开平方除之
Kaibao (period) 开宝
Kaibao Temple 开宝寺
kaifangshu 开方术

Kaifeng 开封
kailifang 开立方
kaipingfang 开平方
Kaiyuan (reign) 开元
Kaiyuan Gate 开远门
Kang Xi, emperor 康熙
ke (division of time) 刻
Ke Shangqian 柯尚迁
kesi (silk fabric with large, stand-
  out jacquard patterns) 缂丝
kewu 渴乌
Kin (dynasty) 金
Kong Pingzhong 孔平仲
Kong Ting 孔挺
Kong Yingda 孔颖达
kongqing (malachite) 空青
Kou Zongshi 寇宗奭
kouqi (lacquerware with inlay) 扣器
kui 葵
Kun (225°) 坤
kun (reed) 捆
Kun Wu 昆吾
Kunlun Mountains 昆仑山
Kunming Lake 昆明湖
Kuqa (county) 库车

L

lai (barley) 来
Laiwu 莱芜
lajian (waxed paper) 蜡笺
Lang kiln 郎窑
langdu 狼毒
Lanshan 蓝山
Lanzhou 兰州
lao (continuous cropping) 劳
Lao Zi 老子
laolong (old wells) 老窿

Leigutai 擂鼓台
Leizu 嫘祖
Lengdao 冷道
lengjin (cold gold) 冷金
Lengshui River 冷水
Leshan 乐山
li (hard) 力
li (measure of distance) 里
Li (nationality) 黎
li (ward) 里
Li Bing 李冰
Li Chun 李春
Li Chunfeng 李淳风
Li Daoyuan 郦道元
Li Gao 李皋
Li Huang 李璜
Li Jie 李诫
Li Lan 李兰
Li Longji (prince) 李隆基
Li Rui 李锐
Li Shizhen 李时珍
Li Xu 李诩
Li Ye 李冶
Li Yuheng 李豫亨
Li Zhi 李治
lian (side) 廉
lian (transformation of a dry substance by roasting) 炼
lian (vanity case) 奁
liang (measure of weight) 两
Liang (dynasty) 梁
Liang Lingzan 梁令瓒
Liangyi Hall 两仪殿
Liangzhou 凉州
Liao 辽
Liaodong 辽东
Liaoning (province) 辽宁
Liaoyang 辽阳

Lichun 立春
Lidong 立冬
Lijiang River 漓江
lin (rinsing) 淋
Lin Hong 林洪
Lin Qing 麟庆
Lin Zhiyuan 林知元
Linde Calendar 麟德历
Linde Hall 麟德殿
ling (damask) 绫
Ling Mengchu 凌濛初
Ling Ruxiang 凌汝享
Ling Yingchu 凌赢初
Lingqu Canal 灵渠
Lingshan 灵山
Linhe (county) 临河
Linhu (tribe) 林胡
Linqing 临清
Lintao (county) 临洮
Linxi River 林溪河
Linyi (county) 临沂
Liqiu 立秋
Liquan 礼泉
Lirenshi 利人市
Lishan 骊山
Liu An 刘安
Liu Bei, emperor 刘备
Liu Che (Emperor Wu Di) 刘彻
Liu Hong 刘洪
Liu Hui 刘徽
Liu Meng 刘蒙
Liu Sheng 刘胜
Liu Song (dynasty) 刘宋
Liu Tong 刘侗
Liu Xi 刘熙
Liu Xiang 刘向
Liu Xianting 刘献廷
Liu Yi 刘益

Liu Yin 刘歆
Liu Zhen 刘珍
Liu Zhuo 刘焯
liuhuangjin (sulphur "gold") 硫黄金
liuhuangyin (sulphur "silver") 硫黄
　银
Liujiagang 刘家港
liujin (gildings) 鎏金
Lixia 立夏
liyuan (beginning of calendar) 历元
liyunyi 立运仪
long gu shui che 龙骨水车
Longchuan 隆川
Longjiangguan 龙江关
Longjing 龙井
Longqing (reign) 隆庆
Longquan 龙泉
Longshan 龙山
Longshou Canal 龙首渠
Lou Shu 楼璹
louche (tower carriage) 楼车
Louchuan (turreted ship) 楼船
Loufan 楼烦
lounian (towered carriage) 楼辇
Louying 漏影
Lü Buwei 吕不韦
Lu Daolong 卢道隆
Lu Guimeng 陆龟蒙
Lu Kuisheng 卢葵生
Lü Li 吕礼
Lu (state) 鲁
Lu Xun 卢循
Lu Yu 陆羽
Lü Zuqian 吕祖谦
Lu'an (Anhui) 六安
Lu'an (Shanxi) 潞安
Luding Bridge 泸定桥
lüfan (green vitriol) 绿矾

luganshi (calamine) 炉甘石
Lugou Bridge 芦沟桥
Luhe (county) 六合
luo (leno) 罗
Luo Hongxian 罗洪先
luo pan zhen lu 罗盘针路
Luo Shilin 罗士琳
Luo Tianpeng 罗天鹏
Luocheng 罗城
luodian (mother-of-pearl inlay) 螺钿
luojingpan 罗经盘
Luoqianxi 罗乾溪
Luoshui River 洛水
luowen (woven paper) 罗纹
Luoxia Hong 落下宏
Luoyang 洛阳
lusang 鲁桑
Lushan 卢山
luxian (salt) 卤咸
Luya 露芽

M

Ma Jun 马钧
mafeisan (anaesthetic) 麻沸散
mai (wheat) 麦
Mancheng 满城
Mangzhong 芒种
Mao (90°) 卯
mao (winter peach) 旄
Mao Bowen 毛伯温
Mao Yuanyi 茅元仪
Mao feng 毛峰
Maojian 毛尖
Matteo Ricci 利玛窦
Mawangdui 马王堆
Mei Biao 梅彪
Meng Yuanlao 孟元老

Mengding 蒙顶
menghanyao (narcotic) 蒙汉药
Mengjin 孟津
mi (cross-section) 幂
Mianchi (county) 渑池
Miaodigou 庙底沟
Min Qiji 闵齐级
Minfeng 民丰
ming (tea) 茗
Ming Antu 明安图
Ming (dynasty) 明
Mingde Gate 明德门
mingfanshi (alum) 明矾石
minggu 冥谷
Mingtian Calendar 明天历
Minjiang River 岷江
Minxian (county) 岷县
mise 秘色
moche (mill-cart) 磨车
mofan (pattern, model) 模范
Mou 牟
mou he fang gai 牟合方盖
Mount Emei 峨嵋山
Mount Huashan 华山
Mount Taishan 泰山
mu 亩

N

Nangong Yue 南宫说
Nanhuizui 南汇嘴
Nanjing 南京
Nanling Mountains 南岭
Nanping 南平
Nanyang 南阳
naosha (sal ammoniac) 硇砂
neidan (physiological alchemy) 内
    丹

Neiqiu 内丘
Neixiang 内乡
Ni Changxi 倪长犀
niang (fermentation) 酿
niaochuan ("bird ship") 鸟船
niela (wax-kneading method) 捏蜡
Nieti (Muphrid) 摄提
nijin (sprinkled gold) 泥金
Ningwuguan 宁武关
Ningxia 宁夏
Ningxiang 宁乡
Ningyuan 宁远
Niya 尼雅
nong jia yue ling 农家月令
North Dynasties 北朝
Northern Liang (kingdom) 北梁
Northern Song (dynasty) 北宋
Northern Wei (dynasty) 北魏
Northern Zhou (dynasty) 北周

O

Olong (tea) 乌龙

P

paituo (row of bags) 排橐
Pan Geng, king 盘庚
Pan Jixun 潘季驯
panchi 盘池
Pei Songzhi 裴松之
Pei Xiu 裴秀
Peiling 涪陵
Peng Junbao 彭君宝
pengsha (borax) 蓬砂
Pi-Shi-Hang 淠史杭
Pianguan 偏关

pianti (sideways lifter) 偏提
piaoci (pale-green porcelain) 縹瓷
Pingcheng 平城
Pinglu (county) 平陆
Pingshui Gunpowder 平水珠茶
pingtuo (lacquer object) 平脱
Pinggu (county) 平谷
Pingzhunju 平准局
pishuang (arsenolite) 砒霜
Pochengzi 破城子
pu (garden or plot) 圃
Pu Yuan 蒲元
Pucheng (county) 蒲城
Pu'er tea 普洱茶
Purple Mountain 紫金山
puxiao (sodium sulphate) 朴硝
Puyang (county) 濮阳

Q

qi 棋
qi (damask) 绮
Qi (dynasty) 齐
Qi (state) 齐
Qi Dezhi 齐德之
Qi Shaoweng 齐少翁
qian (measure of weight) 钱
qian (copper coins) 钱
Qian Baocong 钱宝琮
qian feng cui se 千峰翠色
Qian Lezhi 钱乐之
Qian Long, emperor 乾隆
qianbai (white lead) 铅白
qiandan (minium) 铅丹
qiandu (right triangular prism) 堑堵
Qianniu (lunar mansion) 牵牛
Qianshanyang 钱山漾
Qiantang (river) 钱塘江

Qianxiang Calendar 乾象历
Qianyang (county) 千阳
qiaotou (leeboard) 橇头
Qiaowancheng 桥湾城
Qicheng 崎城
Qidan (Khitan) 契丹
Qilian Mountains 祁连山
Qimen 祁门
Qin (dynasty) 秦
qin (lute) 琴
Qin Jiushao (Ch'in Chiu-shao) 秦九韶
Qin Keda 秦可大
Qin Shi Huang (Shi Huang Di, First Emperor of Qin) 秦始皇
Qing (dynasty) 清
qing hua you li hong 青花釉里红
Qing Xu Zi 清虚子
Qingdu 青犊
qingfen (calomel) 轻粉
Qinglong Temple 青龙寺
Qingming 清明
Qingming Canal 清明渠
Qingning (reign) 清宁
Qingshuihe 清水河
Qingxing Palace 庆兴宫
Qinshi (the people of Qin) 秦氏
Qinzhou 秦州
Qiongzhou 琼州
qishu (odd number) 奇数
Qiu Chuji 邱处机
qiu lian bei yuan 求廉本源
Qiufen 秋分
Qiwu Huaiwen 綦毋怀文
Qixia Gate 启夏门
Qiyun 齐云
Qu Yuan 屈原
Qu Zhongrong 瞿中溶

Quanzhou 泉州
Qujiang 曲江
qujidu 去极度

R

ren (man) 人
"Ren Di" 任地
Ren Zong , emperor 仁宗
Rong (tribe) 戎
rong (velvet) 绒
rong (fusing) 熔
rongyan (table salt) 戎盐
rongzhu (smelting) 熔铸
rou (soft) 柔
Rouyuan Gate 柔远门
Ru kiln 汝窑
ruxiudu 入宿度

S

Sanchahe 三汊河
sanchenyi 三辰仪
Sanhe (county) 三河
Sanji Calendar 三纪历
Sanjiang 三江
Santong Calendar 三统历
sanxiandan (mercuric oxide) 三仙丹
sanyuanshu 三元术
Schall von Bell 汤若望
se (zither) 瑟
sha (gauze) 纱
sha-luo 纱罗
"Sha Qing" 杀青
Shaanxi (province) 陕西
shachuan ("sand ship") 沙船
Shan Zhaoming 闪昭明
Shandong (province) 山东

shang (root) 商
Shang (dynasty) 商
"Shang Nong" 上农
Shang Yang 商鞅
Shangcai 上蔡
shangfangling 尚方令
Shanghai 上海
shanglian 上廉
Shangsiwu 上四坞
Shangyan Mountain 商颜山
Shangyu 上虞
shangyuan 上元
Shanhaiguan 山海关
Shanidi 沙泥地
Shanshui Lake 山水池
Shanxi (province) 山西
Shaoguan 韶关
Shaopi 苟陂
Shaoxi (reign) 绍熙
Shaoxing 绍兴
Shaozhou 韶州
Shazhou 沙州
she (tea) 荼
Shen (star) 参
Shen (240°) 申
shen huo fei ya 神火飞鸦
Shen Jiaben 沈家本
Shen Kuo 沈括
Shen Qi 沈启
Shen Zhong 沈重
Shen Zong, emperor 神宗
shendan (elixir) 神丹
shenggong (corrosive sublimate) 升汞
Shengmeiting 胜梅亭
shenshi (magical chamber) 神室
Shenshui River 深水
shi (hour) 时
shi (area of the rectangle) 实

su steel 苏钢
Su Yijian 苏易简
Subutun 苏埠屯
Sui (dynasty) 隋
Sun Quan, emperor 孙权
Sun Simiao 孙思邈
Sun Wu 孙武
Sun Yanxian 孙彦先
Sun Zhilu 孙之骈
Sun Zi 孙子
Suzhou (in Jiangsu Province) 苏州
Suzhou (now Jiuquan) 肃州

T

tai (the absolute) 太
tai yin xuan jing (selenite) 太阴玄精
Tai Zong, emperor 太宗
Tai Zu, emperor 太祖
Taibai (people's commune) 太白
Taicang (county) 太仓
Taichu Calendar 太初历
Taiguan Garden 太官园
Taiji Hall 太极殿
Taiji Palace 太极宫
Taiping 太平
Taiping Heavenly Kingdom 太平天国
Taiwan (province) 台湾
Taiwei 太微
Taiyuan 太原
Taijiao 太角
tan tree-bark paper 檀皮纸
Tancheng 郯城
Tang (dynasty) 唐
Tang Fu 唐福
Tanglang Mountain 螳螂山
Tangshan 唐山

Tanshan 滩山
Tao Hongjing 陶弘景
Tao Zongyi 陶宗仪
taoye (mould) 陶冶
Teng Yi 滕揖
Three Kingdoms 三国
tian ("heaven") 天
Tian Shou Shan 天寿山
tianbai (sweet white) 甜白
Tianbao (reign) 天宝
Tiancang (commune) 天仓
tianerong (swan velvet) 天鹅绒
Tianjin 天津
tianqi 填漆
Tianshan Mountains 天山
Tianshi 天市
Tianxian Bridge 天仙桥
tianyuanshu 天元术
tianyuanyi 天元一
Tianzhen 天镇
tiejiao (iron-armoured car) 铁轿
Tielian Mountain 铁镰山
Tieshenggou 铁生沟
tihong (carved red lacquer) 剔红
Tonghua Gate 通化门
tongjiangjun (gun) 铜将军
tongjin (copper "gold") 铜金
Tonglü Mountain 铜绿山
Tongshan (county) 铜山
tongshu 通数
Tongtian Calendar 统天历
tongyin (copper "silver") 铜银
Tongzi 潼梓
toushi (copper and zinc) 鍮石
toushijin (brass "gold") 鍮石金
toutai 投胎
tu (tea) 茶
tuhuoqiang (fire-spitting lance) 突火

枪

Tunxi 屯溪

tuo (bag) 橐

tuo (measure of length) 托

tuo yuan qiu zhou shu (method for finding the circumstance of an eclipse) 椭圆求周术

Turpan 吐鲁番

tuzi 土子

W

Wafangzhuang 瓦房庄

waidan (laboratory alchemy) 外丹

Wan Bridge 卍桥

Wan Nian Ru Yi Jin 万年如意锦

Wang Bao 王褒

Wang Bing 王冰

Wang Chong 王充

Wang Fan 王蕃

Wang Guan 王观

Wang Ji 汪机

Wang Jian 王建

Wang Kentang 王肯堂

Wang Lai 汪莱

Wang Nongwan 汪弄丸

Wang Shuhe (Wang Shu Khu) 王叔和

Wang Tao 王焘

Wang Weiyi 王惟一

Wang Xiaotong 王孝通

Wang Yi 王逸

Wang Youhuai 王又槐

Wang Yu 王与

Wang Zhen 王祯

Wang Zhenduo 王振铎

Wanli (period) 万历

Wanquan (county) 万泉

Wanrong 万荣

Wanshou Bridge 万寿桥

Warring States Period 战国

Wei (210°) 未

Wei Boyang 魏伯阳

Wei (dynasty) 魏

Wei (state) 魏

Wei Yilin 危亦林

Wei Yuan 魏源

Weichuan 渭川

weijilu (alchemist furnace) 未济炉

weishu (minute numbers) 微数

Weishui River 洧水

Wen Di, emperor 文帝

Wen, king 文王

Wenhe River 汶河

Wenxian (county) 温县

Wenzhou 温州

Western Han (dynasty) 西汉

Western Jin (dynasty) 西晋

Western Zhou (dynasty) 西周

woduan (Japanese satin) 倭缎

Wu (180°) 午

Wu (kingdom) 吴

wu (matter) 物

Wu (state) 吴

Wu Deren 吴德仁

Wu Di, emperor 武帝

Wu Pu 吴普

Wu Qi 吴起

Wu Wu 吴惧

Wu Xian 巫咸

Wu Yue (state) 吴越

Wu Zimu 吴自牧

Wuchang 武昌

Wuding River 无定河

Wudu 武都

Wuji Calendar 五纪历

Wujiang (county) 吴江
Wujiang River 吴江
Wuling Mountains 五岭
wumingyi (pyrolusite) 无名异
Wunijing 乌泥泾
Wusong River 吴淞江
Wuwei 武威
Wuxi 无锡
Wuxian (county) 吴县
Wuxing 无兴
Wuyin Calendar 戊寅历
Wuzhuhou 五诸侯

X

xi (to lie fallow) 息
Xi Zhong 奚仲
Xia (dynasty) 夏
Xia (state) 夏
Xiadu 下都
xialian 下廉
Xiamen 厦门
Xi'an 西安
xian ("god") 仙
xianchu (wedge) 羡除
xianchu theorem 羡除公式
Xiang Mingda 项明达
Xiangjiang River 湘江
Xiangzi Bridge 湘子桥
Xianyang 咸阳
Xiangyang Mountain 象羊山
Xianyun (tribe) 猃狁
Xiao Zong, emperor 孝宗
Xiaoeryuan 小儿原
Xiaohan 小寒
Xiaohe River 洨河
Xiaoman 小满
xiaoshi (hour) 小时

xiaoshi (potassium nitrate) 硝石
xiaoshi (stone dissolvent) 消石
Xiaoshu 小暑
Xiaoshui River 潇水
Xiaotianping 小天平
Xiaotun 小屯
Xiaoxili 小溪里
Xiaoxue 小雪
Xiaoyan Pagoda 小雁塔
Xiasiwu 下四坞
Xiaxian (county) 夏县
Xiazhi 夏至
Xiezhou 解州
Xinan 新安
Xinchang Fang 新昌坊
Xincun 新村
xing ("shape") 形
Xing Bridge 荇桥
Xing kiln 邢窑
Xing Yunlu 邢云路
Xingan (county) 兴安
Xingan Gate 兴安门
Xinglong 兴隆
Xingping (county) 兴平
Xingqing Palace 兴庆宫
Xingshan Temple 兴善寺
Xing Zong, emperor 兴宗
Xinjiang 新疆
Xintian 新田
Xinxiang (county) 新乡
Xinyang 信阳
Xinzheng 新郑
Xiongnu (Huns) 匈奴
xionghuang (realgar) 雄黄
Xiyin Village 西阴村
Xizhaoxian Village 西招贤村
Xu Fu 徐市
Xu Guangqi 徐光启

Xu Jian 徐坚
Xu Jing 徐兢
Xu Shang 许商
Xu Xiake 徐霞客
Xu Xinlu 徐心鲁
Xu Yue 徐岳
xuan (hypotenuse) 弦
Xuan Di, emperor 宣帝
Xuan Zong, emperor 玄宗
Xuande (period) 宣德
Xuandu Monastery 玄都观
Xuanfu 宣府
Xuanhua 宣化
xuan ming long gao 玄明龙膏
xuantaiding (suspended inner vessel) 悬胎鼎
Xuanwu Gate 玄武门
Xuanyuan 轩辕
Xuanzheng Hall 宣政殿
Xue (clan) 薛
Xue Jingshi 薛景石
Xuecheng 薛城
Xueshui River 泜水
Xunyi 浚仪

Y

Yacheng 崖城
yahua (calendered pattern) 砑花
Yan (state) 燕
yan (table salt) 盐
Yan Du 延笃
Yan Su 燕肃
Yanan 延安
Yanchi (county) 盐池
Yang (positive) 阳
yang (simmering) 养
Yang Di, emperor 炀帝

Yang Er 杨尔
Yang Guozhu 杨国柱
Yang Hui 杨辉
Yang Jian 杨坚
Yang Jizhou 杨继州
Yang Mao 杨茂
Yang Wei 杨伟
Yang Xiong 杨雄
Yang Yao 杨么
Yang Zhongfu 杨忠辅
yangma (pyramid) 阳马
yangqishi (actinolite) 阳起石
Yangshao 仰韶
yangtao (*Actindia chinensis*) 杨桃
yangtaoteng (*Actindia chinensis planch*) 杨桃藤
Yangzhou 扬州
Yanhe River 延河
Yanmen 雁门
Yanmenguan 雁门关
Yanping Gate 延平门
Yanshi (county) 偃师
Yanshou (county) 延寿
yanweiqiang (swallow-tailed arrow) 燕尾箭
Yanxi Gate 延禧门
Yanxing Gate 延兴门
Yanying Hall 延英殿
Yanyuan (reign) 延元
Yao, emperor 尧
Yao Shunfu 姚舜辅
yaobian 窑变
Yazhou 崖州
Yecheng 邺城
Yellow River (Huanghe) 黄河
Yeting Palace 掖庭宫
Yi (about 105°) 乙
Yi Xing 一行

Yichang 宜昌

Yijiang Gate 挹江门

yijishu 益积术

Yin (negative) 阴

Yin Shitong 尹世同

Yin (time) 殷

yinbai (jade-white) 莹白

Yinchuan 银川

yingbuzu 盈不足

Yingpu 营浦

yingqing (moon-white) 影青

Yingxian (county) 应县

Yingzhou 应州

Yingqueshan 银雀山

Yinshan Mountains 阴山

yinzhu (vermilion) 银硃

yiwofeng (hornet nest) 一窝峰

Yixing 宜兴

yiyu 益隅

Yong Le, emperor 永乐

Yong Ning Si 永宁寺

Yong Zheng, emperor 雍正

Yongan Canal 永安渠

Yongding River 永定河

Yongkang (reign) 永康

Yongchu (reign) 永初

yu (corner) 隅

yu (carriage box) 輿

yu guo tian qing 雨过天晴

Yu Hao 喻皓

Yu the Great 大禹

Yu Xi 虞喜

Yu Yan 俞琰

Yu Zongben 俞宗本

yuan (the primary) 元

yuan (shaft) 辕

Yuan (dynasty) 元

Yuan Hanqing 袁翰青

Yuan Yi 袁宜

Yuan Shen 袁慎

Yuanjia Calendar 元嘉历

Yuanshi Calendar 元始历

yuanshu 元数

Yue Huan 乐隈

Yue kiln 越窑

Yuejun (county) 越巂

Yulin 榆林

Yulucha 玉露茶

Yumen (county) 玉门

Yumenguan 玉门关

Yuncheng 运城

yunmu 云母

Yunnan (province) 云南

Yunyang 云阳

yushi (mispickel) 礜石

Yushui 雨水

yusuan 隅算

Yuwen Kai 宇文恺

Yuxian (county) 蔚县

Yuyao 余姚

yuyuliang (haematite) 禹余粮

Z

zao (dry) 燥

Zeng Gongliang 曾公亮

Zeng Sanyi 曾三异

Zeng Shenzhai 曾慎斋

zeng cheng kai fang fa 增乘开方法

zengcheng 增乘

zengqing (chalcanthite) 曾青

zengqingyin (chalcanthite "silver") 曾青银

zhacai 榨菜

Zhai Jinsheng 翟金生

zhang (measure of length) 丈
Zhang Cang 张苍
Zhang Cheng 张成
Zhang Degang 张德刚
Zhang *duan* 漳缎
Zhang Heng 张衡
Zhang Hong 张泓
Zhang Hua 张华
Zhang Qian 张骞
Zhang *rong* 漳绒
Zhang Shinan 张世南
Zhang Tuxin 张徒信
Zhang Yi 张揖
Zhang Yingwen 张应文
Zhang Yuzhe (Y.C. Chang) 张钰哲
Zhang Zeduan 张择端
Zhang Zhan 张湛
Zhang Zhihe 张志和
Zhang Zhongjing 张仲景
Zhang Zhongyan 张中彦
Zhang Zhouxuan 张胄玄
Zhang Zihe 张子和
Zhang Zixin 张子信
Zhangjiakou 张家口
Zhangye 张掖
Zhangzhai 张寨
Zhangzhou 漳州
Zhanlin (tribe) 澹林
Zhanqu 毡曲
Zhao Fei 赵瞰
Zhao Guo 赵过
Zhao Jianzi 赵简子
Zhao Shuang 赵爽
Zhao (state) 赵
Zhao Xihu 赵希鹄
Zhao Yizhai 赵遗斋
Zhao Youqin 赵友钦
Zhao Yurong 赵与衮

Zhao Zhiwei 赵知微
Zhaoshigang 赵士岗
Zhaoxian (county) 赵县
Zhaoxiang, king 昭襄王
Zhaozhou 赵州
Zhejiang Province 浙江省
zhen (irrigation ditch) 甽
Zhen Luan 甄鸾
zheng (positive) 正
Zheng Guo 郑国
Zheng He 郑和
Zheng Ke 郑克
Zheng (state) 郑
Zheng Xingyi 郑兴裔
Zhengde (reign) 正德
Zhengzhou 郑州
Zhenjiang (city) 镇江
Zhenping 镇平
zhentianlei 震天雷
zhi (baking) 炙
Zhi (characters of the duodenary cycle) 支
Zhi Cong 知聪
zhicheng (woven fabric) 织成
Zhide Gate 至德门
zhinanyu (south-pointing fish) 指南鱼
zhinanzhen (south-pointing needle) 指南针
zhishiying (amethyst) 紫石英
Zhiyuan (reign) 至元
zhiyao (paper drug) 纸药
Zhongbaocun Village 中保村
Zhongmou (county) 中牟
Zhongnan Mountains 终南山
Zhongping (reign) 中平
zhongqi 中气
Zhongquan 重泉

Zhongshan Park 中山公园
zhongsui (focus) 中燧
Zhongxinglu 中兴路
Zhongyan (production team) 中颜
Zhou Chu 周处
Zhou Daguan 周达观
Zhou (dynasty) 周
Zhou Enlai, premier 周恩来
Zhou Shihou 周师厚
Zhou Zong 周琮
zhu (reed) 杼
zhu (boiling) 煮
zhu (measure of weight) 铢
Zhu Jifang 朱继芳
Zhu Shijie 朱世杰
Zhu Siben 朱思本
Zhu Xi 朱熹
Zhu Yan 朱琰
Zhu Yu 朱彧
Zhu Yunming 祝允明
Zhu Zundu 朱遵度
Zhuang Yuanchen 庄元臣
Zhuang Zhou 庄周

Zhuang Zhuo 庄绰
Zhuanxu Calendar 颛顼历
Zhucheng (county) 诸城
Zhuge Liang 诸葛亮
Zhujiang (Pearl) River 珠江
Zhuozheng Garden 拙政园
Zhupu Bridge 珠浦桥
Zhuque Gate 朱雀门
zhusi (rami silk) 纻丝
zhuzi (wine pipette) 注子
Zichen Hall 紫宸殿
Zicheng 子城
ziheche 紫河车
Zijingguan 紫荆关
zike (bullets) 子窠
Ziliujing 自流井
Ziwei 紫微
ziwu (meridian) 子午
zu (sacrificial utensil) 俎
Zu Chongzhi 祖冲之
Zu Geng 祖暅
zun (wine vessel) 尊

General Index

Postscript

The articles that have been presented here are selected from *Achievements in Science and Technology in Ancient China*, published in Chinese by the China Youth Publishing House in March 1978. Certain mistakes and inaccuracies in the original were discovered during translation and have been corrected by the authors.

The translation of a book of this type is a challenging task given the diversity of fields covered and the classical Chinese language (including technical terms peculiar to ancient Chinese science and technology) used to describe them. We are deeply grateful to Mao Yisheng, Vice-President of the China Science and Technology Association and Director of the Railway Research Academy, who revised and approved the Preface and showed sincere concern for the translation of the book; and to Chen Yousheng, who edited the Chinese original for English-speaking readers; to Liu Zuwei, Du Youliang, Li Tiansheng, Liu Naiyuan and Fu Zhengyuan, who translated the articles into English; and to Tang Bowen, Betty Chandler and Zhao Shuhan, who edited the English text.

It is our hope that by informing readers abroad on ancient China's culture, science and technology the present book will in its way contribute towards promoting cultural exchange in science and technology between China and other countries.

The book has been compiled by the Institute of the History of Natural Sciences, Chinese Academy of Sciences. Among those responsible for the work are Du Shiran, Fan Chuyu, Jin Qiupeng, Chen Meidong and He Shaogeng.

> The Institute of the History of Natural Sciences, Chinese Academy of Sciences